Aerospace Engineering Handbook

Aerospace Engineering Handbook

Edited by **Ted Dunham**

CLANRYE INTERNATIONAL

New Jersey

Published by Clanrye International,
55 Van Reypen Street,
Jersey City, NJ 07306, USA
www.clanryeinternational.com

Aerospace Engineering Handbook
Edited by Ted Dunham

© 2015 Clanrye International

International Standard Book Number: 978-1-63240-058-1 (Hardback)

Printed in the United States of America.

Contents

Preface

Aerospace engineering is a branch of engineering that focuses on the study of development processes, construction, tests, scientific aspects, and technological aspects of aircrafts and space-crafts. Developing and manufacturing a modern flight vehicle is a highly complex process. A lot of factors need to be kept in consideration such as design, available technology and costs. Aerospace engineering deals with all these aspects so that the design can be reliably built and maintained. New technologies are also constantly being developed by aerospace engineers for use in aviation, defence systems, and space.

If we turn the pages of history world over, the first aviation pioneers can be traced to the late 19th to early 20th centuries. One of the most prominent names in the history of aeronautics is of Sir George Cayley, who contributed significantly to this field from the last decade of the 18th to mid-19th century. The first definition of aerospace engineering appeared in the month of February in the year 1958.

There are two main branches of this field of engineering, commonly known as aeronautical engineering and astronautical engineering. There is a vast difference between these two sub branches, as aeronautical engineering deals with aircrafts that are operational in Earth's atmosphere, and astronautical engineering studies about the space-craft which are basically operational outside the Earth's atmosphere. Aerospace engineering also studies about the aerodynamic characteristics of a spacecraft.

Each chapter includes an extensive review of literature, as well as current information and trends on aerospace engineering. We have been fortunate to have an outstanding group of aeronautic specialists from all over the world, who have contributed to this publication. As the latest trends have been illuminated through various perspectives, it allows the reader to understand different approaches to the same problem and then decide his/her own interests.

Editor

Ultrasonic Characterization of the Fiber-Matrix Interfacial Bond in Aerospace Composites

D. G. Aggelis,[1,2] D. Kleitsa,[1] and T. E. Matikas[1]

[1] *Department of Mechanics of Materials and Constructions, Vrije Universiteit Brussel, Pleinlaan 2, 1050 Brussels, Belgium*
[2] *Department of Materials Science and Engineering, University of Ioannina, 45110 Ioannina, Greece*

Correspondence should be addressed to D. G. Aggelis; daggelis@cc.uoi.gr

Academic Editors: K.-M. Chung, E. E. Imrak, and A. F. B. A. Prado

The properties of advanced composites rely on the quality of the fiber-matrix bonding. Service-induced damage results in deterioration of bonding quality, seriously compromising the load-bearing capacity of the structure. While traditional methods to assess bonding are destructive, herein a nondestructive methodology based on shear wave reflection is numerically investigated. Reflection relies on the bonding quality and results in discernable changes in the received waveform. The key element is the "interphase" model material with varying stiffness. The study is an example of how computational methods enhance the understanding of delicate features concerning the nondestructive evaluation of materials used in advanced structures.

1. Introduction

Reinforcement of a bulk material with fibers is commonly applied in order to upgrade its properties in terms of stiffness, strength, and durability. Fiber composites are applicable in any type of material, like steel-fiber-reinforced concrete, polymer, or ceramic composites as well as metal matrix composite materials [1–3]. In most cases, the fibers exhibit higher mechanical properties than the matrix to improve its behavior. However, in order to take full advantage of the fiber potential, the bonding between fiber and matrix is of primary importance. Efficient stress transfer is desirable, and this is the reason that in certain cases chemical treatment of the fibers is applied in order to enhance bonding [4]. The chemical reaction between the matrix and fiber results in an "interphase" zone with properties different than the ones of the constituent phases (see Figure 1). This interphase zone may be very thin but it plays a crucial role in the mechanical performance of the medium. This is the zone through which stress is transferred, and therefore, all important mechanical properties of the composite like its strength and toughness heavily depend on the quality of the interphase [5, 6]. However, environmental and stress effects degrade the quality of the interphase compromising the structural capacity of the

whole composite. It is understandable that the initial bonding conditions between fiber and matrix should be optimized in order to minimize the effect of service-induced deterioration. Assessing the quality of initial bonding is a task that can be conducted by certain mechanical tests like pull-out or push-in [7, 8]. However, these are destructive and focus on the strength, totally neglecting the elastic properties of the interphase. Therefore, a fast, nondestructive, and simple methodology of assessing the quality of the interphase is desirable. This should result in a quantifiable parameter related to the stiffness of the interphase and could be applied after manufacturing to assess the initial bonding condition in single fiber specimens or, under suitable circumstances, after service for assessment of bonding degradation. A reliable test will assist the design of the composite material by evaluating the stiffness of the interphase in terms of the different constituent materials' elastic and thermal properties, as well as regarding the suitable conditions to achieve optimal bonding, that is, temperature and pressure.

In the present paper, a numerical study of stress wave reflection is described in order to estimate the potential for characterization of the bonding and guide relevant experimental efforts [9]. The suitability of stress waves has been pointed out concerning characterization of surfaces and

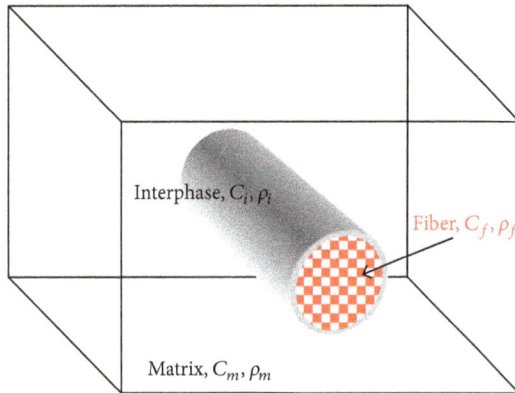

FIGURE 1: A typical part of the microstructure of the composite.

thin layers [10]. Stress waves employ infinitesimally small displacements and are influenced by the elastic properties of the materials. Therefore, their study reveals information on the stiffness of the materials and not directly their strength. Particularly, the immersion pulser-receiver technique is targeted. According to this technique, a pressure pulse is emitted inside a liquid (normally water). This pulse propagates with the sound velocity of water towards the material under test. When it meets an interface, it is partially transmitted and partially reflected through the medium. The number of reflections recorded by the sensor and their delay define the possible defects and their depth below the surface of the tested object. In the specific case described herein, the direction of the pulse relatively to the materials' surface is dictated by the shear critical angle so that only shear waves propagate in the medium, which are arguably more sensitive to the existence and quality of interphases [8, 11, 12]. The shear wave interacts with the embedded fiber and the reflection is eventually recorded by the receiver. Analysis of the recorded waveform sheds light into the condition of the interphase since the reflection depends on the relative mismatch of mechanical properties. When the bonding is inadequate due to incompatible materials or has worn out, there is essentially no contact between the materials. Therefore, a strong reflection is bound to occur since the fiber volume acts as a void in terms of wave propagation. In the case of a single scatterer, although dispersive effects are not expected, the analysis is based on the reflection on the scatterer in a pulser-receiver mode. On the other hand when two materials of similar stiffness are in good contact, the reflection will be minimized. The different possible conditions of bonding are simulated herein by an elastic "interphase" material with varying stiffness. This interphase should not be confused with the "interface" which is a boundary between the matrix and the fiber. This interphase is used to model the elastic properties of the interphasial zone between the matrix and fibers resulting from the chemical reaction between the two materials. Hence its behavior is governed by a user-defined varying elastic wave velocity to simulate different degrees of stiffness. This study includes the exact geometry of the fiber as an advancement of the analytical solution that was provided in [9] for the reflection on an inclined straight line instead of the circular

cross-section of the fiber. The material system targeted herein is a metal matrix composite material [9]. Specifically, the matrix of the targeted material is Ti-6Al-4V reinforced with continuous SCS-6 fibers, a material widely used in aerospace. High strength titanium alloys, as well as fiber-reinforced metal matrix composite materials, are suitable for a number of highly demanding applications because of their improved mechanical properties in high temperature conditions. In applications where dynamic loading is expected and where life management is required, consideration must be given to the behavior of the material in the sensitive area of the interphase between the matrix and the fiber in order to verify the best possible performance of the material.

2. Numerical Simulations

2.1. Model. Numerical simulations are generally used to expand to cases that cannot be experimentally tested due to cost, geometry, or other limitations and also to increase the physical understanding in specific problems. Wave simulation studies enable also the recognition of wave modes and reflections inside a whole waveform. In the specific case, two-dimensional simulations were conducted on a cross-section of the geometry as is explained below.

The fundamental equation of two-dimensional propagation of elastic waves in an elastic medium neglecting viscosity is

$$\rho \frac{\partial^2 \underline{u}}{\partial t^2} = \mu \nabla^2 \underline{u} + (\lambda + \mu) \nabla \nabla \cdot \underline{u}, \qquad (1)$$

where $\underline{u} = u(x, y, t)$ is the displacement vector as a function of time, t, ρ is the material density, and λ and μ are the first and second Lame constants, respectively. These parameters are related to the wave propagation velocities with the following equations:

$$\begin{aligned} C_L &= \sqrt{\frac{\lambda + 2\mu}{\rho}}, \\ C_S &= \sqrt{\frac{\mu}{\rho}}, \end{aligned} \qquad (2)$$

where C_L is the longitudinal and C_S is the shear wave velocities respectively.

The simulations were conducted with commercially available software [13] that solves the above equation with respect to the boundary conditions of the object and the initial conditions [14]. The solution is in time domain with the finite difference method in the plane strain field. The excitation pulse has a defined displacement-time function and is applied at specified nodes of the geometry that simulate the "pulser." Continuity equations must be fulfilled at the interfaces between different entities. In the present analysis, individual materials are included in the geometry, and therefore propagation is solved in each distinct phase, while the continuity conditions for stresses and strains must be satisfied on the interfaces.

In the case described herein, the propagation of a stress wave after excitation in water is simulated. The geometry is

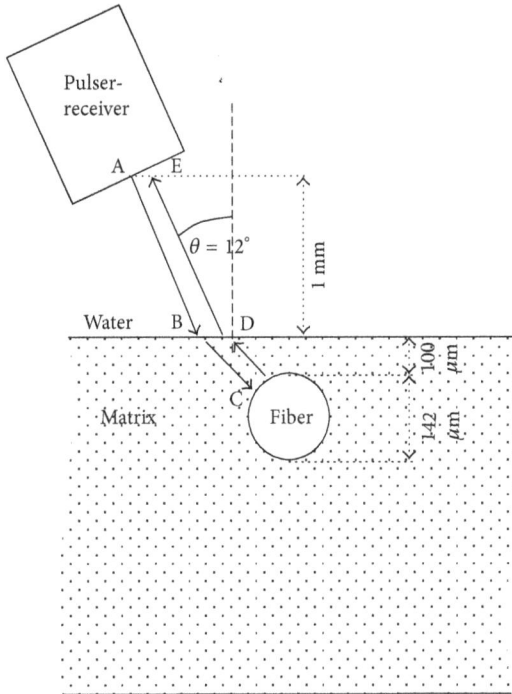

FIGURE 2: Geometry of the simulated test including wave directions.

TABLE 1: Basic properties for material modeling.

	Water	Matrix (Ti-6Al-4V)	Fiber (SCS-6)
λ (GPa)	2.25	25.9	61.9
μ (GPa)	10^{-4}	32.8	177.0
ρ (kg/m^3)	1000	2580	3000
C_L (m/s)	1500	5954	11774
C_S (m/s)	10	3566	7681

shown in Figure 2 and the wave path of interest is indicated, while it is discussed in more detail in Section 3. The wave impinges on the matrix under the shear critical angle, thus allowing only shear waves to propagate into the matrix. The shear wave interacts with the fiber and a part is reflected back. After being refracted from the matrix/water interface, a longitudinal wave propagates through water back to the receiver (same as pulser, see Figure 2).

The "source" is placed at a specific angle, θ relatively to the vertical axis, equal to the critical shear angle of this horizontal liquid/solid interface. In the specific case, the angle is 12°, as calculated based on Snell's law and the mechanical properties of water and the titanium matrix [9]. The pulser introduces one cycle of different frequencies in the longitudinal mode. The applied frequencies were 1 MHz, 5 MHz, 10 MHz, 25 MHz, and 50 MHz.

The employed materials were considered elastic without viscosity. The basic properties of all the materials except the interphase are seen in Table 1. Both matrix and fiber materials are quite stiff with the fiber exhibiting approximately twice the longitudinal and shear wave velocities of the matrix. As already mentioned, the interphase obtained different values of stiffness expressed by the corresponding longitudinal wave velocities. This is a key parameter of the study and a practical way to simulate different contact levels between the matrix material and the fiber [11, 15]. Specifically, the lowest value was 300 m/s (case of loose interphase similar to air), and the maximum 11770 m/s which is the longitudinal wave velocity of the fiber. In between, the values were incremented by 1000 m/s, for example, 1000 m/s, 2000 m/s, 3000 m/s, and so forth. This includes the possible range of equivalent stiffness values that could be obtained by the interphase layer. The

diameter of the fiber is 142 μm, and it is embedded 100 μm below the surface, (see Figure 2). Since there was no physical insight for the thickness of the actual interphase layer, it was set to 50 μm. In similar cases, it has been shown that the thickness of the interphase does not make critical difference in the results [11]. The vertical distance of the pulser was indicatively set to 1 mm above the surface of the specimen, while it can be adjusted to suit the relevant experimental geometry each time.

As in any simulation study, here also certain conditions must apply in order to ensure reliable and repeatable results. The mesh size is a crucial parameter since if it is defined to a relatively large value, the outcome will not be accurate but on the other hand there are computational power and time restrictions that prevent from applying an infinitesimally small value. Restrictions on the computational power do not always allow to use several elements per wavelength. In any case, since the study employs four materials (water, matrix, fiber, and interphase), there is no standard wavelength to adjust the element size accordingly. Therefore, another holistic approach was followed; different values of mesh sizes were tested; namely, from 0.4 mm down to 70 μm and the resulted waveforms were compared. Figure 3(a) shows the time window when the first part of the reflection (case of a loose interphase) is recorded for the frequency of 25 MHz for some indicative mesh sizes. Simulations with mesh sizes larger than 0.1 mm (specifically 0.3 and 0.4 mm in Figure 3(a)) result in quite different waveforms compared to the finer meshes and were not further considered. From the mesh size of 100 μm and finer, the waveforms converge in shape. In order to quantify the comparison of these cases, a threshold was chosen, namely, −0.012 units of amplitude (u.a.) in order to deterministically define the onset of the reflection (see Figure 3(a) and compare between different cases). As the mesh becomes finer, the calculated onset times changed and can be seen in Figure 3(b). The finer mesh tested (70 μm) resulted in an onset of 0.90067 μs but it was extremely time consuming. The simulations were conducted with the mesh of 80 μm which resulted in transit time of 0.89959 μs being 0.12% away from the result of the finest mesh applied. This was considered a suitable approximation due to the limited amount of error relatively to the specific available computational power. As an indication, a full simulation of a case in a computer with RAM of 3 GB and processor of 2.1 GHz lasted about 1 hr. Concerning the time step resolution, it resulted in 0.00034 μs, which even for the highest frequency (50 MHz with period of 0.02 μs) contains

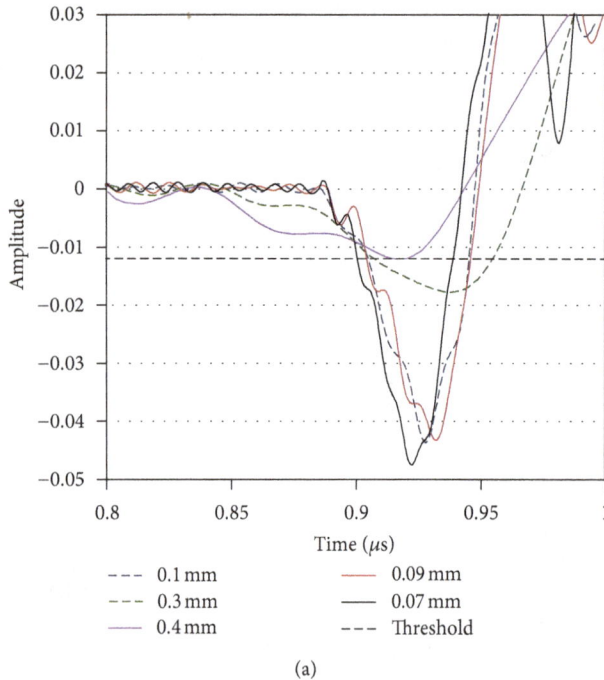

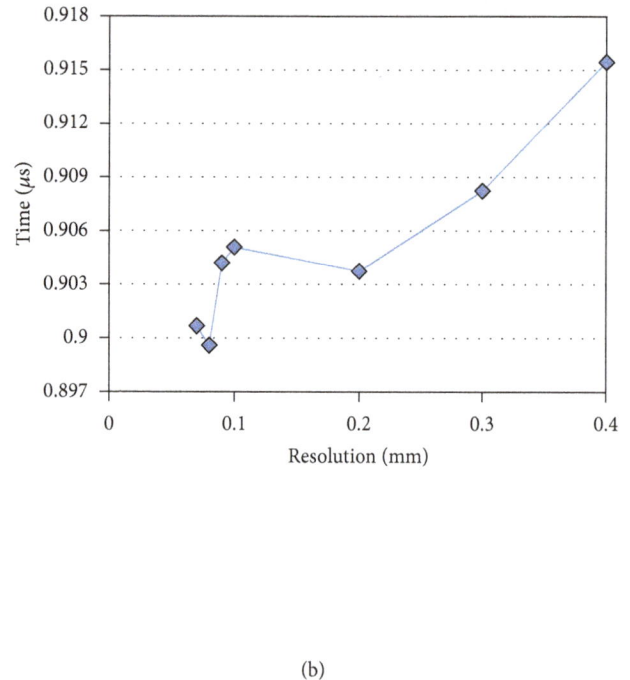

FIGURE 3: (a) Onset of the reflection in the received waveform for different resolutions. (b) Arrival time to the receiver for different resolutions (measured from the threshold crossing at 0.012 u.a.).

approximately 55–60 points in a cycle and is considered more than adequate sampling in similar cases [16].

3. Results

The longitudinal wave pulse is emitted by the pulser (point A in Figure 2). This pulse propagates initially through water and hits the water/matrix interface under the shear angle, θ, as has been discussed above (point B). The shear wave is transmitted through the matrix and reaches the fiber (C). Reasonably one part is reflected and another is transmitted past the fiber. The amount of energy reflected will depend on the shear wave impedance (product of shear velocity and density) mismatch of the two materials. The matrix impedance is of the level of 9 MRayl, while the fiber which is stiffer exhibits impedance of 23 MRayl. Therefore, in any case a reflection is expected when the materials are in perfect contact. If, on the contrary, the fiber is totally debonded from the matrix, the reflection will be stronger since the impedance of air is negligible compared to that of the matrix. It is reasonable that for any intermediate condition of bonding quality the reflection will be in between the above-mentioned extreme cases. This role (quality of bonding) is played by the "interphase" material, which in our analysis obtains variable values of stiffness, as expressed by the different longitudinal wave velocities. The wave reflected by the fiber, which now may again include longitudinal components after the reflection on the circular surface, propagates back to the surface of the matrix (D), and a part is refracted within water as longitudinal wave following the opposite direction of the initial incident pulse. This wave reaches the sensor as shown in Figure 2, point E. A typical

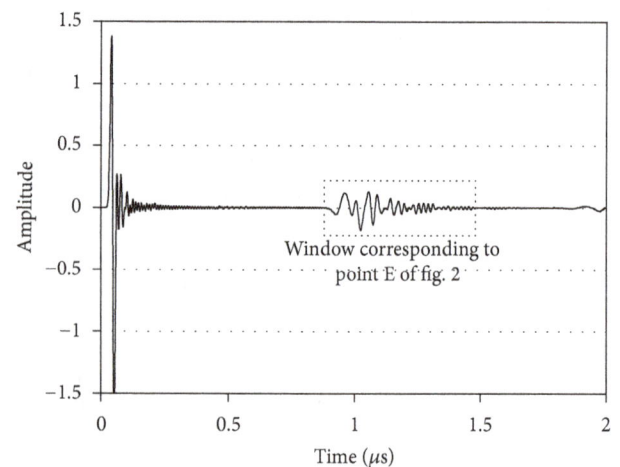

FIGURE 4: Typical waveform after excitation of 25 MHz.

waveform is seen in Figure 4 where the initial pulse and the reflection (window corresponding to point E of Figure 2) are shown, and in this part of the wave any analysis and evaluation should be focused to characterize the quality of the interphase.

Figure 5 shows some indicative views of the displacement field for the frequency of 25 MHz and for the stiff interphase with pulse velocity of 11770 m/s. In the first case (a), the wave is propagating through water, while in Figure 5(b) the shear wave starts to be refracted in the matrix traveling on a higher speed than the wave in water. In the last case of Figure 5(c), the clear reflection can be seen in water (see arrow) while

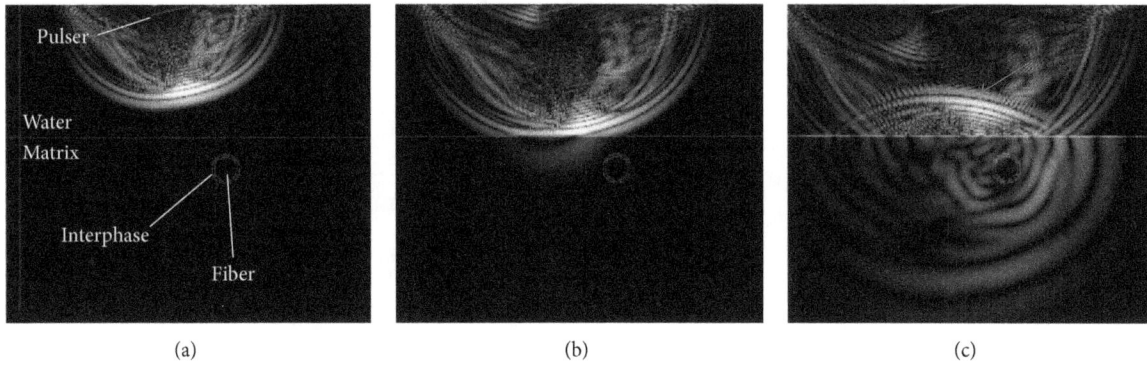

(a)   (b)   (c)

FIGURE 5: Consecutive snapshots of the displacement field for the case of stiff interphase.

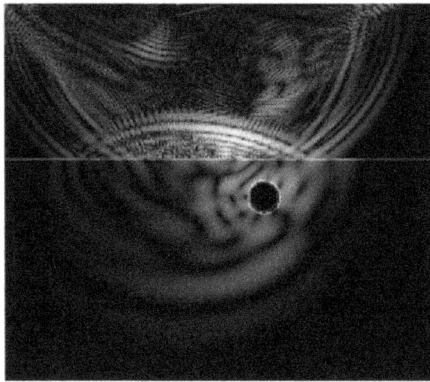

FIGURE 6: Snapshot of the displacement field for the case of loose interphase.

the refracted wave propagates deep in the matrix. Figure 6 shows the field at approximately the same time but with loose interphase. It is obvious that no wave is transmitted through the fiber, while the reflection traveling back to the receiver is similar to the previous case. However, it contains critical differences that make characterization of the different interphases possible, as discussed next.

Figure 7(a) shows the reflections (corresponding to window E of Figure 4) as recorded by the receiver for two extreme cases of interphase stiffness values, namely, equivalent to air (C_i = 330 m/s) and fiber (C_i = 11770 m/s). The waveforms are identical up to 1 μs, since the initial part of the waveform is due to the direct reflection on the water/solid interface which is not influenced by the fiber. The wave packet of the reflection between the matrix/fiber interphase arrives slightly later since the fiber is at a depth of 100 μm from the surface. Therefore, some discrepancies are visible after the time of 1 μs, with the waveform from the loose interphase exhibiting higher amplitude attributed more likely to the higher reflection coefficient. In order to focus on the differences between the two waveforms, they are subtracted and the resulted waveform is seen in Figure 7(b). Quite detectable discrepancies are noted after 1 μs. The result of the subtraction is a wave of similar amplitude mainly because the reflections from a less stiff second material are of opposite

phase. The discrepancy can be quantified by the area of the signal envelope (measured area under the rectified signal envelope, see Figure 7(c)) denoted as "energy," which is a parameter widely used in waveforms analysis [17, 18]. The reflection from the stiff interphase was maintained as reference and the waveforms obtained for each other stiffness were subtracted by the reference in order to calculate the energy difference. The results are seen in Figure 8. For any of the applied frequencies, this energy indicator increases monotonically as the interphase stiffness decreases from its maximum value down to the value of loose interphase. This is because the reflection from the fiber with a loose interphase is maximum due to the extreme impedance mismatch, as has already been mentioned. Comparing the results derived for different frequencies, the maximum energy difference comes for the frequency of 5 MHz, where its value is more than 100 units, while its lowest for loose interphase comes at 1 MHz. Frequencies of 10, 25, and 50 MHz result in intermediate values of 40 to 55, while specifically 50 MHz exhibits a quite constant rate, being equally sensitive to changes of interphase velocity at any interphase velocity level. On the contrary, the 5 MHz curve is very sensitive to changes at the low level of interphase velocities but is not as sensitive to higher values close to good bonding (i.e., the initial signs of debonding in a real case). Therefore, in actual application the use of higher frequencies (25 MHz or 50 MHz) is suggested for assessment of even slight incompatibility or debonding trends, which corresponds to a drop of interphasial stiffness from 11770 m/s to 10000 m/s or 15%. In the same figure, the experimental values of the reflection coefficient for the two extreme cases (good bonding and simulated debonding) are also included, as measured in [9]. This reflection coefficient was obtained by comparing the FFT of the waveform corresponding to the actual geometry (e.g., with a hole) to the waveform of an angled surface which was considered a reference. This reflection coefficient, which again depends on the mismatch between the two sides of the interface, is much higher for the debonding than the case of regular bond between the fiber and the matrix. The qualitative similarity in the decreasing trend between the experimental and numerical energy-related features as the interphasial stiffness increases shows that the approach is in the right direction, and further study will enable accurate evaluations of the interphase quality.

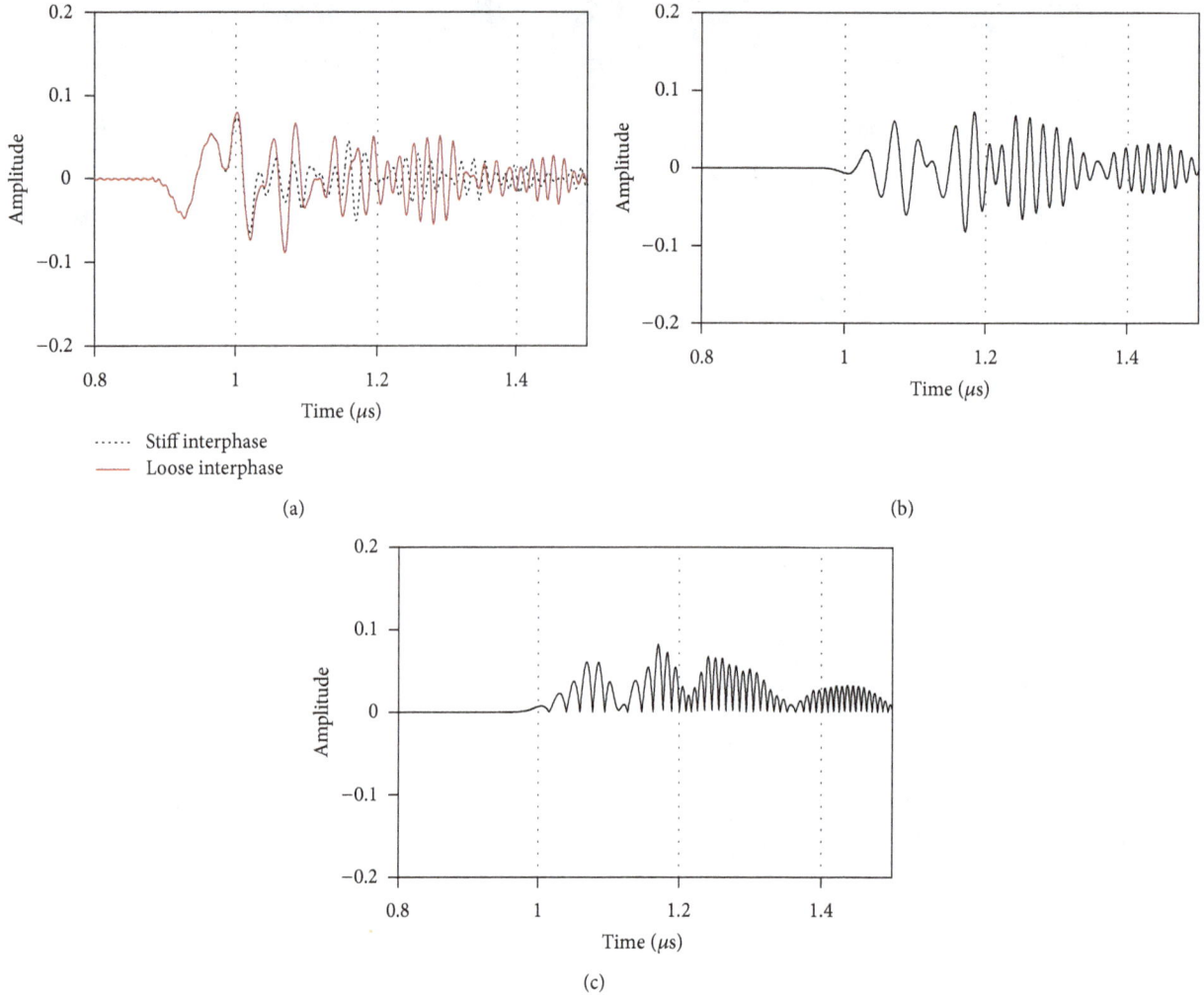

FIGURE 7: (a) Part of the waveform containing the reflection from the fiber for stiff ($C_i = 11770$ m/s) and loose ($C_i = 300$ m/s) interphase. (b) Subtraction of the two waveforms of (a). (c) Rectification of the waveform of (b).

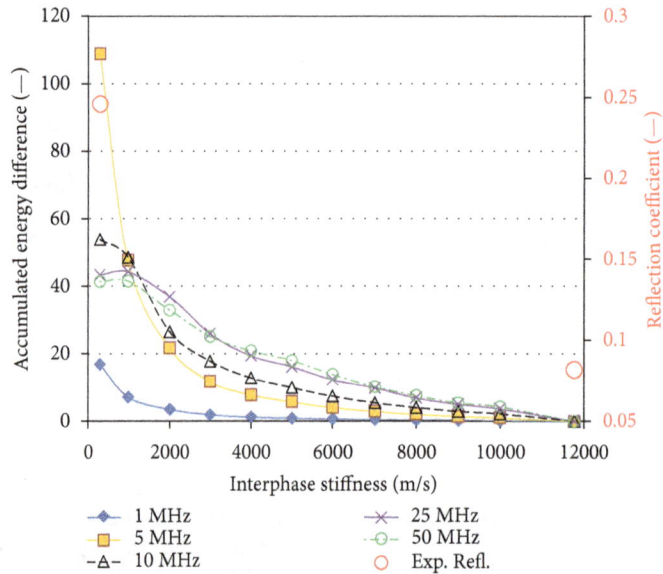

FIGURE 8: Reflection energy difference for several values of interphase stiffness (longitudinal velocities) and different frequencies.

4. Discussion

The results presented above show that ultrasonic reflection parameters exhibit a monotonic trend with respect to the interphase stiffness. This opens the possibilities not only to detect debonding or poor compatibility but to quantify the stiffness of the interphase. This property is handled by the equivalent wave velocity of the modeling material called "interphase" in this study in accordance to the actual layer between the matrix and fiber materials. The values of interphase stiffness are varied from the two extreme cases of similar to air (loose contact) and similar to fiber (strong bonding) including all the possible realistic values in between. Concerning some specific parameters that are encountered towards the experimental application, it should be mentioned that though the measurement is delicate, in a real experiment with the immersion technique, the quality of the acoustic coupling provided by water is constant and therefore, any difference due to even slight reflection changes will be detected. The sensor scans along the longitudinal axis of the fiber enabling characterization of the interphase bonding on its whole length. It should be kept in mind that this test is intended for material design purposes (compatibility of constituents) rather than deterioration assessment. Therefore, the targeted geometry is simple (e.g., single fiber specimen [9]), in order to avoid the interference with neighboring fibers that would occur in the actual material. The simple geometry will enable derivation of accurate information on the fiber-matrix interphase and will act as a guide for the material design process. This way the results from different systems can be compared in order to judge sort their interface compatibility. Additionally, the corresponding "stiffness" of the interphase can be correlated to the results of mechanical tests if they are also performed (i.e., pull-out or push-in). Concerning the fiber alignment, which is crucial for the aforementioned destructive tests, it is not crucial for the proposed ultrasonic reflection technique because the experimental wave beam cross-section is much larger than the fiber diameter.

5. Conclusion

Advanced metal matrix composites for aerospace applications require delicate methods to accurately assess their initial state as well as service-induced damage. This study concerns the nondestructive evaluation of the quality of bonding between fiber and matrix in such composites. The exact fiber geometry is simulated as an advancement of the previous analytic studies on a simplified geometry. The immersion ultrasonic technique is numerically simulated, while shear waves are targeted due to their sensitivity on bonding conditions. Different bonding is modeled by altering the stiffness of the "interphase" material which acquires properties from near-zero, simulating negligible contact up to stiffness similar to the fiber, simulating the stiffest possible bonding. The results indicate that despite the consecutive refractions between the water and matrix, the influence of the de-bonding is distinguishable compared to the case of stiff interphase. This is because the amount of energy reflected depends on the interphase elastic properties which cause small but discernible differences in the received waveform. The study shows how computational methods enhance our understanding and can give direction to the relevant experimental techniques with the aim of providing better characterization of crucial aspects of the material's condition in a nondestructive manner.

References

[1] P. Stähli and J. G. M. van Mier, "Manufacturing, fibre anisotropy and fracture of hybrid fibre concrete," *Engineering Fracture Mechanics*, vol. 74, no. 1-2, pp. 223–242, 2007.

[2] K. G. Dassios and C. Galiotis, "Direct measurement of fiber bridging in notched glass-ceramic-matrix composites," *Journal of Materials Research*, vol. 21, no. 5, pp. 1150–1160, 2006.

[3] U. Ramamurty, "Assessment of load transfer characteristics of a fiber-reinforced titanium-matrix composite," *Composites Science and Technology*, vol. 65, no. 11-12, pp. 1815–1825, 2005.

[4] D. G. Aggelis, D. V. Soulioti, N. M. Barkoula, A. S. Paipetis, and T. E. Matikas, "Influence of fiber chemical coating on the acoustic emission behavior of steel fiber reinforced concrete," *Cement and Concrete Composites*, vol. 34, no. 1, pp. 62–67, 2012.

[5] G. D. Zhang and R. Chen, "Effect of the interfacial bonding strength on the mechanical properties of metal matrix composites," *Composite Interfaces*, vol. 1, no. 4, pp. 337–355, 1993.

[6] H. An, Y. X. Li, M. Li, Y. Z. Gu, Y. N. Liu, and Z. G. Zhang, "Interfacial adhesion and micro-failure phenomena in multi-fiber micro-composites using fragmentation test," *Composite Interfaces*, vol. 19, no. 6, pp. 385–396, 2012.

[7] B. Hassoune-Rhabbour, L. Poussines, and V. Nassiet, "Development of an adhesion test for characterizing the interface fiber/polymer matrix," *Key Engineering Materials*, vol. 498, pp. 210–218, 2012.

[8] M. C. Waterbury, P. Karpur, T. E. Matikas, S. Krishnamurthy, and D. B. Miracle, "In situ observation of the single-fiber fragmentation process in metal-matrix composites by ultrasonic imaging," *Composites Science and Technology*, vol. 52, no. 2, pp. 261–266, 1994.

[9] P. Karpur, T. E. Matikas, and S. Krishnamurthy, "Ultrasonic characterization of the fiber-matrix interphase/interface for mechanics of continuous fiber reinforced metal matrix and ceramic matrix composites," *Composites Engineering*, vol. 5, no. 6, pp. 697–711, 1995.

[10] T. E. Matikas, "Assessment of interface deformation and fracture in metal matrix composites under transverse loading conditions," *Composite Interfaces*, vol. 15, no. 6, pp. 589–609, 2008.

[11] D. Kleitsa, K. Kawai, T. Shiotani, and D. G. Aggelis, "Assessment of metal strand wire pre-stress in anchor head by ultrasonics," *NDT and E International*, vol. 43, no. 7, pp. 547–554, 2010.

[12] P.-N. Tzounis, L. N. Gergidis, T. E. Matikas, and A. Charalambopoulos, "Mathematical investigation of interfacial property in fiber reinforced model composites," *Composites B*, vol. 43, no. 6, pp. 2605–2612, 2012.

[13] Wave2000, Cyber-Logic, New York, NY, USA, http://www.cyberlogic.org/.

[14] J. Kaufman, G. Luo, and R. Siffert, "Ultrasound simulation in bone," *IEEE Transactions on Ultrasonics, Ferroelectrics, and Frequency Control*, vol. 55, no. 6, pp. 1205–1218, 2008.

[15] C. Pecorari, "Scattering of a Rayleigh wave by a surface-breaking crack with faces in partial contact," *Wave Motion*, vol. 33, no. 3, pp. 259–270, 2001.

[16] F. Moser, L. J. Jacobs, and J. Qu, "Modeling elastic wave propagation in waveguides with the finite element method," *NDT and E International*, vol. 32, no. 4, pp. 225–234, 1999.

[17] D. G. Aggelis, "Classification of cracking mode in concrete by acoustic emission parameters," *Mechanics Research Communications*, vol. 38, no. 3, pp. 153–157, 2011.

[18] A. Anastasopoulos, D. Kourousis, S. Botten, and G. Wang, "Acoustic emission monitoring for detecting structural defects in vessels and offshore structures," *Ships and Offshore Structures*, vol. 4, no. 4, pp. 363–372, 2009.

2

Performance Prediction of a Synchronization Link for Distributed Aerospace Wireless Systems

Wen-Qin Wang and Huaizong Shao

School of Communication & Information Engineering, University of Electronic Science and Technology of China, Chengdu, China

Correspondence should be addressed to Wen-Qin Wang; wqwang@uestc.edu.cn

Academic Editors: P.-C. Chen, J. Y. Fu, K. M. Isaac, and M. G. Perhinschi

For reasons of stealth and other operational advantages, distributed aerospace wireless systems have received much attention in recent years. In a distributed aerospace wireless system, since the transmitter and receiver placed on separated platforms which use independent master oscillators, there is no cancellation of low-frequency phase noise as in the monostatic cases. Thus, high accurate time and frequency synchronization techniques are required for distributed wireless systems. The use of a dedicated synchronization link to quantify and compensate oscillator frequency instability is investigated in this paper. With the mathematical statistical models of phase noise, closed-form analytic expressions for the synchronization link performance are derived. The possible error contributions including oscillator, phase-locked loop, and receiver noise are quantified. The link synchronization performance is predicted by utilizing the knowledge of the statistical models, system error contributions, and sampling considerations. Simulation results show that effective synchronization error compensation can be achieved by using this dedicated synchronization link.

1. Introduction

Distributed aerospace wireless systems have attained more and more interests over the last years as they are seen as a potential means of countering vulnerability to electronics countermeasure [1–7], especially in directional responsive jamming, and avoiding physical attack to the communication platforms [8]. Furthermore, distributed configuration allows a passive receiver teamed with a transmitter at a safe standoff distance. Distributed aerospace wireless systems can be used in many different applications, for example, wireless communications, wireless sensor networks and distributed radars [9–11]. Without loss of generality, this paper considers mainly radar-related applications, especially for the distributed synthetic aperture radar (SAR) imaging. The proposed method is also effective for other distributed aerospace wireless systems.

In distributed radar systems, to measure the echo pulses coherently, the phase information of the transmitted pulse has to be preserved. For a monostatic radar system, in which the colocated transmitter and receiver use the same oscillator, the phase decorrelates over a very short period of time. In contrast, a distributed radar system uses a receiver that is spatially displaced from the transmitter, and hence, the independent phase noise of the transmitter and receiver oscillators does not cancel out. This superimposed phase noise corrupts the received signal over the whole coherent integration time, and therewith severely compromises the subsequent radar performance.

Although the feasibility of distributed radar system concept was already demonstrated by experimental investigations in [12–15], the time and frequency synchronization aspects are still impediments to distributed radar system development [16–20]. The requirement of phase stability for distributed radar system was discussed in [16]. The impact of limited oscillator stability in bi- and multistatic SAR was discussed in [21], which pointed out that uncompensated phase noise may cause a time variant shift, spurious sidelobes, and a deterioration of the impulse response, as well as a low-frequency phase modulation on the received signal. An estimation of oscillator's phase offset in bistatic interferometry SAR was investigated in [22]. In practice, time synchronization is also required for data acquisition. The linear and random time synchronization errors were discussed in [23]; a conclusion was made that linear frequency synchronization errors would lead to a lower imaging resolution and a movement of the target image.

In [24], we investigated a direct-path signal-based technique to compensate the oscillator phase noise for distributed radar systems. The direct-path signal of the transmitter is received with an appropriative antenna divided into two channels. One is passed through an envelope detector and used to synchronize the sampling clock, and the other one is down-converted and used to compensate the phase synchronization errors. However, this approach can be applied only in a limited observation region. The use of continuous duplex intersatellite links for oscillator drift is proposed in [25] and further investigated in [26]. However, this approach destroys the passive characteristic of the receiver and increases its vulnerability, which greatly limits the application scope of the distributed radar system. To get around this disadvantage, we extend the approach to general distributed radar system. The use of a dedicated synchronization link to quantify and compensate the carrier frequency instability is proposed. With the analytical models of phase noise, the possible synchronization accuracy, which may be impacted by oscillator, phase-locked loop, and receiver noise, is quantified. This work can provide a reference to develop practical time and frequency synchronization for distributed radar systems.

The remaining sections of this paper are organized as follows. The time and frequency synchronization scheme via a dedicated microwave communication link is proposed in Section 2. With the analytical models described in Section 3, the time and frequency synchronization accuracy is predicated in Section 4. Finally, Section 5 concludes the whole paper.

2. Synchronization Schemes

Depending on the hardware and affordable synchronization system complexity, various hardware configurations can be employed to establish the synchronization link. As mentioned previously, the duplex intersatellite link [25] demolishes the passive characteristic of receiver and increases its vulnerability, and the direct-path signal-based approach [24] limits the observation region. To overcome these disadvantages, in this paper, we investigate a monodirection synchronization link, as shown in Figure 1.

According to the synchronization schemes, the transmitter repeatedly transmits a synchronization signal, which is a linearly frequency modulated (LFM) signal. The frequency of the oscillator in the transmitter at the start of data take t_0 is $f_i = f_0 + \Delta f_i$, with a nominal frequency f_0 and a frequency offset Δf_i. The phase $\phi_T(t)$ at time t is the integration over frequency [26]:

$$\phi_T(t) = 2\pi \int_{t_0}^{t} f_T(t)\, dt + \varphi_{Ti} + n_{\varphi T}(t), \quad (1)$$

where φ_{Ti} is the initial phase and $n_{\varphi T}(t)$ is the oscillator phase noise.

The receiver receives the signal after a delay τ_i corresponding to the time it takes the signal to travel the

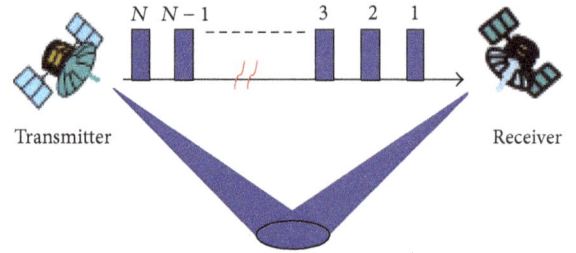

FIGURE 1: Model of the time and frequency synchronization link.

transmitter-to-receiver distance r. At the receive instance $t + \tau_i$, the phase $\phi_R(t + \tau_i)$ of the oscillator in receiver is

$$\phi_R(t + \tau_i) = 2\pi \int_{t_0}^{t+\tau_i} f_R(t)\, dt + \varphi_{Ri} + n_{\varphi R}(t + \tau_i). \quad (2)$$

The demodulated phase available at receiver is the difference between (1) and (2) after including the system and path contributions. This phase difference can be used to obtain the compensation phase. Note that, here, t_0 can be set to zero without restricting generality.

A practical problem is to decide the synchronization repeatedly frequency f_{syn}, carrier frequency f_0, and pulse duration T_p. Additionally, the changes of propagation conditions will result in amplitude and phase fluctuations. Furthermore, an estimate of the time-continuous compensation phase must be recovered from the discrete samples (e.g., *sinc* interpolation). Therefore, the synchronization accuracy must be predicted, and its feasibility must be evaluated prior to developing this synchronization system. In the following, we focus on deriving quantitative estimations for predicting the performance of this synchronization link.

3. Modeling Frequency Instability in Distributed Radar Systems

3.1. Model of Reference Oscillator Frequency Instability. Generally, the performance of oscillator instability is evaluated with Allan variance [27] in time domain or phase noise power spectral density (PSD) $S_\varphi(f)$ in frequency domain [28]. Although an oscillator's phase noise is a complex interaction of variables, ranging from its atomic composition to the physical environment of the oscillator, in the condition that the phase fluctuations occurring at rates f and are small compared with one radian, a good approximation is [29]

$$S_\varphi(f) = a \cdot f^{-4} + b \cdot f^{-3} + c \cdot f^{-2} + d \cdot f^{-1} + e, \quad (3)$$

where the coefficients a to e describe the contributions from (a) random walk frequency noise, (b) frequency flicker noise, (c) white frequency noise, (d) flicker phase noise, and (e) white phase noise, respectively.

One cannot foresee to predict the synchronization accuracy without a model of phase instability. Unfortunately, the frequency-domain expression $S_\varphi(f)$ cannot be directly used in distributed radar systems. The white noise model cannot describe the statistical process of phase noise. The

Wiener noise model [30] cannot describe the low-frequency phase noises which are of great interest for distributed radar system. Hence, we use a time-domain analytical model of reference oscillator phase noise. This model may represent the output signal of a hypothetical filter with impulse response $h(t)$ receiving an input signal $x(t)$.

The spectral density of the output signal is given by the product $S_x(f)|H_\varphi(f)|^2$, where the filter transfer function $H_\varphi(f)$ is the Fourier transform of $h_\varphi(t)$. Note that, here, the $|H_\varphi(f)|^2$ must be satisfied with

$$|H_\varphi(f)|^2 = \begin{cases} S_\varphi(f), & f_l \le |f| \le f_h, \\ S_\varphi(f_l), & |f| \le f_l, \\ 0, & \text{otherwise}, \end{cases} \qquad (4)$$

where a sharp-up cutoff frequency f_h and a sharp-down cutoff frequency f_l are introduced. Note that time domain stability measures sometimes depend on f_h and f_l which should be given to obtain numerical results. In this paper, $f_h = 5\,\text{kHz}$ and $f_l = 0.01\,\text{Hz}$ are assumed.

The phase noise in time domain can then be represented by

$$\varphi_{\text{osc}}(t) = \sqrt{K} x(t) \otimes h(t), \qquad (5)$$

where K is a constant, $\varphi(t)$ denotes the phase noise sequence in time domain and $\otimes$ denotes the convolution operator.

3.2. Model of Phase-Locked Loop (PLL) Phase Noise. Figure 2 shows a fairly general PLL arrangement with a phase detector (PD), a low-pass loop filter $H_L(s)$, a voltage controlled oscillator (VCO) in the forward path and a mixer, an intermediate frequency (IF) filter $H_M(s)$, and a divider ($\div N$). Additionally, a divider ($\div Q$) and a multiplier ($\times N$) are also placed. Since all the noises generated or added in individual PLL blocks are small compared with the useful signals, the small signal theory makes it possible to use the Laplace transform to find the output noise of the considered PLL system or, more exactly, the, respectively, power spectral densities.

According to Figure 2, we can get that [31]

$$N_{\text{PLL}}(s)$$

$$= \left[N_{\text{in}}(s) \left(M + \frac{N}{Q}\frac{1}{F_M(s)} \right) \right.$$

$$+ \left(N_{\text{DQ}}(s) - N_{\text{dn}}(s) + \frac{V_{\text{PDn}}(s) + V_{\text{Fn}}(s)}{K_d} \right) \frac{N}{F_M(s)}$$

$$\left. + N_{\text{mu}}(s) - N_{\text{mi}}(s) \right] \cdot H(s) + N_{\text{osc}}\left[1 - H(s)\right], \qquad (6)$$

where the effective loop transfer function $H(s)$ is

$$H(s) = \frac{K_0 K_d F(s)}{s + K_0 K_d F(s)}, \qquad (7)$$

with

$$F(s) = \frac{1 + s\tau_2}{s\tau_1}. \qquad (8)$$

The τ_1 and τ_2 are the loop low-pass filter parameters. The details can be found in [32]. Note that all the other variables are illustrated in Figure 2. Since most of the noise components are random and uncorrelated, the power spectral density of the PLL output phase noise is

$$S_{\varphi,\text{PLL}}(f)$$

$$= \left\{ S_{\varphi,\text{in}}(f) \left(M + \frac{N}{Q} \right)^2 \right.$$

$$+ \left[S_{\varphi,\text{DQ}}(f) + S_{\varphi,\text{dn}}(f) + \frac{S_{\varphi,\text{PDn}}(f) + S_{\varphi,\text{Fn}}(f)}{K_d^2} \right]$$

$$\times N^2 + S_{\varphi,\text{mu}}(f) + S_{\varphi,\text{mi}}(f) \right\}$$

$$\left. \cdot |H(jf)|^2 + S_{\varphi,\text{osc}}(f) |1 - H(jf)|^2. \right. \qquad (9)$$

We see that the first term in the brace of (9) is inevitable since it is merely a multiplied reference oscillator noise. The second term includes the divider noise, phase detector noise, and loop filter noise, all multiplied by the division ratio N. Finally, with the third term, the multiplier and mixer noises are added; generally, they are small compared with the second term [33]. Hence, all the additive noises, due to the phase detector, loop frequency divider, loop amplifiers, and loop filters are required to quantify prior to predicting the synchronization accuracy.

3.2.1. Phase Detector and Mixer. There are both theoretical and experimental lines of evidences that additive noise due to the mixers is quite small and of the order of the loading circuit noise. Experimental results show that the best phase detector is a double-balanced mixer [28]. Measurements reveal that the phase noise in a double-balanced mixer can be approximated as [34]

$$S_{\varphi,\text{PDn}}(f) \approx \frac{10^{-14\pm 1}}{f} + 10^{-17} \qquad (10)$$

and for phase detector based on CMOS logic family is

$$S_{\varphi,\text{PDn}}(f) \approx \frac{10^{-12.7}}{f} + 10^{-16.2}. \qquad (11)$$

3.2.2. Frequency Divider. Theoretically, the division process reduces the input PSD in proportion to the square of the division factor N^2. However, investigation of the divider output phase noise performed by Kroupa [35] reveals that the output phase noise is

$$S_{\varphi,\text{dn}}(f) \approx \frac{S_{\varphi,\text{dn,in}}(f)}{N^2} + \frac{10^{-10\pm 1} + 10^{-27\pm 1} f_0^2}{f}$$

$$+ 10^{-16\pm 1} + 10^{-22\pm 1} f_0. \qquad (12)$$

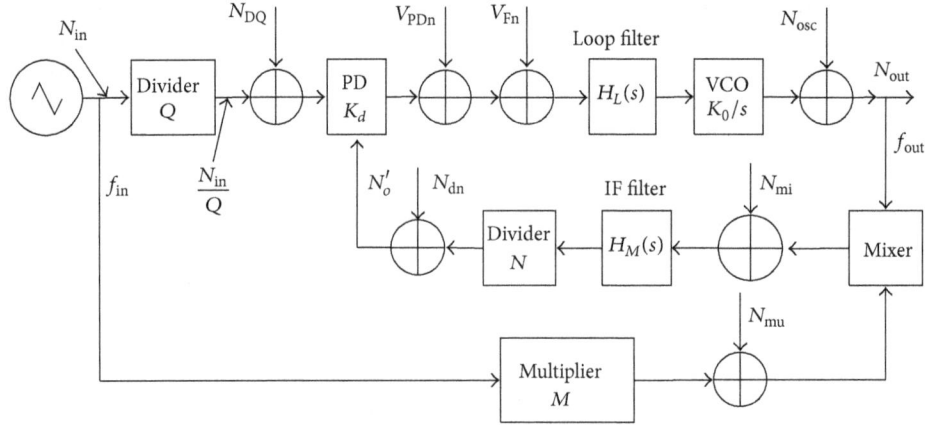

FIGURE 2: Model of a general PLL with additive noise.

3.2.3. Frequency Multiplier.
The phase noise PSD at the output of a frequency multiplier is equal to its input multiplied by the square of multiplication factor plus an additive term, that is,

$$S_{\varphi,\text{mu}}(f) \approx S_{\varphi,\text{mu,in}}(f) \cdot N^2 + \frac{10^{-13\pm2}}{f} + 10^{-16\pm1}. \quad (13)$$

3.2.4. Amplifier.
The output phase noise in the low-frequency operational amplifier implemented with GaAs/GaAlAs heterojunction bipolar transistors is [36]

$$S_{\varphi,\text{amp,IF}}(f) \approx \frac{10^{-13}}{f}. \quad (14)$$

For radio frequency (RF) amplifier noise, generally, only a narrow bandwidth around the carrier is considered; one-half of the thermal white noise contributes to the amplitude noise modulation and the other half to the phase noise modulation. Hence, a theory limit of the phase noise at the RF amplifier output is [36]

$$S_{\varphi,\text{amp,RF}}(f) \approx 4\frac{kTR}{V_{\text{rms}}^2}, \quad (15)$$

where k and T are Boltzmann's constant and temperature in Kelvin, respectively.

3.2.5. VCO.
The VCO phase noise improves as it goes to farther offsets from the carrier. Although there could be more regions with different slopes to the phase noise, a reasonable model for this is to divide this noise into three regions. A fairly general VCO phase noise equation is [37]

$$S_{\varphi,\text{vco}}(f) \approx \frac{f_0^2 \cdot 10^{-11.6}}{f^3 \cdot Q_L^2} + \frac{f_0^2 \cdot 10^{-15.6}}{f^2 \cdot Q_L^2}$$

$$+ \frac{10^{-11}}{f \cdot Q_L^2} + 10^{-15}, \quad (16)$$

where Q_L is the loaded quality factor of the oscillator.

3.2.6. Loop Filter.
As one of the most important parts in the PLL synthesizer, loop filter has various topologies. For distributed radar systems, passive filters are generally recommended, because they have the advantages of lower cost and no active devices to add noise. Moreover, to reduce spur levels, a fourth-order filter is used in this paper, because fourth order and higher-order filters become more practical when the spurs to be filtered are at least 20 times the loop bandwidth [32]. As all resistors create thermal noise, there are two major sources of noise, namely, some types of capacitors and resistors. Typically, the contribution from this resistor noise within the loop bandwidth is negligible. In the case of a resistor, this noise voltage is the thermal noise generated by the resistor. We then have

$$R_{\text{noise}}(R) = \sqrt{4TkR}, \quad (17)$$

where the units are $V/\sqrt{\text{Hz}}$. Since phase noise is normalized to a 1 Hz bandwidth, one can disregard the denominator and consider the units to be in Volts.

3.3. Model of Receiver Noise.
The receiver noise, consisting of thermal noise and the noise collected by the antenna, will introduce both amplitude and phase fluctuation to the synchronization signal. Here, only phase fluctuation is considered; its influence on the signal phase is described by the receiver phase noise spectral density function $S_{\varphi,\text{SNR}}(f)$. For band-limited white Gaussian noise, the PSD of phase noise is related to the SNR (signal-to-noise ratio) through [38]

$$S_{\varphi,\text{SNR}}(f) \approx \frac{1}{2B_w \cdot \text{SNR}}, \quad (18)$$

with the receiver (noise) bandwidth B_w. Note that, the following matched filtering will further improve the receiver SNR.

3.4. Model of Distributed Radar System Frequency Instability.
Since only phase noise is of great interest, the modulation waveform used for range resolution can be ignored, and the distributed radar system can be simplified to an "azimuth

only" system [16]. Suppose that the transmitted signal is sinusoid whose phase argument is

$$s_T(t) = 2\pi f_T t + \varphi_T(t). \tag{19}$$

The first term is the carrier frequency, and the second represents the phase deviations from the error-free carrier which includes the sum of phase noises discussed previously, that is,

$$\varphi_T(t) = \varphi_{\text{sum}}(t) = \varphi_{\text{osc}}(t) + \varphi_{\text{PLL}}(t) + \varphi_{\text{SNR}}(t). \tag{20}$$

After reflection from a target, the received signal phase is that of the transmitted signal delayed by the round-trip time τ. The receiver output signal phase $\widehat{s}(t)$ results from demodulating the received signal with the receiver oscillator which has the same form as the transmitter oscillator:

$$s_R(t) = 2\pi f_R t + \varphi_R(t). \tag{21}$$

Hence, we have

$$\widehat{s}(t) = 2\pi(f_R - f_T)t + 2\pi f_T \tau + \varphi_R(t) - \varphi_T(t). \tag{22}$$

The first term is a frequency offset arising from the non-identical oscillator frequencies. It is not important and can be ignored. The second term forms the usual Doppler term as round-trip time to the target varies it should be preserved. The last two terms represent the frequency synchronization errors which are of interest for extracting the frequency synchronization errors; hence, the phase errors in distributed radar system can be modeled as

$$\phi_B(t) = \varphi_T(t) - \varphi_R(t). \tag{23}$$

It is assumed that $\varphi_T(t)$ and $\varphi_R(t)$ are independent random variables having identical PSD $S_{\varphi,\text{sum}}(f)$; then, the phase synchronization error PSD in distributed radar system is

$$S_{\varphi_B}(f) = 2S_{\varphi,\text{sum}}(f), \tag{24}$$

where the factor 2 arises from the use of two independent oscillators.

4. Prediction of Link Synchronization Accuracy

The synchronization signals must be sufficiently decoupled from the radar signals; otherwise, they may cause problems when using the same carrier frequency. Hence, we suppose the phase arguments of synchronization signal and normal radar signal are given, respectively, by

$$
\begin{aligned}
s_{T,\text{syn}}(t) &= 2\pi f_{T,\text{syn}} t + \varphi_{T,\text{syn}}(t), \\
s_{T,\text{sar}}(t) &= 2\pi f_{T,\text{sar}} t + \varphi_{T,\text{sar}}(t).
\end{aligned}
\tag{25}
$$

Similarly, the first term is the error-free carrier frequency, and the second represents the phase deviations from the error-free carrier. In the same manner like (23), we get that

$$
\begin{aligned}
\phi_{\text{syn}}(t) &= \varphi_{T,\text{syn}}(t) - \varphi_{R,\text{syn}}(t), \\
\phi_{\text{sar}}(t) &= \varphi_{T,\text{sar}}(t) - \varphi_{R,\text{sar}}(t).
\end{aligned}
\tag{26}
$$

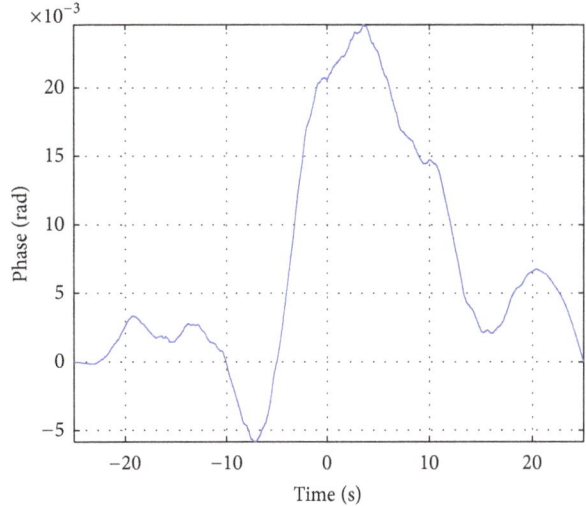

FIGURE 3: Prediction of phase noise in a typical 10 MHz crystal oscillator.

TABLE 1: Phase noise parameters of one typical crystal oscillator.

Frequency offset (Hz)	1	10	100	1 k	10 k
Phase noise PSD (dBc/Hz)	−80	−100	−130	−150	−160

With the proposed synchronization compensation method, the phase compensation term is $\phi_{\text{syn}}(t) \cdot f_{T,\text{sar}}/f_{T,\text{syn}}$. Thus, the synchronization accuracy is decided by

$$\phi_{\text{err}}(t) = \phi_{\text{sar}}(t) - \phi_{\text{syn}}(t) \cdot \frac{f_{T,\text{sar}}}{f_{T,\text{syn}}}. \tag{27}$$

To derive quantitative estimation, we take a typical crystal oscillator, shown in Table 1, as an example. With the model of reference oscillator frequency instability, simulated phase errors in this 10 MHz oscillator are shown in Figure 3 for a time interval of 50 seconds. From its Allan variance shown in Figure 4, we can see that this oscillator can be regarded as a representative example for the ultrastable oscillator of current aerospace radar systems. Note that Figure 3 is the result of one typical simulation, and Figure 4 is the statistical result of many simulations.

From the models of phase noise described in the previous sections, we get the comparative results of the output phase noise between the radar channel and synchronization channel, shown in Figure 5. The observation could be made that, within the loop bandwidth, the PLL phase detector is typically the dominant noise source, and outside the loop bandwidth, the VCO noise is often the dominant noise source. Hence, the performance of the synchronization link will be impacted by the common misconception that the phase noise will vary with $20\log(f_0/f_{\text{osc}})$ with f_{osc} being the oscillator frequency.

Additionally, the phase of the synchronization link may be influenced by receiver noise, analog-to-digital convertor (ADC), and data interpolation. As we have mentioned previously, the receiver noise determined by SNR is of special interest. Furthermore, the synchronization phase is sampled, which requires a later interpolation of the compensation

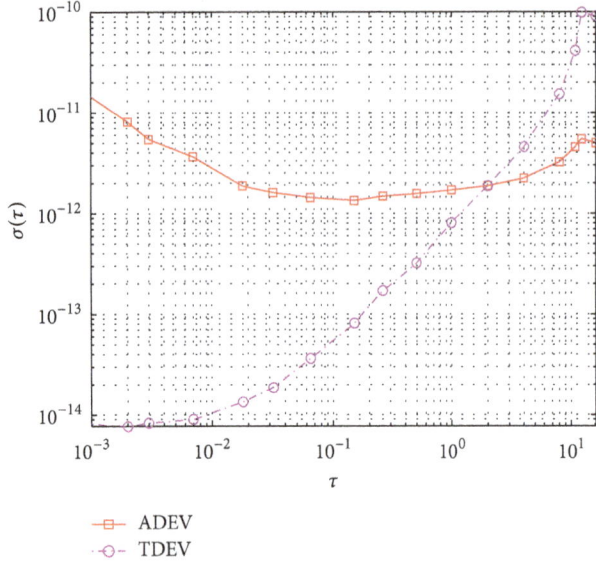

FIGURE 4: Prediction of frequency instability, ADEV: overlapping Allan standard deviations, TDEV: Allan time standard deviations.

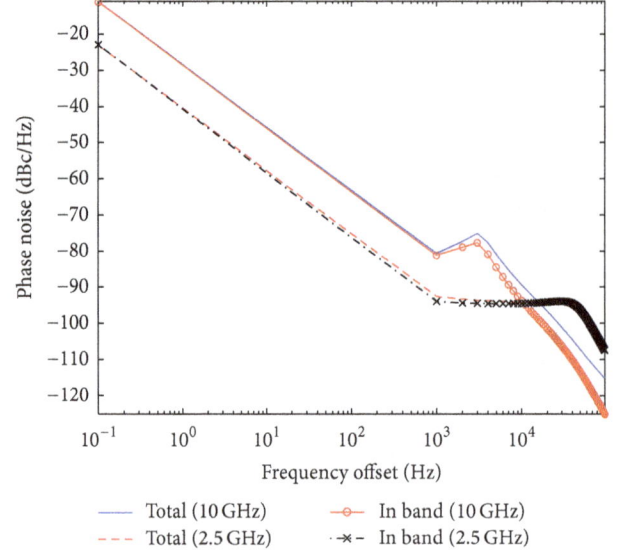

FIGURE 5: Comparative results of output phase noise between synchronization channel and radar channel.

phase. We may choose to filter the compensation phase with an arbitrary transfer function $H_{syn}(f)$ like [26]. Note that, if distributed SAR imaging is considered, the compensated SAR phase (SAR phase after subtracting the compensation phase) is filtered through azimuth compression. This filter is described by the transfer function $H_{az}(f)$ and is dependent on the azimuth processing. The impact of receiver noise on synchronization phase is [26]

$$\sigma_{SNR}^2 = \frac{1}{2 f_{syn} \cdot SNR}$$
$$\times \int_{-f_{syn}/2}^{f_{syn}/2} S_{\varphi,SNR}(f) \left| H_{syn}(f) H_{az}(f) \right|^2 df, \tag{28}$$

where f_{syn} represents the synchronization repeatedly frequency rate.

In the case of digital-to-analog convertor (DAC), the quantization errors result in what appears to be a white noise floor but is actually a "sea" of very finely spaced discrete spurs. For a N-bit DAC, the phase errors due to quantization errors are determined by [39]

$$\varphi_{max} \approx \arctan\left(\frac{1}{2^N - 1}\right). \tag{29}$$

Note that $N = 12$ is assumed in the following simulation.

The interpolation error is because frequency components outside the range $-f_{syn}/2 < f < f_{syn}/2$ are lost due to the sampling and hence cannot be reconstructed. The interpolation variance is [21]

$$\varphi_{int}^2 = 2\left(\frac{f_0}{f_{osc}}\right)^2 \int_{f_{syn}/2}^{\infty} \left| H_{az}(f) \right|^2 df. \tag{30}$$

The factor 2 is due to the use of two independent oscillators, and the scaling factor in the parentheses is due to the

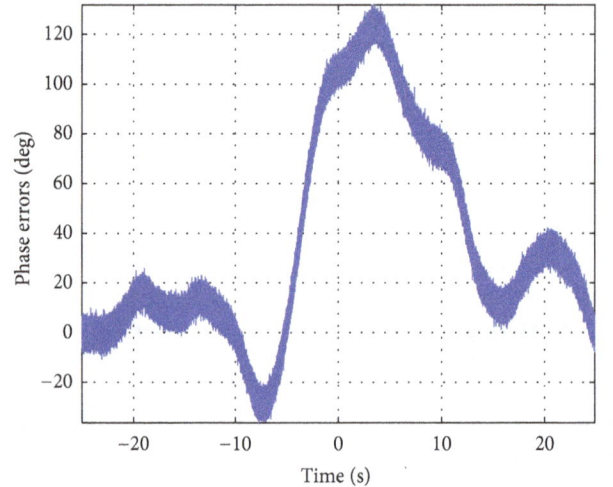

FIGURE 6: Prediction of phase errors via the dedicated synchronization link.

multiplication of the oscillator frequency f_{osc} with (f_0/f_{osc}) to obtain the radar signal with carrier frequency f_0.

Finally, further suppose that the signal bandwidths of radar signal and synchronization signal are 100 MHz and 1 GHz, respectively. Figure 6 shows the prediction of total phase errors contributed by the synchronization link. The prediction of standard deviation (STD) of the phase synchronization error contributed by synchronization link versus synchronization rate is shown in Figure 7. Note that Figure 7 is a statistical result with twenty realizations of the stochastic process described previously. From Figure 7, we can see that successful synchronization error compensation is possible by using this dedicated synchronization link with enough synchronization rate.

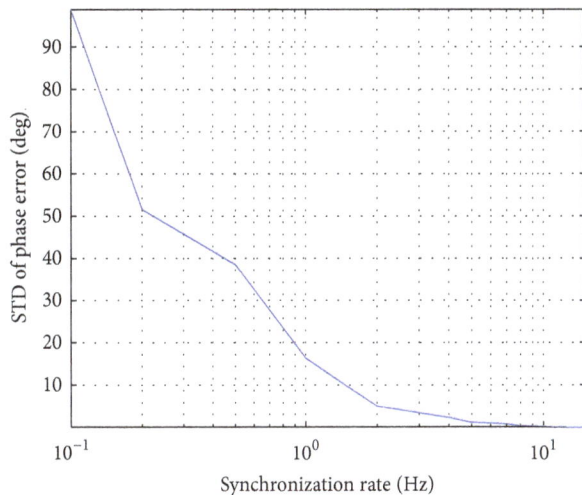

FIGURE 7: Prediction of phase synchronization accuracy versus synchronization repeatedly frequency rate.

5. Conclusion

A dedicated synchronization link is a solution to avert the performance degradation due to oscillator frequency instability in distributed radar system. Hence, the use of a dedicated synchronization link to quantify and compensate oscillator frequency instability is investigated in this paper. With analytical models of phase noise, closed analytic expressions for the link performance are derived. We utilize the knowledge of statistical models, system error contributions, sampling considerations, and signal processing parameters to investigate the residual phase error after synchronization, and the possible error contributions including oscillator, PLL, and receiver noise are quantified. Simulation results show that effective synchronization error compensation is possible by using this dedicated synchronization link. Note that this paper considers mainly radar-related synchronization applications, but the presented method and analysis results are also effective for other distributed wireless systems.

Acknowledgments

This work was supported in part by the National Natural Science Foundation of China under Grant no. 41101317, the Fundamental Research Funds for the Central Universities under Grant no. ZYGX2010J001, and the Program for New Century Excellent Talents in University under Grant no. NCET-12-0095.

References

[1] A. Ozgur, O. Leveque, and D. Tse, "Spatial degrees of freedom of large distributed MIMO systems and wireless Ad Hoc networks," IEEE Journal on Selected Areas in Communications, vol. 31, no. 2, pp. 202–214, 2013.

[2] J. A. Jackson, B. D. Rigling, and R. L. Moses, "Canonical scattering feature models for 3D and bistatic SAR," IEEE Transactions on Aerospace and Electronic Systems, vol. 46, no. 2, pp. 525–541, 2010.

[3] V. T. Muralidharan and B. S. Rajan, "Distributed space time coding for wireless two-way relaying," IEEE Transactions on Signal Processing, vol. 61, no. 4, pp. 980–991, 2013.

[4] W. Q. Wang, "GPS-based time & phase synchronization processing for distributed SAR," IEEE Transactions on Aerospace and Electronic Systems, vol. 45, no. 3, pp. 1040–1051, 2009.

[5] Y. F. Chen, Y. Nijsure, C. Yuen, Y. H. Chew, Z. G. Ding, and B. Said, "Adaptive distributed MIMO radar waveform optimization based on mutual information," IEEE Transactions on Aerospace and Electronic Systems, vol. 49, no. 2, pp. 1374–1385, 2013.

[6] W. Q. Wang, "Distributed passive radar sensor networks with near-space vehicle-borne receivers," IET Wireless Sensor Systems, vol. 2, no. 3, pp. 183–190, 2012.

[7] C. C. Liu and W. D. Chen, "Sparse self-calibration imaging via iterative MAP in FM-based distributed passive radar," IEEE Geoscience and Remote Sensing Letters, vol. 10, no. 3, pp. 538–542, 2013.

[8] W. Wang and J. Cai, "A technique for jamming Bi- And multistatic SAR systems," IEEE Geoscience and Remote Sensing Letters, vol. 4, no. 1, pp. 80–82, 2007.

[9] Y. Wang and G. Noubir, "Distributed cooperation and diversity for hybrid wireless networks," IEEE Transactions on Mobile Computing, vol. 12, no. 3, pp. 596–608, 2013.

[10] W. L. Zhang, Q. Y. Yin, H. Y. Chen, F. F. Gao, and N. Ansari, "Distributed angle estimation for localization in wireless sensor networks," IEEE Transactions on Wireless Communications, vol. 12, no. 2, pp. 527–537, 2013.

[11] F. De Rango, F. Guerriero, and P. Fazio, "Link-stability and energy aware routing protocol in distributed wireless networks," IEEE Transactions on Parallel and Distributed Systems, vol. 23, no. 4, pp. 713–726, 2012.

[12] M. Wendler, "Results of a bistatic airborne SAR experiment," in Proceeding of the International Radar Conference, pp. 247–253, September 2003.

[13] T. Espeter, I. Walterscheid, J. Klare, A. R. Brenner, and J. H. G. Ender, "Bistatic forward-looking SAR: results of a spaceborne-airborne experiment," IEEE Geoscience and Remote Sensing Letters, vol. 8, no. 4, pp. 765–768, 2011.

[14] I. Walterscheid, T. Espeter, A. R. Brenner et al., "Bistatic SAR experiments with PAMIR and TerraSAR-X-setup, processing, and image results," IEEE Transactions on Geoscience and Remote Sensing, vol. 48, no. 8, pp. 3268–3279, 2010.

[15] A. S. Goh, M. Preiss, N. J. S. Stacy, and D. A. Gray, "Bistatic SAR experiment with the Ingara imaging radar," IET Radar, Sonar and Navigation, vol. 4, no. 3, pp. 426–437, 2010.

[16] J. L. Auterman, "Phase stability requirements for a bistatic SAR," in Proceedings of the IEEE National Radar Conference, pp. 48–52, May 1984.

[17] M. Weiß, "Time and phase synchronization aspects for bistatic SAR systems," in Proceedings of the European Synthetic Aperture Radar Conference, pp. 395–398, 2004.

[18] W. Q. Wang, Multi-Antenna Synthetic Aperture Radar, CRC Press, London, UK, 2013.

[19] B. J. Choi, H. Liang, X. M. Shen, and W. H. Zhuang, "DCS: distributed asynchronous clock synchronization in delay tolerant networks," IEEE Transactions on Parallel and Distributed Systems, vol. 23, no. 3, pp. 491–504, 2012.

[20] M. Leng and Y. C. Wu, "Distributed clock synchronization for wireless sensor networks using belief propagation," IEEE Transactions on Signal Processing, vol. 59, no. 11, pp. 5404–5414, 2011.

[21] G. Krieger, M. R. Cassola, M. Younis, and R. Metzig, "Impact of oscillator noise in bistatic and multistatic SAR," in *Proceedings of the IEEE International Geoscience and Remote Sensing Symposium (IGARSS '05)*, pp. 1043–1046, July 2005.

[22] P. Ubolkosold, S. Knedlik, and O. Loffeld, "Estimation of oscillator's phase offset, frequency offset and rate of change for bistatic interferometric SAR," in *Proceedings of the European Synthetic Aperture Radar Conference*, pp. 1–4, 2006.

[23] X. Zhang, H. Li, and J. Wang, "The analysis of time synchronization error in bistatic SAR system," in *Proceedings of the IEEE International Geoscience and Remote Sensing Symposium (IGARSS '05)*, pp. 4619–4622, July 2005.

[24] W. Q. Wang, C. B. Ding, and X. D. Liang, "Time and phase synchronisation via direct-path signal for bistatic synthetic aperture radar systems," *IET Radar, Sonar and Navigation*, vol. 2, no. 1, pp. 1–11, 2008.

[25] M. Eineder, "Ocillator Clock Drift Compensation in Bistatic Interferometric SAR," in *Proceedings of the IEEE International Geoscience and Remote Sensing Symposium (IGARSS '03)*, pp. 1449–1451, July 2003.

[26] M. Younis, R. Metzig, and G. Krieger, "Performance prediction of a phase synchronization link for bistatic SAR," *IEEE Geoscience and Remote Sensing Letters*, vol. 3, no. 3, pp. 429–433, 2006.

[27] D. W. Allan, "Statistics of atomic frequency standards," *Proceedings of the IEEE*, vol. 54, no. 2, pp. 221–230, 1966.

[28] W. Q. Wang, J. Y. Cai, and Y. W. Yang, "Extracting phase noise of microwave and millimetre-wave signals by deconvolution," *IEE Proceedings*, vol. 153, no. 1, pp. 7–12, 2006.

[29] J. Rutman, "Characterization of phase and frequency instabilities in precision frequency sources: fifteen years of progress," *Proceedings of the IEEE*, vol. 66, no. 9, pp. 1048–1073, 1978.

[30] G. J. Foschini and G. Vannucci, "Characterizing filtered light waves corrupted by phase noise," *IEEE Transactions on Information Theory*, vol. 34, no. 6, pp. 1437–1448, 1988.

[31] V. F. Kroupa, "Noise properties of PLL systems," *IEEE Transactions on Communications*, vol. 30, no. 10, pp. 2244–2252, 1982.

[32] D. Banerjee, *PLL Performance, Simulation, and Design*, National Semiconductor, 2006.

[33] W. Cheng, A. J. Annema, J. A. Croon, and B. Nauta, "Noise and nonlinearity modeling of active mixers for fast and accurate estimation," *IEEE Transactions on Circuits and Systems I*, vol. 58, no. 2, pp. 276–289, 2011.

[34] D. G. Meyer, "A test set for the accurate measurement of phase noise on high-quality signal sources," *IEEE Transactions on Instrumentation and Measurement*, vol. IM-19, no. 4, pp. 215–227, 1970.

[35] V. F. Kroupa, "Jitter and phase noise in frequency dividers," *IEEE Transactions on Instrumentation and Measurement*, vol. 50, no. 5, pp. 1241–1243, 2001.

[36] V. F. Kroupa, *Direct Digital Frequency Synthesizers*, IEEE Press, New York, NY, USA, 1998.

[37] V. F. Kroupa, "Low-noise microwave-frequency synthesisers design principles," *IEE Proceedings H: Microwaves Optics and Antennas*, vol. 130, no. 7, pp. 483–488, 1983.

[38] F. G. Stremler, *Introduction to Communication Systems*, Addison-Wesley, Reading, Mass, USA, 1982.

[39] M. M. Abousetta and D. C. Cooper, "Noise analysis of digitized FMCW radar waveforms," *IEE Proceedings: Radar, Sonar and Navigation*, vol. 145, no. 4, pp. 209–215, 1998.

3

CIB: An Improved Communication Architecture for Real-Time Monitoring of Aerospace Materials, Instruments, and Sensors on the ISS

Michael J. Krasowski,[1] **Norman F. Prokop,**[1] **Joseph M. Flatico,**[2] **Lawrence C. Greer,**[1] **Phillip P. Jenkins,**[3] **Philip G. Neudeck,**[1] **Liangyu Chen,**[2] **and Danny C. Spina**[4]

[1] *NASA Glenn Research Center, 21000 Brookpark Road, Cleveland, OH 44135, USA*
[2] *Ohio Aerospace Institute, NASA Glenn Research Center, 21000 Brookpark Road, Cleveland, OH 44135, USA*
[3] *U. S. Naval Research Laboratory, 4555 Overlook Avenue SW, Washington, DC 20375, USA*
[4] *Jacobs Technology, NASA Glenn Research Center, 21000 Brookpark Road, Cleveland, OH 44135, USA*

Correspondence should be addressed to Norman F. Prokop; norman.f.prokop@nasa.gov

Academic Editors: T. E. Matikas and M. R. Woike

The Communications Interface Board (CIB) is an improved communications architecture that was demonstrated on the International Space Station (ISS). ISS communication interfaces allowing for real-time telemetry and health monitoring require a significant amount of development. The CIB simplifies the communications interface to the ISS for real-time health monitoring, telemetry, and control of resident sensors or experiments. With a simpler interface available to the telemetry bus, more sensors or experiments may be flown. The CIB accomplishes this by acting as a bridge between the ISS MIL-STD-1553 low-rate telemetry (LRT) bus and the sensors allowing for two-way command and telemetry data transfer. The CIB was designed to be highly reliable and radiation hard for an extended flight in low Earth orbit (LEO) and has been proven with over 40 months of flight operation on the outside of ISS supporting two sets of flight experiments. Since the CIB is currently operating in flight on the ISS, recent results of operations will be provided. Additionally, as a vehicle health monitoring enabling technology, an overview and results from two experiments enabled by the CIB will be provided. Future applications for vehicle health monitoring utilizing the CIB architecture will also be discussed.

1. Introduction

The International Space Station is a unique space vehicle in that it is currently the largest artificial satellite in orbit around the Earth. The U S portion of the ISS has been designated as a national laboratory by the Congress. The ISS provides a unique environment of extreme hot-cold thermal cycling, cosmic radiation exposure [1], atomic oxygen presence, vacuum, and microgravity. This allows for long duration experiments and space testing of devices and structures. While testing and experiments take advantage of this unique environment, facility equipment must operate reliably in it. Electronic components and integrated circuits (IC) are especially susceptible to radiation effects from the environment in LEO. These effects can present themselves in two ways: long-term

dose damage associated with total ionizing dose (TID) or through single event effects (SEE). Long-term or TID results in permanently damaging an IC by altering the crystal lattice of the semiconductor, which can result in changing bias voltages and currents which affect circuit operation. Where single event upsets (SEU) are transient, energy is transferred from ionizing particles to the IC. This energy transfer is localized, so individual transistors on an IC are affected. Single events caused by radiation may only result in the flip of a single bit or the corruption of an analog signal. If an affected bit is part of an instruction for microcontroller or processor, the result could result in operational failure. Further, an SEE-induced phenomenon known as a single event latchup (SEL) could also result in a loss of data but, in extreme cases, may cause a hard destructive failure which could result in the permanent loss of a circuit component. With this in mind, care must be taken in the design of any electronic system expected to operate reliably on the ISS for extended period of time and be tolerant of the expected radiation environment.

To support science payloads, the ISS as a facility provides three telemetry communications interfaces [2] for resident experiments with its associated physical layer/protocol: low rate data link: MIL-STD-1553; medium rate data link: Ethernet; and high rate data link: fiber optic. Each interface can provide telemetry to the ground. Increased bandwidth comes with increased cost in development to meet the physical interface requirements. The only interface available throughout the ISS is the MIL-STD-1553 bus. In addition to telemetry data, the Mil-STD-1553 bus performs the command, health, and status data transfer. This is done for safety reasons, so that health and status data are transmitted on the most reliable communication bus.

The MIL-STD-1553 [3] "Aircraft Internal Time Division Command/Response Multiplex Data Bus" defines a physical layer as well as a bus protocol. The MIL-STD-1553 bus is highly reliable and robust in that it is a deterministic command and response protocol. The ISS adds additional layers to those of the MIL-STD-1553 standard, in effect making the ISS MIL-STD-1553 implementation a superset of the military standard. Which allows ISS MIL-STD-1553 hardware to interface with other space and aircraft platforms, but not necessarily the converse. The MIL-STD-1553 bus is deployed on numerous U S military aircrafts including the F-16 Falcon, F/A-18 Hornet, AH-64 Apache, and P-3C Orion and has also been adopted by North Atlantic Treaty Organization (NATO). The MIL-STD-1553 bus transmits data at 1 Mbit/second. The command and response protocol of the bus adds overhead, reducing the effective bit rate for data transfer. In addition, the ISS adds overhead to the bus transfers. One layer added by ISS to the MIL-STD-1553 telemetry is the Consultative Committee for Space Data Systems (CCSDS) headers, which allow for data telemetry routing in space. Each node in the telemetry system is provided with a unique application identifier (APID), which is part of the CCSDS header, to enable routing. These APIDs allow a user on the ground to receive telemetry and command their node from anywhere with internet connectivity. The ISS allots 12.8 kBytes of telemetry data per second, some of which is used for overhead. The ISS allows commands from the ground to be routed to APIDs residing on the MIL-STD-1553 bus.

Onboard the ISS, there is an Express Logistics Carrier (ELC) facility which is primarily designed to store ISS replacement parts but also has two science payload slots per platform [4]. Each science payload site has 28 V and 120 V power available. A MIL-STD-1553 communications repeater (bus controller) link to the ISS is also provided by the ELC, allowing telemetry data to pass through.

The CIB was needed to be the communications backbone for a permanent testbed on the ELC of the ISS which would utilize the ISS telemetry of the MIL-STD-1553 bus. The CIB needed to provide simpler communication interface to experimenters while maintaining the reliability expected of a space flight system on the ISS. The CIB needed to be designed with radiation tolerance and reliability as primary considerations. The CIB was designed by the NASA Glenn Research Center (GRC) Mobile and Remote Sensing (MaRS) Laboratory. The first experiment set to utilize the testbed was the Materials on the International Space Station Seven (MISSE7) [5]. MISSE7 was comprised of numerous passive experiments and over 21 active experiments using the command and telemetry capabilities provided by the CIB. MISSE7 used two Extra-Vehicle Activity (EVA) deployable suitcase-like containers called a Passive Experiment Container (PEC) in which numerous experiments are contained. These PECs are carried by astronauts, opened, and mounted on the ELC. When the experiment was completed, an additional EVA was used to retrieve the PECs and install a new one for the follow-on experiment set, MISSE8. This exchange required a physical communications link which was swappable. The need for on-orbit exchange of PECs required a physical communications link robust enough to withstand potential damage encountered during the swap.

The CIB was installed during Space Transportation System-129 (STS-129) and is currently flying on the exterior of the International Space Station mounted on the Express Logistics Carrier 2 (ELC-2). The CIB provides a simple, RS-485-based communications interface between experiments, instruments, or sensor systems and the ELC's MIL-STD-1553 bus. This allows developers to design on simple platforms without having to confront the difficulty of integrating their experiments directly into the ELC or the ISS. The CIB currently provides serial communication supporting twenty active systems. Figure 1 shows the CIB within the MISSE7 communication architecture including the interface to the ISS telemetry through the ELC. Follow-on experiments (MISSE8) leave the CIB in place and swap out one or both of the allowed PECs.

This paper will discuss in detail the technical development and design of the CIB hardware and architecture. Recent results from current operations aboard the ISS will be provided. Specific experiments supported by the CIB like a Silicon Carbide Junction Field Effect Transistor (SiC JFET) health monitoring and a solar cell health monitoring will be discussed. Future health monitoring applications to include node to node schemes will also be briefly discussed.

2. CIB Design

The CIB is designed to be a highly reliable and radiation hard communications bridge from the ISS/ELC MIL-STD-1553 to onboard experiments, sensors, and health monitoring systems. The CIB is constructed from components designed or known to be radiation tolerant of LEO conditions for over 20 years of operation. Further, the CIB is designed to be tolerant of the electrostatic discharge (ESD) anticipated to occur during removal and insertion of new payload systems over multiple missions.

Communications over each RS-485 bus was limited to 9600 baud, which is 9600 bits of data per second for eight bit words. Though theoretically capable of at least ten times· this bandwidth, the CIB was designed for 9600 baud as a consequence of some experiments in MISSE7 being only capable of that rate due to hardware or software constraints. Thus,

CIB: An Improved Communication Architecture for Real-Time Monitoring of Aerospace Materials, Instruments, and Sensors on the ISS

19

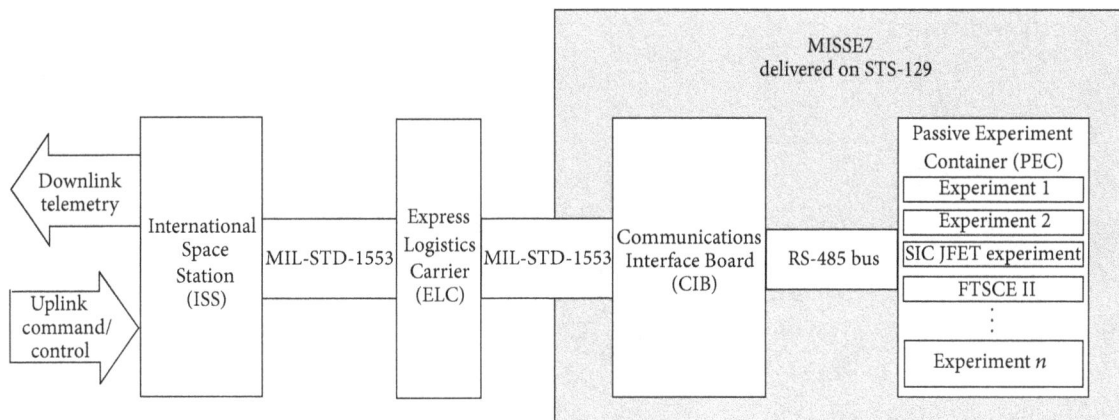

FIGURE 1: Diagram of the MISSE7 communication architecture and interface to the International Space Station telemetry through the Express Logistics Carrier.

the lowest common denominator dictated bus speed. A reprogrammed CIB could support higher baud rates and could dynamically change baud rate to accommodate each experiment. A block diagram of the CIB interfaced to a string of experiment systems is given below in Figure 2. Note that all systems interface to the CIB via a full duplex RS-485 bus system and each system has associated with it a unique hardware enable line.

3. Hardware Interface

The RS-485 standard specifies a multidrop serial bus which can be full or half-duplex. The CIB implements a full duplex multidrop bus, where each experiment or sensor is a stub or drop on the bus, and reception and transmission happen on separate lines. Since MISSE7 was to support two physically separate experiment containers, the designers provided two RS-485 buses from the CIB, one for each container. The multidrop bus provides a concern for reliability as one errant transceiver can corrupt the bus. To mitigate against this risk, the CIB implements a hardware transmission enable, as well as requiring a software enable function on the experiment side.

Experimenters must provide a hardware handshaking interface as shown in Figure 3, wherein the enable line from the CIB is shown schematically to provide one half of a signal set necessary to enable a system's transmit hardware. The second half of the signal set is provided by the system through response to an initiating packet transfer from the CIB on the RS-485 bus containing the system's APID. This interface prevents experiments from transmitting without being selected and communicated to by the CIB. The interface must be implemented in hardware so that the experiment cannot interfere with the RS-485 bus during communications slots not associated with it. The cause of this out of order bus use could be the result of a software failure, for example. The CIB maintains a permission bitmap that can be overwritten by a command from the ground. If a given experiment is locked out via the bitmap (for any number of scheduling reasons or after detection of system failure), its enable is never asserted and is prevented from communicating.

This transmission scheme allows for the full duplex hardware Universal Asynchronous Receiver Transmitter (UART) common to most microcontrollers/microcomputers/computers or which can be easily configured into programmable logic or as a software UART routine on microcontrollers which allow for interrupt on transition. Thus, systems may be easily configured using simple and available hardware and design tools.

The byte format is 8 bit, one start bit, stop bit, and no parity. The baud rate is 9600 which simplifies requirements for the simpler hardware and offer forgiveness in timing and deskew.

The CIB rotates around through its list of 20 APIDS, polling each experiment one at a time, in order, as long as power is applied. This method is chosen as the desired way to maximize bandwidth for the experiments. If a system needs to pass more than one packet set of data, it merely waits until the CIB returns to it at a later time.

In addition to the transmission lines (TX), receive lines (RX), +5 Volt power, and ground are the enable lines for the experiments. Within the example experiment hardware is the logic to enable failsafe transmission on the shared RS-485 bus as required by the CIB Interface Control Document [6]. The separate enable wire provided to each experiment is detailed in the following illustration in Figure 4.

An enable line is pulled low for a particular system prior to the CIB transmitting to that system. A 10 Volt transient absorber (dual redundant 5 volt transient absorbers) and a $1\,k\Omega$ resistor occupy the output of each enable line as shown in Figure 4 and are present to mitigate against electrostatic discharge and also to a short to +28 condition should it occur at an experiment. As noted earlier, the enable wire is used by the system to enable its transmitter hardware if it has also received a valid packet from the CIB. Commencement of the packet transfer from CIB occurs no less than 250 ms from the assertion of enable. This enable signal removes the obligation of the systems from having to monitor all the traffic on the TX line while listening for transmissions dedicated to themselves.

Also, this active low enable signal can be used to locally enable a pass element to provide the system power. Thus, an experiment may come alive upon a powerup initiated by this

Dual redundant MIL-STD-1553 bus to ISS

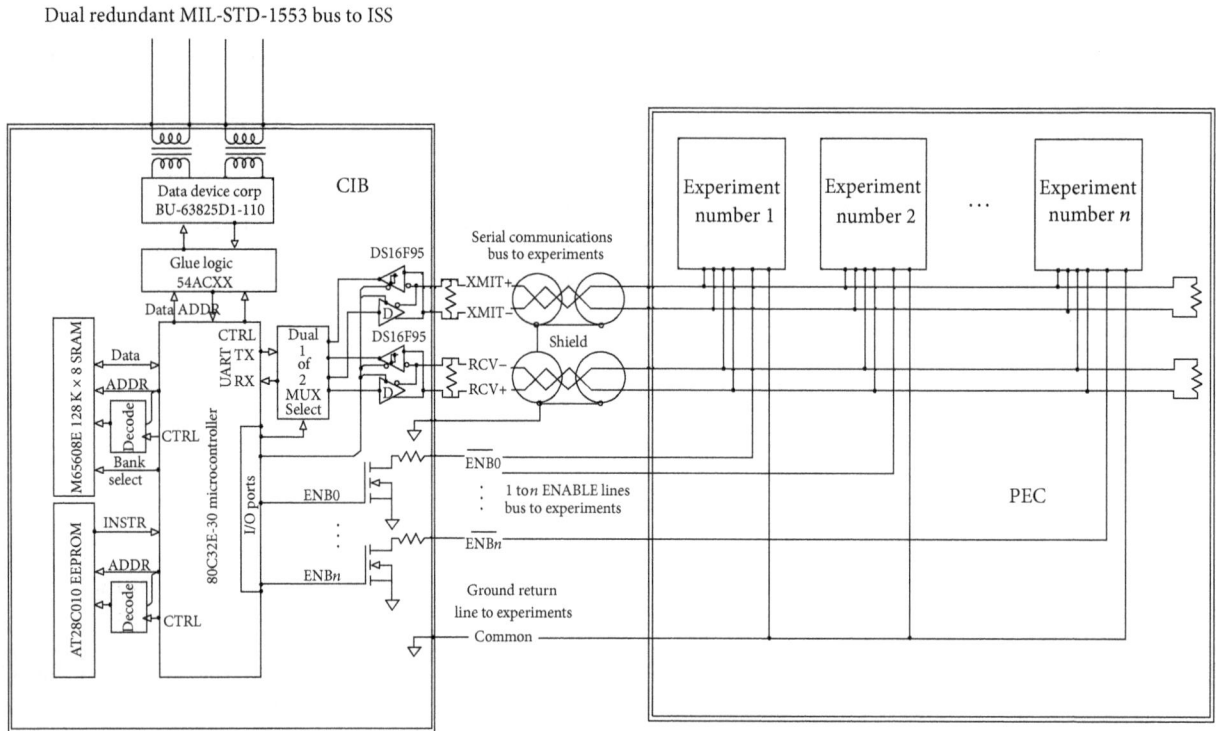

FIGURE 2: A block diagram of the Communications Interface Board (CIB) interfacing between the International Space Station (ISS) MIL-STD-1553 bus and the experiments residing on the Passive Experiment Container (PEC) (image courtesy of NASA).

An example schematic for experiment n stub onto the PEC RS485 bus and to its ENABLE

using the DS16F95

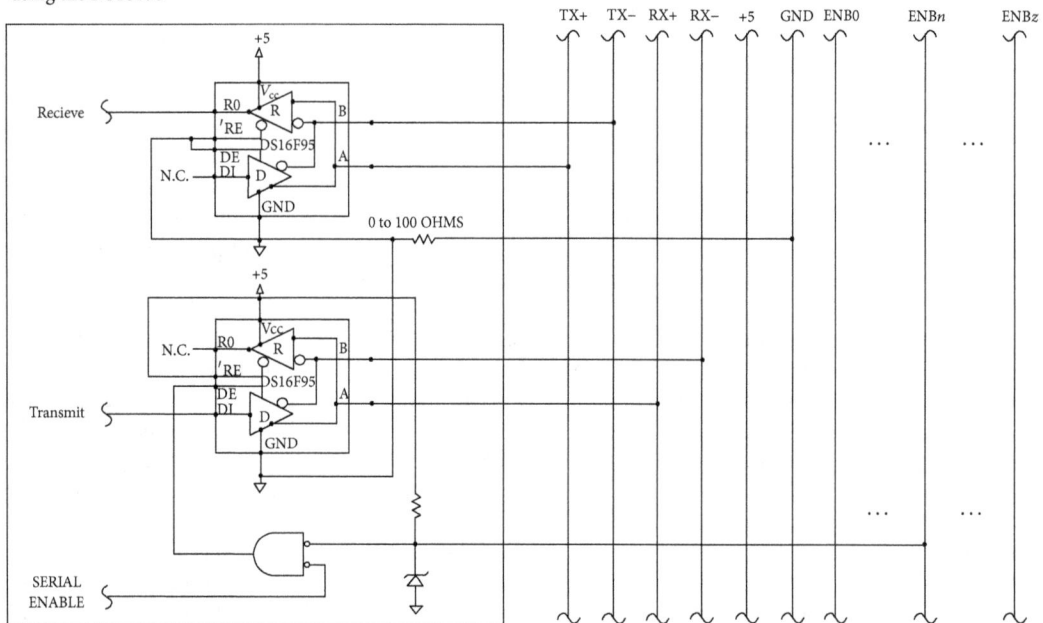

Note: the above NOR gate represents a NOR function

and does not necessarily specify the use of a NOR gate

SERIAL ENABLE is generated by experiment n upon valid

handshake with CIB after assertion of ENBn

NASA GRC flight electronics lab		
Stubbing in		
Krasowski	Rev 1.0 1/8/2008	1 of 1

FIGURE 3: Schematic diagram of the RS-485 bus interface provided to experiments by Communications Interface Board (CIB) (image courtesy of NASA).

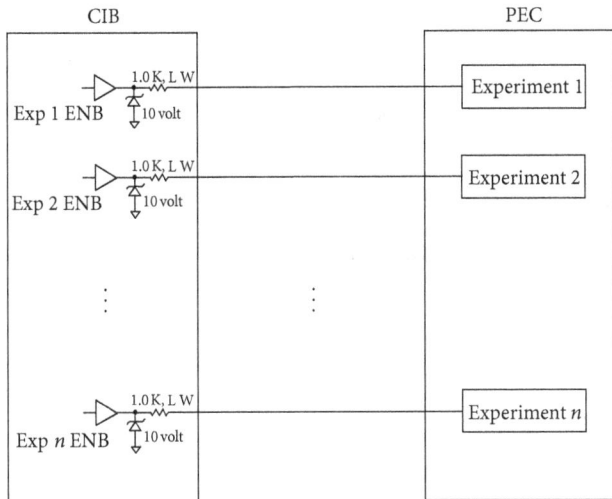

FIGURE 4: Schematic of the CIB enable lines for experiments onboard the Passive Experiment Containers (PECs). Current limiting resistors as well as voltage limiting Zender diodes to protect CIB enable circuitry from experimental failure are shown within the CIB (image courtesy of NASA).

signal. The systems are expected to be capable of accepting a packet from the CIB 250 milliseconds from the assertion of the enable, and so boot-up must occur consistent with this delay. Failure to receive this command and reply after 250 ms results in the CIB disasserting the enable line and moving on to the next experiment. This is of consequence to a system which uses the time between sequential acquisitions of itself by the CIB to perform its tasks. After completion of task and data transfer with the CIB, a system may turn itself off. Removal of power when not operating reduces some radiation total dose effects and can also remove the conditions for and thus clear a nondestructive latchup.

Each transfer starts with a packet from the CIB containing the experiments APID, and so theoretically experiments do not need to consider the state of an enable line but merely listen for their APID. Given this fact and since the two communications busses are RS-485 links, with proper wiring considerations, over thirty experiments per RS-485 bus are possible with a reprogrammed CIB.

4. Software Description

The CIB firmware is built on a commercially available real-time kernel. The firmware consists of serial input/output, CCSDS packet building, time keeping, and MIL-STD-1553 support modules in addition to top-level modules that manage data handling and poll experiments.

Data handling is straightforward. Experimenters have only to implement a very simple protocol to enable communication. All packets have a fixed length. Command packets are 116 bytes long, data packets are 1274 bytes long, and acknowledgment packets are 10 bytes long [6]. The CIB will query experiments one at a time. Each query has the form in Box 1.

The firmware packs experiment data into CCSDS packets for transmission and unpacks commands received from the

ground. This relieves developers from the task of implementing CCSDS packet handling.

Experiments can be controlled through commands of up to 104 bytes in length. No processing is done on commands by the CIB; they are simply passed to the experiment. Commands may contain any kind of data including firmware updates. Experiments are polled in round-robin fashion, each transaction requiring roughly 2.5 seconds. Experiments can transmit up to 2 MB/day. An experiment may receive a command from the ground or transmit data during a single transaction. Data from the experiment is wrapped in a CCSDS packet and buffered in the MIL-STD-1553 transceiver.

MIL-STD-1553 transactions occur in frames. The minor frames contain different transactions. A major frame consists of minor frames. The major frames are repeated. The ELC uses ten 100 ms minor frames for each major frame. The result is that data packets received from experiments are buffered a maximum of one second before being transmitted on the MIL-STD-1553 bus. In addition to passing data, the MIL-STD-1553 module initializes the transceiver, updates the time, and transmits the health and status packets of Table 1 once per major frame.

5. CIB Hardware

The core of the CIB is the 80C32E-5962-0051801QQC radiation tolerant 8 Bit ROMless microcontroller. This performs all processing functions within the system. The 80C32E has the following environmental operating specifications [7]: temperature range is military ($-55°$C to $125°$C), no single event latch-up (SEL) below a linear energy transfer (LET) threshold of 80 MeV/mg/cm^2, and is tested up to a total dose of 30 krads (Si) according to MIL-STD-883 method 1019.

Program memory for the CIB is stored in the AT28C010-12DK-MQ-128Kx8 parallel EEPROM which is reprogrammable in circuit at the bench but which is not reprogrammable on-orbit. The AT28C010 has the following environmental operating specifications [8]: temperature range is military ($-55°$C to $125°$C), no SEL below an LET threshold of 80 MeV/mg/cm^2, and is tested up to a total cose of (according to MIL-STD-883 method 1019): 10 krad (Si) read-only mode when biased and 30 krad (Si) read-only mode when unbiased.

Dynamic memory for the CIB is embodied in an M65608E-5962-8959818MZC radiation tolerant 128Kx8 very low-power Complementary Metal Oxide Semiconductor Static Random Access Memory (CMOS SRAM). The M65608M has the following environmental operating specifications [9]: military temperature range is ($-55°$C to $+125°$C), no SEL below a LET threshold of 80 MeV/mg/cm^2 @ $125°$C, and is tested up to a total dose of 30 krad (Si) according to MIL-STD-883 method 1019.

MIL-STD-1553 communications are effected through the BU-63825D1-300 Space Advanced Communication Engine (Sp'ACE II) BC/RT/MT interface module. The Sp'ACE II has the following environmental operating specifications [10]: total gamma dose immunity of 1×10^6 Rad, LET threshold of 63 MeV/mg/cm^2, and a soft error rate of 2.56×10^{-5} errors/device-day.

/ENBn for experiment *n* is asserted

After 250 milliseconds the CIB will transmit:
SOP   CRC   Address   Timestamp   Packet Type   Command or Padding

If the above packet is valid the experiment will transmit:
SOP   CRC   Address   Timestamp   Packet Type   Data and/or Padding

If the CIB receives a valid packet from the experiment the CIB will acknowledge the packet:
SOP   CRC   Address   Timestamp   Packet Type = ACK

Box 1

TABLE 1: Health and status packet format transmitted once per major frame (once per second).

Type	Variable	Description
Unsigned char	opto[2]	State of optoisolators that determine whether the AO and FTSCE experiments are enabled
Unsigned long	last_poll_time	Beginning of the current polling cycle
Unsigned short	poll_cycles	Number of completed polling cycles
Unsigned char	poll_map[20]	Determines which experiments are to be polled
Unsigned short	sequence[20]	Packet sequence number of each of the experiments
Unsigned char	valid_crc[20]	0×FF = valid
Unsigned long	last_cmd_time[20]	Time last experiment command received
Unsigned long	last_cib_cmd_time	Time last CIB command received
Unsigned char	valid_ack[20]	0×FF = valid
Unsigned short	cmd_cnt[20]	Count of commands sent to experiments
Unsigned short	cib_cmd_cnt	Count of commands sent to CIB
Unsigned short	err_cmd	Command for nonpolled APID. Error = APID. 0×FFFF = no error
Unsigned short	err_poll	Invalid polling table received. Unused
Unsigned char	last_cmd[106]	Last command transmitted to experiments
Unsigned char	last_cib_cmd[106]	Last command transmitted to the CIB
Unsigned char	last_ftsce_cmd[100]	Last command sent to FTSCE
Unsigned char	ftsce_data[124]	Data transmitted by FTSCE
Unsigned char	reserved[64]	

The core components of the CIB hardware are listed in Table 2 along with their relevant environmental specifications.

MIL-STD-1553 coupling transformers and the two clock crystals are screened for space applications. All other active components are deemed radiation hard to this mission through consultation with the customer. A photograph of the CIB is given in Figure 5. The radiation environment in LEO consists primarily of trapped protons with some galactic cosmic rays and solar particles. The expected and agreed to limit for total ionizing dose for the CIB was less than 300 rad(si) per year (behind shielding). The CIB componentry is accepted to be SEU tolerant and SEL hard to the ISS LEO radiation environment. As such, the CIB was designed for long duration survival to the LEO radiation environment.

6. Results

The Communications Interface Board, or CIB, embodies the first demonstration of a permanent communications and control interface for deployable experiments to characterize materials, systems, and components exposed to LEO. The CIB was and is the core of two of the earliest science payloads on the ELC affixed to the external structure of the ISS as part of an Express Payload Adapter (ExPA). These two experiments, deployed by astronauts during EVA, are MISSE7 and -8. MISSE7 was deployed, had a successful mission, and was subsequently returned to Earth. MISSE8 remains on orbit at the time of this writing. The CIB was delivered to the ISS on STS-129 in November 2009 along with MISSE7. Table 3 lists experiments which use command and telemetry data provided by the CIB. MISSE7 then returned when MISSE8 was delivered in May 2011, with the CIB remaining aboard to provide the simplified telemetry interface for MISSE8. Table 4 lists MISSE8 experiments utilizing command and telemetry data enabled by the CIB. MISSE8 is scheduled for return in March 2014.

The original intent for the CIB was to embody the communications component of a MISSE-specific infrastructure capable of supporting two PECs with up to 20 separate experiments with power, uplink, and downlink capabilities. A PEC is a suitcase-like metal container which, when opened

CIB: An Improved Communication Architecture for Real-Time Monitoring of Aerospace Materials, Instruments, and Sensors on the ISS

23

TABLE 2: CIB core system components and respective environmental operating specifications.

Function	Part number	Temperature range	Single event latchup (SEL) threshold	Total radiation dose
Microcontroller	80C32E	−55°C to 125°C	>80 MeV/mg/cm^2	30 krad (Si)
Program memory	AT28C010	−55°C to 125°C	>80 MeV/mg/cm^2	10 krad (Si) under bias 30 krad (Si) when unbiased
Dynamic memory	M65608E	−55°C to 125°C	>80 MeV/mg/cm^2	30 krad (Si)
MIL-STD-1553 interface module	BU-63825		Immune	1 Mrad

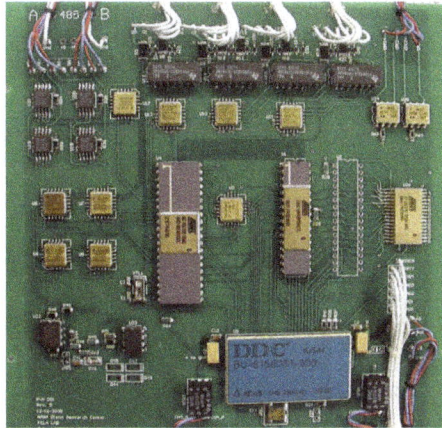

FIGURE 5: Photograph of the flight Communications Interface Board (CIB) circuit board. This image was taken prior to delivery, during functional testing of the circuit board, and prior to the insertion of the flight MIL-STD-1553 transceiver (image courtesy of NASA).

and deployed, presents two opposing experiment surfaces. Two PECs can thus provide zenith and nadir along with ram and wake presentations. This infrastructure greatly reduces the ISS interface complexity for future MISSE experiments. Experimenters would thus have a well-defined power and communication interface to the ISS to build to as well as an EVA compatible mechanical structure for deploying two MISSE style PECs. Further, continuous monitoring of components and samples while on orbit removes the requirement of an experiment returning to Earth for postmission analysis. As such, at end of mission, experiments may be disposed of via deorbiting as will be the case for part of MISSE8.

The decommissioning of the space shuttle brought with it a loss of EVA deployable experiments, and as such the PEC structure is inconsistent with new requirements for robotic deployment of science payloads and environmental sensors for insertion onto the ELCs. However, the CIB arguably represents a communication and control interface for a system consistent with current specifications.

While the CIB hardware currently residing on ELC2 onboard the ISS may not be utilized after the completion of MISSE8, it should be recognized that the architecture is suitable for future use in vehicle health monitoring applications. The current hardware design has been proven to be highly reliable in the demanding space flight environment. The simple telemetry interface provided by the CIB enabled

the experiments listed in Table 3 on MISSE7 and Table 4 on MISSE8, respectively. Broad applications enabled by the CIB should be noted, from processor testing to solar cell health monitoring and from CMOS image sensor testing to a variety of materials testing. Two individual experiments specifically related to vehicle health monitoring will be discussed in the next section.

7. Health Monitoring Enabled by the CIB

7.1. SiC JFET Health Monitoring Experiment. An example of a health monitoring circuit flown on MISSE7 was the silicon carbide (SiC) Junction Field Effect Transistor (JFET) Experiment designed by the NASA GRC mobile and remote sensing laboratory and SiC development group. NASA GRC has a long history in extreme temperature range silicon carbide electronics and packaging development having demonstrated SiC logic circuits operating over a temperature range of −125°C to 500°C [11]. Current long duration extreme temperature testing is performed in laboratory ovens and cold chambers, but future use is anticipated on flight vehicles. In an effort to demonstrate the technology in a flight environment, a health monitoring experiment was designed for SiC JFETs in high temperature packaging which was the first space flight of this technology. The experiment consisted of two SiC JFETs, one in room temperature commercial packaging, the other in high temperature packaging developed by the Ohio Aerospace Institute and NASA GRC [12].

The experiment monitors the current versus voltage transfer characteristics or a curve trace of both transistors during the flight. The transfer characteristics of the transistors show any electrical or physical degradation of the transistors, which is the primary concern of this experiment demonstration. This transfer characteristics are generated with a microcontroller-based curve tracing circuit. The CIB RS-485 protocol includes a timestamp in each transaction, which the SiC JFET uses to determine when to initiate a curve trace. To minimize bandwidth used, the SiC JFET experiment will only run once every hour. When the CIB queries this experiment, if the timestamp does not lie in the first ten minutes of the hour, the experiment will power itself down to wait for the next query. In addition to using the CIB enable line as a safety to lockout transmission on the RS-485 bus, this experiment uses the enable line as a signal to a latch to enable powering up. The CIB assertion of the enable line powers the experiment up, after which the experiment listens for the CIB command packet. The experiment receives the timestamp,

TABLE 3: Materials on the International Space Station Experiment Seven (MISSE7) flown on ISS from November 2009 to May 2011.

Short experiment name	APID	Experiment description	Provider
MCPE	1300	Multicore processor single event upset testing	Naval Research Laboratory
GRC experiment set	1301	SiC transistor health testing [12], H2 sensor, and zenith/nadir AO monitor	NASA GRC
SEUXSE	1302	Xilinx FPGA SEU Testing	Sandia National Laboratory
Not used	1303		
Not used	1304		
CIE	1305	CMOS imager experiment	Assurance technology
FTSCE II	1306	Solar cell health monitoring [14–18]	NRL, NASA GRC, and AFRL
Boeing ram side experiment	1307	Materials testing	Boeing
HyperX	1308	High performance low-power processor, SEU testing [19]	NASA GSFC
SpaceCube "A"	1309	Advanced processor design SEU testing [20, 21]	NASA GSFC
SpaceCube "B"	1310	Advanced processor design SEU testing [20, 21]	NASA GSFC
Wake AO fluence monitor	1311	Wake atomic oxygen fluence monitor/thermal control paints Experiment	NASA GRC
Boeing wake side experiment	1312	Materials testing	Boeing
AFRL wake 1	1313	Tribology measurements	AFRL Dayton, U of Florida
AFRL wake 2	1314	Tribology measurements	AFRL Dayton, U of Florida
AFRL ram 1	1315	Tribology measurements	AFRL Dayton, U of Florida
AFRL ram 2	1316	Tribology measurements	AFRL Dayton, U of Florida
iMESA	1317	Miniaturized electrostatic analyzer [22]	U.S. Air Force Academy
LTESE	1318	Lead-free technology experiment in space environment	MSFC
Ames	1319	Thermal protection systems sensors	NASA ARC, NASA LaRC, NASA JSC, and Boeing
Boeing PICA	1320	Materials testing	Boeing
Ram atomic oxygen fluence monitor/thermal control paints experiment	1321		NASA GRC

determines if the time is within the ten-minute window, and then either proceeds with curve tracing or goes to sleep.

Figure 6 shows experiment data in the form of curve traces from midflight monitoring of the SiC transistor health. Both the room (Figure 6(a)) and high temperature (Figure 6(b)) packaged transistor test data are shown. The graph overlays preflight curves with those 6 months into the flight. These midflight results show no degradation of transistor performance during the 6-month period [12].

The experiment board shown in Figure 7 resided at APID 1301 as shown in Table 3. In addition to the SiC JFET experiment, the circuit board and electronics set also supported two additional experiments the Makel Gas Sensor and the atomic oxygen fluence monitor, neither of these will be discussed. The design of the CIB enables many types of health monitoring to be performed, with the sample rate being the limiting factor.

7.2. Forward Technology Solar Cell Experiment II. Another experiment toward vehicle health monitoring enabled by the CIB is the Forward Technology Solar Cell Experiment II (FTSCE II). The experiment was designed by the U. S. Naval Research Laboratory with instrumentation development by the NASA GRC MaRS Lab. This experiment monitored solar cell health by measuring current versus voltage (I-V) curve characteristics. Space vehicles in Earth orbit rely on

CIB: An Improved Communication Architecture for Real-Time Monitoring of Aerospace Materials, Instruments, and Sensors on the ISS

25

TABLE 4: Materials on the International Space Station Experiment Eight (MISSE8) flown on ISS from May 2011 to present (April 2013).

Short experiment name	APID	Experiment description	Provider
None	1300		
Reflectarray	1301	Characterize performance of components of a flexible, phased array antenna	NASA GRC
SEUXSE	1302	Xilinx FPGA SEU testing	Sandia National Lab
Not used	1303		
Not used	1304		
PASCAL	1305	Primary arcing effects on solar cells at LEO	Lockheed Martin, JAXA, Kyushu Institute of Technology
FTSCE III	1306	Solar cell health monitoring [23–25]	NRL, NASA GRC, and AFRL
Not used	1307		
HyperX	1308	High performance low-power processor, SEU testing [19]	NASA GSFC
SpaceCube "A"	1309	Advanced processor design SEU testing	NASA GSFC
SpaceCube "B"	1310	Advanced processor design SEU testing	NASA GSFC

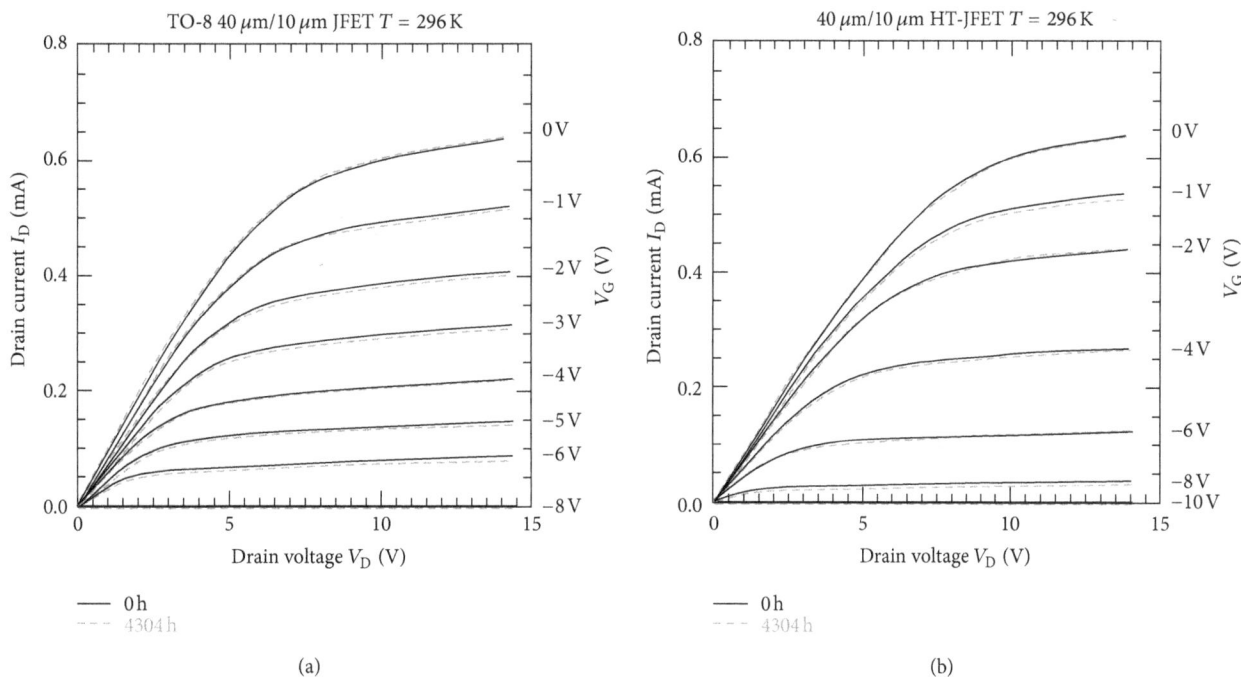

FIGURE 6: Current versus voltage transfer curves for given gate voltages of the two flight silicon carbide (SiC) Junction Field Effect Transistors (JFETs). (a) shows characteristics for room temperature packaged JFETs, for preflight, and after over six months of flight. (b) shows the characteristics for the high temperature packaged SiC JFET (image courtesy of NASA).

photovoltaic or solar cells as their power source. Space environment effects can degrade photovoltaic cell performance providing less power to the vehicle, and thus affecting the health and operating envelope of the space vehicle. While technology can be used to mitigate the degradation, the space vehicle designer and operator should know how the system will degrade and respond to the changing power system output.

Photovoltaic cells have current versus voltage characteristic which can provide insight into the health of the cell. Figure 8 shows an example of a solar cell I-V curve from initial FTSCE II on-orbit data. Performance degradation of the cell will show up as less current output, shifting the curve in Figure 8 downward along the vertical axis as the cell degrades. FTSCE II is comprised of 36 experimental solar cells as shown in Figure 9 along with measurement circuitry

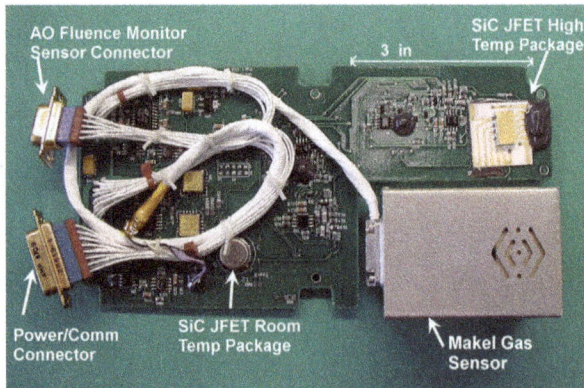

FIGURE 7: Silicon carbide junction field effect transistor (SiC JFET) health monitoring flight experiment circuit board (image courtesy of NASA).

FIGURE 9: Sun looking face of the Passive Experiment Container (PEC) containing passive material samples and cells as flown on Materials International Space Station Experiment 7 (MISSE7). Solar cells can be denoted by the wires to attach to the measurement circuitry of the Forward Technology Solar Cell Experiment II (FTSCE II) (image courtesy of NASA).

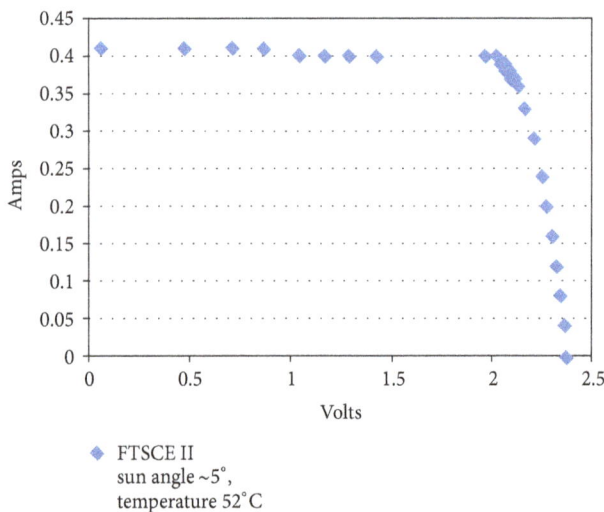

♦ FTSCE II
sun angle ~5°,
temperature 52°C

FIGURE 8: Typical I-V (current versus voltage) curve for a solar cell as measured by the Forward Technology Solar Cell Experiment II (FTSCE II).

to provide I-V curve data along with environment parameters of temperature and sun position in relation to the solar cells. The measurement hardware of FTSCE II flown on MISSE7 is actually a reflight of the same hardware originally designed for and flown as FTSCE I on MISSE 5 [13]. FTSCE II resided at the APID 1306 and would transmit data packets consisting of I-V curve data along with relevant environmental parameters. Future results covering the complete FTSCE II experiment are expected, while selected results from in flight data are presented in [14–17].

8. Future Health Monitoring Applications

The CIB design discussed in this paper is one possible hardware implementation of the architecture. This design was focused on providing reliable operation within a radiation environment. So, the hardware implementation described represents a design to achieve this goal. The architecture of

the CIB may also be realized with a different hardware set to suit other needs. For instance, in a more benign environment, the choices for suitable hardware increase significantly and the design could also be shrunk physically. The hardware could be implemented in a variety of processing platforms which might even include the MIL-STD-1553 interface onboard. One likely hardware target for the CIB architecture is a field programmable gate array (FPGA). This reconfigurable platform would allow the processing unit and communication to exist in single IC, greatly shrinking the physical footprint of the circuitry. A smaller CIB would provide greater access to additional flight platforms.

Another consideration for future use is the reliability of the sensors and instruments which utilize the RS-485 bus. As the sensors mature and are demonstrated to be more reliable in operation, the enable lines provided by the CIB might not be needed. This elimination would reduce the wiring needed as well as the physical size of the CIB.

Hardware changes are not the only possible architecture modifications which would lead to expanded use. Software alterations may also be applied to allow increased use of the architecture. For instance, current CIB protocol might be expanded. The CIB operated with a command/response protocol, where the CIB controlled the RS-485 bus in much the same way as the bus controller in the MIL-STD-1553 bus. One feature possible on a command/response bus is node to node communication. Because the CIB uses packet-based transfers, the software protocol could be expanded to include these node to node transfers, whereby a request for node to node transfer is communicated to the bus controller, which is then granted by the CIB through an additional message. Even in the current configuration, node to node transfer is possible, since a node needs only to actively listen to the bus, for messages encoded in its packet.

Future use of this architecture may be a vehicle health monitoring bus included in aircraft/spacecraft electronic systems. The CIB or future hardware could be the interface between all health monitoring sensors and the crafts central control unit. Used in this manner, the CIB could provide the craft with vehicle health information through a single interface such as the MIL-STD-1553 bus.

CIB: An Improved Communication Architecture for Real-Time Monitoring of Aerospace Materials, Instruments, and Sensors on the ISS

27

9. Conclusion

This paper demonstrates that the CIB is an improved architecture enabling simple vehicle health monitoring on the ISS. While the ISS has in place a communications architecture which enables health monitoring, telemetry, and control, the CIB improves and adds to this infrastructure by providing a simpler interface. This is accomplished by interfacing with the ISS MIL-STD-1553 bus and providing two multidrop RS-485 busses for experiments, sensors, and health monitoring systems to communicate on. The RS-485 busses make available the MIL-STD-1553 telemetry, command, and control services without the complex development required for MIL-STD-1553. This simpler interface enables sensor and instrumentation development for vehicle health monitoring applications.

The CIB provides this simpler communication interface while maintaining the reliability expected of a space flight system on the ISS. The CIB was developed with radiation tolerance and reliability as the primary design considerations. With over three years of successful operation on the ISS over two missions, the CIB has proven to be highly reliable and tolerant to the LEO radiation environment while providing simpler vehicle health monitoring communication support.

The CIB also enabled future materials and device development which lead to further use in health monitoring systems. This paper gave two examples of a health monitoring experiment flown on the external of the ISS for over a year. The SiC JFET is a high temperature component which has uses throughout a vehicle to include health monitoring in extremely hot environments. The flight on MISSE7 was the first space flight of this technology and was only possible with the simple telemetry interface provided by the CIB. FTSCE II demonstrated solar cell health monitoring on the ISS with real-time telemetry enabled by the CIB.

The CIB design discussed is one implementation of the architecture. This architecture might be expanded on or implemented differently for a different platform. Another processing technology might be utilized in the architecture in a more benign environment such as an aircraft. Future software development may use the CIB protocol to support node to node transfer, allowing sensor fusion techniques for vehicle health monitoring.

Acknowledgments

The authors would like to acknowledge George Y. Baaklini, branch chief of the Optical Instrumentation and NDE Branch at the NASA Glenn Research Center, and Robert Walters of the Naval Research Laboratory for support of this work. This work was funded by the U. S. Naval Research Laboratory.

References

[1] "Space station ionizing radiation design environment," Tech. Rep. NASA SSP-30512 Revision C, June 1994.

[2] "International standard payload rack to international space station, software interface control document part 1," Tech. Rep. NASA SSP, 52050 Revision G, 2007.

[3] "Aircraft internal time division command/response multiplex data bus," Tech. Rep. U.S. Department of Defense MIL-STD-1553B, 1996.

[4] "EXPRESS Logistics Carrier (ELC) development specification," Tech. Rep. NASA SSP-52055 Revision C, 2006.

[5] P. P. Jenkins, R. J. Walters, M. J. Krasowski et al., "MISSE7: Building a permanent environmental testbed for the international space station," in *Proceedings of the 9th International Conference on Protection of Materials and Structures From Space Environment (ICPMSE '08)*, vol. 1087, pp. 273–276, May 2008.

[6] "Communications interface board serial interface specification," Tech. Rep. NASA/NRL ICD Document, Revision 9, 2009.

[7] Atmel Corporation, "Rad. Tolerant 8-bit ROMless Microcontroller 80C32E Datasheet," 2007.

[8] Atmel Corporation, "Space 1-MBit (128K x 8) Paged Parallel EEPROMs AT28C010-12DK Datasheet," 2011.

[9] Atmel Corporation, "Rad. Tolerant 128K x 8, 5-Volt Very Low Power CMOS SRAM M65608E Datasheet," 2008.

[10] Data Device Corporation, "BU-63285 Space Level MIL-STD-1553 BC/RT/MT Advanced Communication Engine (SP'ACE II) Terminal Datasheet," 2005.

[11] P. G. Neudeck, M. J. Krasowski, L.-Y. Chen, and N. F. Prokop, "Characterization of 6H-SiC JFET integrated circuits over a broad temperature range from −150°C to +500°C," *Materials Science Forum*, vol. 645–648, pp. 1135–1138, 2010.

[12] P. G. Neudeck, N. F. Prokop, L. C. Greer III, L.-Y. Chen, and M. J. Krasowski, "Low earth orbit space environment testing of extreme temperature 6H-SiC JFETs on the international space station," *Materials Science Forum*, vol. 679-680, pp. 579–582, 2011.

[13] M. J. Krasowski, L. C. Greer, J. M. Flatico, P. P. Jenkins, and D. C. Spina, "Big science, small-budget space experiment package aka MISSE-5: a hardware and software perspective," in *Proceedings of the 19th Space Photovoltaic Research and Technology Conference*, September 2005.

[14] P. P. Jenkins, R. J. Walters, M. Gonzalez et al., "Initial results from the second forward technology solar cell experiment," in *Proceedings of the 35th IEEE Photovoltaic Specialists Conference (PVSC '10)*, pp. 1124–1127, June 2010.

[15] K. M. Edmondson, A. Howard, P. Hausgen et al., "Initial on-orbit performance analysis of Inverted Metamorphic (IMM3J) solar cells on MISSE-7," in *Proceedings of the 37th IEEE Photovoltaic Specialists Conference (PVSC '11)*, pp. 3719–3723, June 2011.

[16] T. Stern and A. Reid, "Modular solar panels using components engineered for producibility," in *Proceedings of the 37th IEEE Photovoltaic Specialists Conference (PVSC '11)*, pp. 1626–1629, June 2011.

[17] A. D. Howard, D. M. Wilt, P. P. Jenkins, K. M. Trautz, P. Hausgen, and J. M. Merrill, "Selected on-orbit data from the FTSCE II aboard the MISSE 7 testbed," in *Proceedings of the 38th IEEE Photovoltaic Specialists Conference (PVSC '12)*, pp. 3275–3280, June 2012.

[18] T. D. Sahlstrom, P. E. Hausgen, J. Guerrero, A. D. Howard, and N. A. Snyder, "Ultraviolet degradation testing of space protective coatings for photovoltaic cells," in *Proceedings of the 33rd IEEE Photovoltaic Specialists Conference*, pp. 1–5, May 2008.

[19] A. S. Keys, J. H. Adams, R. E. Ray, M. A. Johnson, and J. D. Cressler, "Advanced avionics and processor systems for space and lunar exploration," in *AIAA Space 2009 Conference and Exposition*, Anaheim, Calif, USA, September 2009, AIAA 2010-8783.

[20] D. Petrick, "SpaceCube: current missions and ongoing platform advancements," The MAPLD Workshop, 2009, https://nepp .nasa.gov/mapld_2009/.

[21] K. M. Zick, C.-C. Yu, J. P. Walters, and M. French, "Silent data corruption and embedded processing with NASA's spacecube," *IEEE Embedded Systems Letters*, vol. 4, no. 2, pp. 33–36, 2012.

[22] R. Balthazor, M. G. McHarg, C. L. Enloe et al., "Sensitivity of ionospheric specifications to in situ plasma density observations obtained from electrostatic analyzers onboard of a constellation of small satellites," in *Proceedings of the AIAA/USU Conference on Small Satellites, SSC12-IV-1*, Logan, Utah, USA, 2012.

[23] B. Cho, R. Lutz, J. Pappan et al., "IMM experimentation in the next frontier: Emcore's participation in the MISSE-8 program," in *Proceedings of the 35th IEEE Photovoltaic Specialists Conference (PVSC '10)*, pp. 110–112, June 2010.

[24] T. D. Sahlstrom, P. E. Hausgen, D. M. Wilt, A. D. Howard, M. D. Anderson Jr., and N. A. Snyder, "Space flight experiment: advanced solar cells and protective materials on the ISS exterior," in *Proceedings of the 35th IEEE Photovoltaic Specialists Conference (PVSC '10)*, pp. 2610–2615, June 2010.

[25] S. Gasner, M. Tresemer, S. Billets, D. Bhatt, and P. Wallis, "Design & fabrication of the lockheed martin solar cell demonstration experiment for the ISS Forward Technology Solar Cell Experiment II," in *Proceedings of the 34th IEEE Photovoltaic Specialists Conference (PVSC '09)*, pp. 791–793, June 2009.

4

Imaging Tasks Scheduling for High-Altitude Airship in Emergency Condition Based on Energy-Aware Strategy

Li Zhimeng, He Chuan, Qiu Dishan, Liu Jin, and Ma Manhao

Science and Technology on Information Systems Engineering Laboratory, National University of Defense Technology, Changsha 410073, China

Correspondence should be addressed to He Chuan; chuanhe@nudt.edu.cn

Academic Editors: C. Bigongiari and B. Vrsnak

Aiming to the imaging tasks scheduling problem on high-altitude airship in emergency condition, the programming models are constructed by analyzing the main constraints, which take the maximum task benefit and the minimum energy consumption as two optimization objectives. Firstly, the hierarchy architecture is adopted to convert this scheduling problem into three subproblems, that is, the task ranking, value task detecting, and energy conservation optimization. Then, the algorithms are designed for the subproblems, and the solving results are corresponding to feasible solution, efficient solution, and optimization solution of original problem, respectively. This paper makes detailed introduction to the energy-aware optimization strategy, which can rationally adjust airship's cruising speed based on the distribution of task's deadline, so as to decrease the total energy consumption caused by cruising activities. Finally, the application results and comparison analysis show that the proposed strategy and algorithm are effective and feasible.

1. Introduction

One of the most significant features of emergency scheduling problem (ESP) is timeliness; that is, the execution of task must be completed in its deadline. Otherwise, the task will lose its executive value or become invalid [1–3]. Under the emergency condition, the imaging task has its observation slot to reflect the requirements on the execution timing interval. For example, the emergency tasks, such as the observations on the targets about moving missile system, massing troops, and cruising battleship, generally need the responding agencies to scout timely in order to rapidly analyze the situation and to plan the operational activity.

Over the last decade, many military groups such as the US army have been devoted to development of the emergency imaging technology and improvement of the quick response ability of the reconnaissance system by incorporating multiple platforms. High-altitude airship is a promising solution for the emergency observation platform in the near-space [4, 5]. Unlike conventional heavier-than-air (HTA) aircraft, high-altitude airship is a lighter-than-air (LTA) aircraft equipped with steering and propulsion systems, and it generates lift force through the buoyancy instead of aerodynamics

[6]. At present, high-altitude airship located in the near-space has attracted wide attention in many countries, and it is well known that some projects have been studied, for example, the HARV and HAA projects [7] in USA, Sky cat and CL-160 projects [8] in the European Union, ETRI [9] in South Korea, and Sky Net in Japan [10]. The scientists and engineers in China have conducted corresponding researches since the last century, and the verification airship has completed its low-altitude flight experiment in 2003.

As a new application platform, the high-altitude airship has many advantages in reconnaissance activities. For instance, it has a long duration, and a great deal of load carrying, and can achieve the fixed-point successive observation, and so forth [11]. In comparison with the traditional unmanned aircraft vehicle (UAV) [12, 13], the high-altitude airship can be operated continuously for several months, even for more than one year in the assigned airspace. It is also easy to acquire data and information uninterruptedly in a long period. Due to the fuel restriction, UAV has to implement the aerial refueling or return to the base frequently, so it is impossible to achieve the long-term and continuous monitoring at a lower cost. The capsule of the high-altitude airship is usually made from the nonmetallic materials with less electromagnetism and heat reflecting, which makes it hard to be captured by radar. In addition, the high-altitude airship is invulnerable to be attacked and intercepted by many air-defense missiles, due to the operational height which is out of their fire range. Compared with the imaging reconnaissance satellite [14–16], the high-altitude airship has stronger ability of rapid response. Generally speaking, the ground support equipments for launching a high-altitude airship are fewer in number and have shorter period of launching preparation. Therefore, the theater reconnaissance, surveillance, and warning system can be established by high-altitude airship in a few hours, and the mass deployment can be rapidly implemented with its strong maneuverability. In terms of the efficiency-cost ratio, the in-orbit time of a high-altitude airship is nearly equal with that of an imaging reconnaissance satellite, but the usage cost is far less than the latter. In addition, the satellite is restricted by the fixed orbit in use and only can observe the targets in a certain time slot. On the contrary, the high-altitude airship can achieve the long-term and continuous observation on the target in the hover-and-stare way. Due to the previous advantages, the high-altitude airship has huge application potential in the emergency activities, such as the antiterrorism, disaster relief, and regional battles. The aforementioned advantages have turned high-altitude airship into an ideal imaging observation platform.

Existing studies on high-altitude airship are scattered over a range of journals, conferences, books, and reports. Rao et al. [17] presented a mission path following controller for the airship by employing artificial neural network (ANN). Tan et al. [18] introduce some methods and techniques to realize lightweight structure and present a review of current research on high-altitude airship with lightweight structures. Bessert and Frederich [19] investigated the aerodynamics behavior of high-altitude airship and presented a novel technology to test the aerodynamics on the structural behavior of airship. Ren et al. [20] analyzed the aerodynamics problems of high-altitude airship while launching, recovering, hovering, and introducing the achievement of airship dynamics research. Especially, there are numerous studies on energy system of high-altitude airship. Wang et al. [21] presented a novel computation method for solar radiation on solar cells of the airship, given the effect of the airship's attitude on the performance of its energy system. Ma and Sun [22] developed a power management framework of high-altitude airship, which can rationally distribute power to subsystems so as to lighten the energy consumption in certain situation. Wang et al. [23] proposed an energy balance method to analyze the regenerative energy system, which can streamline the configuration design of high-altitude airship. In addition, there are also great deal of works focusing on the propulsion system. Chen et al. [24] constructed a simulation model and made some analyses about propeller of high-altitude airship. Jordi et al. [25] discussed the biomimetic principles for the structural design of airship. Various development tests are completed in their research, including wind tunnel testing and flight trials. Unfortunately, there are only a handful of works reported to date in the literature that propose the task planning of high-altitude airship, which greatly degrades the system performance such as task guarantee ratio and energy consumption.

In this paper, we focus on the imaging tasks scheduling problem on the high-altitude airship under the emergency condition. The power-speed model is constructed, which is employed to evaluate the energy consumption during the airship's reconnaissance actives. We convert this scheduling problem to constrain satisfaction problem (CSP), then a heuristic algorithm based on the optimization sequence rule (OSR) is presented to obtain the task ranking scheme, and a value task detecting (VTD) method is provided to detect the key nodes that each airship needs to fly through in sequence. An energy-aware strategy (EAS) is also provided to optimize the task planning by rationally adjusting the cruising speed of airship. The simulation results show the effectiveness of this strategy.

The reminder of this paper is structured as follows. Section 2 makes detailed description on the reconnaissance process of high-altitude airship and establishes the corresponding models and proposed optimization objects. Section 3 converts the original problem into three subproblems by adopting hierarchy architecture and design the solution algorithms, respectively. The simulation experiments and performance analysis are given in Section 4. The final section will conclude this paper and discuss the future research direction.

2. Problem Description and Modeling

The application of high-altitude airship in imaging reconnaissance activity is an asset for other reconnaissance equipments, and it is of great significance to build and improve the reconnaissance network. To facilitate analysis and modeling of this problem, we summarize the main notations used throughout this paper as follows:

$T_p = [t_{\text{start}}, t_{\text{end}}]$: the active period of airship, where t_{start} refers to the starting time and t_{end} refers to the completion time of observation activity;

$Task = \{task_1, task_2, \ldots, task_n\}$: the imaging task set, where the element $task_i$ refers to the ith task, and n refers to the task number;

$sit_i = (x_i, y_i)$: the observation projection position of $task_i$, where x_i, y_i denote the horizontal coordinate and vertical coordinate, respectively;

$sit_0 = (x_0, y_0)$: the projection coordinate of airship at the beginning;

S_{max}: the maximum cruising speed of airship;

td_i: the deadline of $task_i$;

tb_i: the beginning timing instant of $task_i$;

te_i: the completion timing instant of $task_i$;

t_i: the duration time of $task_i$, which includes the system stability time, load switch time, and data storage time;

$S_{i,j}$: the average cruising speed of airship between sit_i and sit_j;

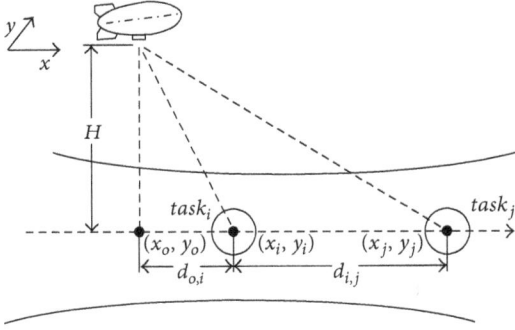

FIGURE 1: The cruising of high-altitude airship.

$d_{i,j}$: the distance between sit_i and sit_j;

$en_{i,j}^{active}$: the energy consumption of cruising from sit_i and sit_j;

en_i^{static}: the energy consumption of balance resistance while airship in fixed-point state.

2.1. The Process of Task Execution.

The imaging payload is usually installed in the cabin of high-altitude airship, which can be tilted or rotated within a certain angle to observe targets on the ground. During the task execution, the high-altitude airship flies according to the predetermined route and hover at a certain observation position, and in this way, the targets can be observed by imaging payload.

As shown in Figure 1, $task_i$ and $task_j$ are located in different positions. After completing $task_i$, the airship moves to another observing position to execute $task_j$. The cruising of airship will take a long time due to the limited speed, which makes it almost impossible to execute $task_j$ timely.

Theorem 1. *If $task_i$ can be observed before its deadline, it is called a value task; otherwise, it is called an invalid task.*

Assume that the current time is T_0, and L is the distance between the airship and $task_i$. If $task_i$ is a value task, the following conditions must be met:

$$T_0 + \frac{L}{S_{max}} + t_i \le td_i. \tag{1}$$

Obviously, the number of value tasks decreases with the time advancement, and this trend is irreversible. Considering that the imaging targets are widely distributed in the battle area, it is nearly impossible to ensure all tasks to be observed timely. Therefore, it is necessary to choose a reasonable task set and allocate the observation time for airship in accordance with various constraint conditions, so as to realize maximum efficiency of observing activity.

2.2. The Power-Velocity Model.

The energy system of high-altitude airship converts the solar radiation into electrical energy, thereby providing energy to the entire platform. Assume that the power of the propulsion system is P_t, and the efficiency is η; then, the actual propulsion power is

$$P_r = P_t \times \eta. \tag{2}$$

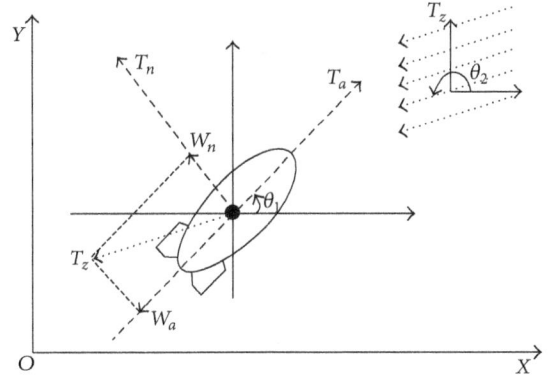

T_a Axial direction
T_n Normal direction
T_z Wind direction

FIGURE 2: The plane motion of high-altitude airship.

The cruising of high-altitude airship is subject to the impact of wind. The magnitude and direction of the propeller thrust are adjusted to balance the wind resistance, so as to realize continuous monitoring at fixed-point position. In this paper, it is assumed that the airship mainly relies on the electric propeller device to provide the thrust, which can be quickly adjusted in accordance with the wind direction and task position [26]. Since the working height of the airship is maintained, we only need to consider its horizontal movement as shown in Figure 2.

Normally, the wind field in near-space is stable; hence, the wind speed can be decomposed along the axial direction and normal direction of the airship. Consider

$$W_a = W \cos(\theta_2 - \theta_1),$$
$$W_n = W \sin(\theta_2 - \theta_1), \tag{3}$$

where W is the velocity of wind, θ_1 is the cruising direction of airship, and θ_2 is the direction of wind field.

The typical power-velocity model of airship is [24, 27, 28]:

$$P_a = \frac{1}{2}\rho V^{2/3}(S - W_a)^3 C_a$$
$$P_n = \frac{1}{2}\rho V^{2/3} W_n^3 C_n, \tag{4}$$

where ρ is the air density, $V^{2/3}$ is the characteristic area of airship based on its volume V, S is the airship's cruising speed relative to the ground, and C_a and C_n refer to the aerodynamic coefficients.

2.3. The Optimization Objectives.

Let $X = (x_1, x_2, \ldots, x_n)$, $Y = [y_{i,j}]_{n \times n}$, and $S = [S_{i,j}]_{(n+1) \times (n+1)}$ be the decision variables. If $task_i$ is effectively performed, then it is a value task (assume that $x_i = 1$); otherwise, let $x_i = 0$. If $x_i x_j = 1$ and $task_i$ is the preceding task of $task_j$, then let $y_{i,j} = 1$; otherwise, let $y_{i,j} = 0$.

The primary optimization objective of the tasks scheduling problem on the high-altitude airship is to maximize the guarantee ration $ER(X, Y, S)$:

$$ER(X, Y, S) = \frac{1}{n} \sum_{i=1}^{n} x_i. \qquad (5)$$

The total energy consumption $En_{\text{total}}(X, Y, S)$ should be minimized on the base of the maximization of $ER(X, Y, S)$. $En_{\text{total}}(X, Y, S)$ includes two parts; $En_{\text{active}}(X, Y, S)$ and $En_{\text{static}}(X, Y, S)$ are the energy consumptions caused by cruising and balance resistance in the suspension position, respectively.

According to (2)–(4), the energy consumption of airship's cruising from sit_i to sit_j is

$$en_{i,j}^{\text{active}} = \frac{\rho V^{2/3} C_n d_{i,j}}{2\eta S_{i,j}} \left| W \sin(\theta_2 - \theta_1) \right|^3$$
$$+ \frac{\rho V^{2/3} C_a d_{i,j}}{2\eta S_{i,j}} \left| S_{i,j} - W \cos(\theta_2 - \theta_1) \right|^3. \qquad (6)$$

If the high-altitude airship locate in the observation position of $task_i$, the energy consumption caused by resisting the effect of wind is

$$en_i^{\text{static}} = \frac{\rho V^{2/3} C_a t_i}{2\eta} \left| W \cos(\theta_2 - \theta_1) \right|^3$$
$$+ \frac{\rho V^{2/3} C_n t_i}{2\eta} \left| W \sin(\theta_2 - \theta_1) \right|^3. \qquad (7)$$

According to (6) and (7), the total energy consumption of airship during execution of the tasks is

$$En_{\text{total}}(X, Y, S) = En_{\text{active}}(X, Y, S) + En_{\text{static}}(X, Y, S)$$
$$= \sum_{i=1}^{n} \sum_{j=1}^{n} y_{i,j} en_{i,j}^{\text{active}} + \sum_{i=1}^{n} x_i en_i^{\text{static}}. \qquad (8)$$

2.4. The Programming Model. In the typical route planning of UAV, it is necessary to simultaneously consider the maximum turning angle, maximum climbing angle, minimum flight altitude, minimum path length, and other constraints. The purpose is to ensure that the cruising path can meet the aircraft's maneuvering characteristics and reduce the probability of damaging the aircraft in the no-fly zone and threatened area. As for the high-altitude airship in this paper, its working space has no spatial constraint, so the no-fly zone is an unnecessary consideration. At the same time, the low speed and slow dynamics provide the airship with a large-angle cornering ability, and it is also unnecessary to consider the minimum path length due to its suspension ability. Moreover, the threatened area of the high-altitude airship can be ignored due to the difficulty of being captured by the radar. However, the main constraints of tasks scheduling on the high-altitude airship are listed as follows.

Constraint 1. The high-altitude airship only executes the observation task within its active period.

Constraint 2. Each task can be executed only once, and it must be completed before its deadline.

Constraint 3. If a task can be executed, the execution time should be no less than the required continuous working time.

Constraint 4. Only one preceding task or one following task of each task is allowable at most.

Constraint 5. The preemptive service in the task execution is prohibited. Once the execution starts, the process cannot be terminated until completion.

Constraint 6. Before performing a new task, the airship needs sufficient time to change the observation position.

Constraint 7. For any two tasks to be executed, the certain priority order exists.

Constraint 8. The moving processes of airship only exist in different observation positions.

Constraint 9. The cruising speed of high-altitude airship in each path segment cannot be higher than the maximum cruising speed.

Let R_1, R_2, and R_2 be the feasible solution space of the decision variables X, Y, and S. In the separate optimization of $ER(X, Y, S)$, their optimal solution spaces are Q_1, Q_2, and Q_3, respectively. The programming model of this scheduling problem is given as follows:

$$Z_1(X, Y, S) = \max_{X \in R_1, Y \in R_2, S \in R_3} \left\{ \frac{1}{n} \sum_{i=1}^{n} x_i \right\},$$

$$Z_2(X, Y, S)$$
$$= \min_{X \in Q_1, Y \in Q_2, S \in Q_3} \left\{ \sum_{i=1}^{n} \sum_{j=1}^{n} y_{i,j} en_{i,j}^{\text{active}} + \sum_{i=1}^{n} x_i en_i^{\text{static}} \right\}$$

$$\text{s.t.} \begin{cases} [tb_i, te_i] \subset [t_{\text{start}}, t_{\text{end}}] \\ \sum_{i=1}^{n} x_i \leq 1, \quad te_i \leq td_i \\ te_i - tb_i \geq t_i, \quad \text{if } x_i = 1 \\ \sum_{i=1}^{n} y_{i,j} \leq 1, \quad \sum_{j=1}^{n} y_{i,j} \leq 1 \\ [tb_i, te_i] \cap [tb_j, te_j] = \emptyset, \quad \text{if } x_i \cdot x_j = 1 \\ te_i + \dfrac{d_{i,j}}{S_{i,j}} - K(1 - y_{i,j}) \leq tb_j \\ y_{i,j} \cdot y_{j,i} = 0 \\ \sum_{i=1}^{n} y_{i,i} = 0 \\ S_{i,j} \leq S_{\max} \\ x_i, y_{i,j} \in \{0, 1\} \\ \forall i, j \in \{1, 2, \ldots, m\}, \quad k \in \{1, 2, \ldots, n\}, \end{cases} \qquad (9)$$

where the former nine inequality formulas correspond to the aforementioned constraints, respectively, and the tenth inequality formula restricts the range of the decision variables.

3. Scheduling Algorithms

There exist numerous constraints in the tasks scheduling problem, so it is difficult to solve this problem directly. Therefore, the hierarchical optimization can be used to convert the original problem to the following sub-problems.

(1) Determine the priority execution order of the tasks, that is, the task sorting problem, which can be solved by the OSR algorithm.

(2) Select the observation tasks for the airship, that is, the value tasks detection problem, which can be solved by the VTD algorithm.

(3) Adjust the planning scheme to reduce energy consumption of the airship, that is, the energy conservation problem, which can be solved by the EAS algorithm.

Firstly, the task guarantee ration is the primary optimization objective of this scheduling problem. Thus, this paper considers that maximum cruising speed should be used by the airship to execute tasks as frequently as possible. Then, the distributions of value tasks are tested, and the initial scheduling scheme of the original problem is obtained. On this basis, energy conservation is regarded as another optimization goal. The cruising speed of airship at each leg is adjusted, and the execution time of all value tasks is updated.

Theorem 2. *The feasible solution of the original problem is expressed with two-tuples $SP = \langle vq, sq \rangle$, where $vq = \{task_{v1}, task_{v2}, \ldots, task_{vn}\}$ is the task set rearranged by priority execution order and $sq = \{S_{0,v1}, S_{v1,v2}, \ldots, S_{v(n-1),vn}\}$ is the speed set of airship at each leg.*

Theorem 3. *For any feasible solutions SP_1 and SP_2, if SP_1 is superior to SP_2, then $SP_1 \rhd SP_2$ or $SP_2 \lhd SP_1$; if SP_1 is inferior to SP_2, then $SP_1 \lhd SP_2$ or $SP_2 \rhd SP_1$; otherwise, $SP_1 = SP_2$.*

The comparison method of any two feasible solutions SP_1 and SP_2 is presented as follow:

$SP_1 \rhd SP_2$, if $ER(SP_1) > ER(SP_2)$,

$SP_1 \rhd SP_2$, if $ER(SP_1) = ER(SP_2)$, $En_{\text{total}}(SP_1) < En_{\text{total}}(SP_2)$,

$SP_1 = SP_2$, if $ER(SP_1) = ER(SP_2)$, $En_{\text{total}}(SP_1) = En_{\text{total}}(SP_2)$,

$SP_1 \lhd SP_2$, otherwise.

Theorem 4. *For any effective solution $SP_1 = \langle vq, sq \rangle$ and feasible solution $SP_2 = \langle vq^*, sq^* \rangle$, if $SP_1 \lhd SP_2$, then SP_2 is an optimal solution of SP_1.*

The conventional methods to obtain effective solutions including (a) improving the priority execution order of tasks

in order to obtain more value tasks; (b) reducing energy consumption by adjusting the cruising speed of airship on the basis of ensuring the quantity of value tasks. The previous methods are realized by the OSR algorithm and EAS algorithm, respectively.

3.1. The OSR Algorithm. The OSR algorithm sorts all elements in *Task* to get the priority execution order of tasks, and the guarantee ration is the primary optimization objective. Although all tasks are ranked, only part of the tasks can be observed timely; that is, the ranking result is only available for value tasks tested by the VTD algorithm, and it shows the observation sequence of executable tasks. The analysis of sorting rules in two tasks is as follows.

Let $Task = \{task_1, task_2\}$ denote the task set, T_0 the current timing instant and sit_0 the initial position of airship. Given two feasible solutions $SP_1 = \langle vq_1, sq \rangle$ and $SP_2 = \langle vq_2, sq \rangle$, where $vq_1 = \{task_1, task_2\}$, $vq_2 = \{task_2, task_1\}$ and $sq = \{S_{\max}, S_{\max}\}$, there is no harm in assuming that $td_1 - d_{0,2}S_{\max}^{-1} \geq td_2 - d_{0,1}S_{\max}^{-1}$, and then, the relationship between $ER(SP_1)$ and $ER(SP_2)$ is discussed as follows.

(a) If $task_1$ and $task_2$ are both value tasks in SP_1, then

$$T_0 + d_{0,1}S_{\max}^{-1} + t_1 \leq td_1,$$
$$T_0 + (d_{0,1} + d_{1,2})S_{\max}^{-1} + t_1 + t_2 \leq td_2. \tag{10}$$

In accordance with the assumption, we can get

$$T_0 + d_{0,2}S_{\max}^{-1} + t_2 \leq td_2,$$
$$T_0 + (d_{0,2} + d_{2,1})S_{\max}^{-1} + t_2 + t_1 \leq td_1. \tag{11}$$

Then, $task_1$ and $task_2$ are both value tasks in SP_2; thereby, $ER(SP_1) = ER(SP_2)$.

(b) As for SP_1, if $task_1$ is a value task while $task_2$ is an invalid task, then:

$$T_0 + d_{0,1}S_{\max}^{-1} + t_1 \leq td_1,$$
$$T_0 + (d_{0,1} + d_{1,2})S_{\max}^{-1} + t_1 + t_2 > td_2. \tag{12}$$

As for SP_2, if $task_2$ is a value task, then $ER(SP_1) \leq ER(SP_2)$; if $task_2$ is an invalid task, we can still ensure that $ER(SP_1) = ER(SP_2)$ due to $T_0 + d_{0,1}S_{\max}^{-1} + t_1 \leq td_1$.

(c) As for SP_1, if $task_1$ is an invalid task while $task_2$ is a value task, then

$$T_0 + d_{0,1}S_{\max}^{-1} + t_1 > td_1,$$
$$T_0 + d_{0,2}S_{\max}^{-1} + t_2 \leq td_2. \tag{13}$$

As for SP_2, $task_2$ is a value task; thereby, $ER(SP_1) \leq ER(SP_2)$.

(d) If $task_1$ and $task_2$ are both invalid tasks in SP_1, then

$$T_0 + d_{0,1}S_{\max}^{-1} + t_1 > td_1,$$
$$T_0 + d_{0,2}S_{\max}^{-1} + t_2 > td_2. \tag{14}$$

There is an equation $ER(SP_1) = ER(SP_2)$.

If $td_1 - d_{0,2}S_{max}^{-1} \leq td_2 - d_{0,1}S_{max}^{-1}$, the similar method can be used to analyze the sorting rules. However, the following conclusions can be acquired.

(i) If $td_1 - d_{0,2}S_{max}^{-1} \geq td_2 - d_{0,1}S_{max}^{-1}$, then $ER(SP_1) \leq ER(SP_2)$.

(ii) If $td_1 - d_{0,2}S_{max}^{-1} \leq td_2 - d_{0,1}S_{max}^{-1}$, then $ER(SP_1) \geq ER(SP_2)$.

According to the previous conclusions, the OSR algorithm is proposed to solve the task sorting problem. The main steps of OSR are presented as in Algorithm 1.

The task sorting problem is a combinatorial optimization problem. At present, there is no algorithm which can be used to obtain the optimal solution within the polynomial time complexity. Similar to the EDF algorithm, the OSR algorithm is a heuristic algorithm, which only generates the optimal scheme instead of a common optimized scheme. The effectiveness of the OSR algorithm will be verified in the subsequent experiments.

3.2. The VTD Algorithm. According to the task sorting result obtained by OSR, the deadline constraints for each element in set *Task* are checked in order. The value tasks in set *Task* are considered as the key nodes, and the cruising path optimization method between the successive key nodes can be learned from [17, 29, 30].

Theorem 5. *If both $task_i$ and $task_j$ are value tasks, and $task_j$ is arranged to be executed just next to $task_i$, then $task_i$ is called the preceding task of $task_j$, and $task_j$ is the following task of $task_i$.*

Consider $task_j$ to be the following task of $task_i$, so the execution time interval of $task_j$ is presented as follows:

$$tb_j = te_i + d_{i,j}S_{i,j}^{-1},$$
$$te_j = tb_j + t_j. \tag{15}$$

Assume that the airship has maximal cruising speed while solving the value tasks detection problem based on the hierarchy architecture. The pseudocode of VTD is shown as Algorithm 2.

3.3. The EAS Algorithm. If $task_i$ is executed, the power consumption of the high-altitude airship caused by cruising in speed S^* to the position sit_j and to observe $task_j$ can be calculated as

$$\text{energy}_{i,j}\left(S^*\right) = en_{i,j}^{active} + en_j^{static}$$

$$= \alpha C_a \left[\left|W\cos\left(\theta_2 - \theta_1\right)\right|^3 t_i \right.$$
$$\left. + \frac{d_{i,j}}{S^*}\left|S^* - W\cos\left(\theta_2 - \theta_1\right)\right|^3 \right] \tag{16}$$
$$+ \alpha C_n \left(\frac{d_{i,j}}{S^*} + t_i\right)\left|W\sin\left(\theta_2 - \theta_1\right)\right|^3,$$

where $\alpha = \rho V^{2/3}(2\eta)^{-1}$ is a constant.

If the cruising speed of airship is limited at $S_{i,j} \in [S_{min}, S_{max}]$, the optimal cruising speed of airship between sit_i and sit_j will be $S^* = \{S^* \mid \text{energy}_{i,j}(S^*) \leq \text{energy}_{i,j}(S), \text{ for all } S^*, S \in [S_{min}, S_{max}]\}$. For the convenience to describe this problem, we define the function of the optimal cruising speed (value range: $[S_{min}, S_{max}]$) between sit_i and sit_j as

$$S^* = \text{OPTE}_{i,j}\left(S_{min}, S_{max}\right). \tag{17}$$

Theorem 6. *If $[S_1, S_2] \supset [S_3, S_4]$, $S_a = \text{OPTE}_{i,j}(S_1, S_2)$, and $S_b = \text{OPTE}_{i,j}(S_3, S_4)$, then one can get $\text{energy}_{i,j}(S_a) \leq \text{energy}_{i,j}(S_b)$.*

Proof. Since $S_a = \text{OPTE}_{i,j}(S_1, S_2)$, we obtain that $\text{energy}_{i,j}(S_a) \leq \text{energy}_{i,j}(S)$, for all $S \in [S_1, S_2]$. For $S_b \in [S_1, S_2]$, $\text{energy}_{i,j}(S_a) \leq \text{energy}_{i,j}(S_b)$ exists. □

Inference 1. If $S_3 \in [S_1, S_2]$, $S_4 = \text{OPTE}_{i,j}(S_1, S_2)$, then $\text{energy}_{i,j}(S_3) \geq \text{energy}_{i,j}(S_4)$ is proven.

Proof. It can be considered that $S_3 = \text{OPTE}_{i,j}(S_3, S_3)$. According to Theorem 6, since $[S_3, S_3] \in [S_1, S_2]$, so Theorem 6 exists. □

Assume that $SP = \langle vq, sq \rangle$ is an efficient solution, where $vq = \{task_{v1}, task_{v2}, \ldots, task_{vg}\}$, and $sq = \{S_{0,v1}, S_{v1,v2}, \ldots, S_{v(g-1),vg}\}$. Let $SP^* = \langle vq, sq^* \rangle$ denote the improved solution of SP, and it is obtained by the EAS listed as follows

Step 1. Let $task_0$ denote a virtual task, and set $t_0 \leftarrow 0$, $tb_0 \leftarrow 0$, $tb_0 \leftarrow 0$, $i \leftarrow 0$.

Step 2. On the premise of ensuring the value tasks, the speed range $[S_{vi,v(i+1)}^{min}, S_{vi,v(i+1)}^{min}]$ of airship cruised from sit_{vi} to $sit_{v(i+1)}$ is calculated.

Step 3. The optimal speed $S_{vi,v(i+1)}^* = \text{OPTE}(S_{vi,v(i+1)}^{min}, S_{vi,v(i+1)}^{min})$ of the airship cruised from sit_{vi} to $sit_{v(i+1)}$ is calculated by (17). Let $S_{vi,v(i+1)} \leftarrow S_{vi,v(i+1)}^*$ and $i \leftarrow i + 1$.

Step 4. The execution period of $task_{vi}$ is updated by (15).

Step 5. If $i \leq g$, go to Step 2; otherwise, the iteration ends.

In the previous algorithm, S_{max} is the maximum cruising speed of airship along each leg. The minimum cruising speed of airship from sit_{vi} to $sit_{v(i+1)}$ is

$$S_{vi,v(i+1)}^{min} = \max_{j=i+1,\ldots,g} \left\{ \frac{\sum_{k=i}^{j-1} d_{vk,v(k+1)}}{td_{vj} - \sum_{k=i}^{j-1} t_{v(k+1)} - te_{vi}} \right\}. \tag{18}$$

Theorem 7. *As for $task_i$ and $task_j$, if $task_i$ is priority to $task_j$, let $task_i \succ task_j$ or $task_j \prec task_i$; otherwise, let $task_i \prec task_j$ or $task_j \succ task_i$.*

Theorem 8. *Assume that $SP = \langle vq, sq \rangle$ is the optimized solution obtained by the EAS algorithm, and then for all $task_{vi} \in vq = \{task_{v1}, task_{v2}, \ldots, task_{vg}\}$, $S_{vi,v(i+1)}^{min} \leq S_{max}$ exists.*

```
(1)  Queue ← Task; NQueue ← NULL; point_0 ← sit_0;
(2)  while Queue ≠ NULL do
(3)    Let set G ← NULL
(4)    for each task task_i in Queue do
(5)      Calculate the distance l_{i,0} beween point_0 and task_i;
(6)      Add g_i = td_i + l_{0,i} S_max^{-1} to set G;
(7)    end for
(8)    g_k = min_{g_i ∈ G} {g_i}; /*Select the minimal member in G*/
(9)    Search the last element task_r in NQueue;/*The set
         NQueue is used to save the execution equence*/
(10)   Let task_k > task_r;/*Assign the order for task_k*/
(11)   Remove task_k from Queue to NQueue, and task_k is the last element in NQueue;
(12)   point_0 ← sit_k;
(13) endwhile
```

ALGORITHM 1: Pseudocode of OSR algorithm.

```
(1)  idle ← 0; VQueue ← NULL; point_0 ← sit_0;/*Initialization*/
(2)  for each task task_i ∈ NQueue in sequence do
(3)    point_1 ← sit_i;/*Search the value task in NQueue*/
(4)    Calculate the distance L_{0,1} between point_0 and point_1;
(5)    if idle + L_{0,1} S_max^{-1} + t_i ≤ t_{di} then/*task_i is a value task*/
(6)      Calculate tb_i and te_i by (15);
(7)      idle ← te_i; point_0 ← point_1;
(8)      Add task_i to the value task set VQueue;
(9)    else
(10)     Reject task_i;/*task_i is a invalid task, which can not been executed in its deadline*/
(11)   end if
(12) end for
(13) for each task task_{gi} ∈ G in sequence do
(14)   Delete task_{gi} from G, and set x_{gi} ← 1; /*Assign value to the decision vector X*/
(15)   if G ≠ NULL then
(16)     Select the first task task_{gj} in G_k; /*task_{gj} is the next value task behind task_{gi}*/
(17)     Let y_{gi,gj} ← 1;
(18)   end if
(19) end for
```

ALGORITHM 2: Pseudocode of VTD algorithm.

Proof. For all $task_{vj} \in vq$, $task_{vj} \prec task_{vi}$, since $task_{vj}$ is a value task, we obtain that

$$te_{vi} + \sum_{k=i}^{j-1} \left(t_{v(k+1)} + d_{vk,v(k+1)} S_{max}^{-1} \right) \leq td_{vj}. \quad (19)$$

We may reach the following conclusion:

$$td_{vj} - \sum_{k=i}^{j-1} t_{v(k+1)} - te_{vi} \geq \sum_{k=i}^{j-1} d_{vk,v(k+1)} S_{max}^{-1}. \quad (20)$$

In other words,

$$\frac{\sum_{k=i}^{j-1} d_{vk,v(k+1)}}{td_{vj} - \sum_{k=i}^{j-1} t_{v(k+1)} - te_{vi}} \leq S_{max}. \quad (21)$$

For the discretion of $task_{vj}$, we may find that

$$S_{vi,v(i+1)}^{min} = \max_{j=i+1,\ldots,g} \left\{ \frac{\sum_{k=i}^{j-1} d_{vk,v(k+1)}}{td_{vj} - \sum_{k=i}^{j-1} t_{v(k+1)} - te_{vi}} \right\} \leq S_{max}. \quad (22)$$

Hereby, Theorem 8 holds. □

Theorem 9. *The optimized solution $SP = \langle vq, sq \rangle$ obtained by EAS algorithm is still an efficient solution.*

Proof. According to Theorem 3, if $SP = \langle vq, sq \rangle$ is still an efficient solution, the following conditions are satisfied: (a) for all $task_{vi} \in vq$ is a value task; (b) for all $S_{vi,v(i+1)} \in sq$ is no larger than S_{max}.

As for all $task_{vi} \in vq$, according to (18), we may find that

$$S_{vi,v(i+1)}^{\min} = \max_{j=i+1,\dots,g} \left\{ \frac{\sum_{k=i}^{j-1} d_{vk,v(k+1)}}{td_{vj} - \sum_{k=i}^{j-1} t_{v(k+1)} - te_{vi}} \right\} \quad (23)$$

$$\geq \frac{d_{vi,v(i+1)}}{td_{v(i+1)} - t_{v(i+1)} - te_{vi}}.$$

In other words,

$$\frac{d_{vi,v(i+1)}}{S_{vi,v(i+1)}^{\min}} \leq td_{v(i+1)} - t_{v(i+1)} - te_{vi}. \quad (24)$$

So

$$te_{vi} + \frac{d_{vi,v(i+1)}}{S_{vi,v(i+1)}} + t_{v(i+1)} \leq te_{vi} + \frac{d_{vi,v(i+1)}}{S_{vi,v(i+1)}^{\min}} + t_{v(i+1)} = td_{v(i+1)}. \quad (25)$$

Thus, it can be regarded that $task_{vi} \in vq$ is a value task. According to Theorem 8, for all $task_{vi} \in vq$, there always exists $S_{vi,v(i+1)}^{\min} \leq S_{\max}$. As for this, there exist $S_{vi,v(i+1)} \in [S_{vi,v(i+1)}^{\min}, S_{\max}]$ and $S_{vi,v(i+1)} \leq S_{\max}$.

In conclusion, Theorem 9 is valid. □

The feasible solution can be converted into an efficient solution by VTD algorithm, so as to figure out the decision variable X, Y; on this basis, the efficient solution can be transformed into an optimal solution by EAS strategy in order to figure out the decision variable S. According to the decision variables, the power consumption of optimized solution SP can be described as follows:

$$En_{\text{total}}(SP) = \sum_{task_i \in Task} \sum_{task_j \in Task} y_{i,j} \text{energy}_{i,j}(S_{i,j}). \quad (26)$$

Conclusion 1. If SP^* is the optimized solution of the efficient solution SP, there exists $ER(SP) = ER(SP^*)$.

Proof. According to Theorem 9, SP^* is still an efficient solution, so that SP and SP^* have the same value tasks that is, $ER(SP) = ER(SP^*)$ exists. □

Theorem 10. *Assume that $SP = \langle vq, sq \rangle$ is an efficient solution and $SP^* = \langle vq, sq^* \rangle$ is an optimized solution obtained with EAS algorithm; as for this, $SP^* \triangleright SP$ exists.*

Proof. For any $S_{vi,v(i+1)}^* \in sq^*$, there is $S_{vi,v(i+1)}^* \in [S_{vi,v(i+1)}^{\min}, S_{\max}]$. According to Theorem 6, $\text{energy}_{vi,v(i+1)}(S_{vi,v(i+1)}^*) \leq \text{energy}_{vi,v(i+1)}(S_{\max})$ is exists. Therefore, we can get

$$En_{\text{total}}(SP^*) = \sum_{task_i \in Task} \sum_{task_j \in Task} y_{i,j} \text{energy}_{i,j}(S_{i,j}^*)$$

$$\leq \sum_{task_i \in Task} \sum_{task_j \in Task} y_{i,j} \text{energy}_{i,j}(S_{\max}) \quad (27)$$

$$= En_{\text{total}}(SP).$$

Based on Conclusion 1, $ER(SP) = ER(SP^*)$; hereby, $SP^* \triangleright SP$ exists. □

TABLE 1: Parameters for high-altitude airship.

Parameters	Value
Active period	[6, 18] h
Maximum speed	90 km/h
Efficiency index	0.5
Duration time	[1, 6] min

TABLE 2: Parameters for environment.

Parameters	Value
Wind direction	30°
Wind speed	5 m/s
Aerodynamic coefficient C_a	0.025

Theorem 10 is actually an authentication to the effectiveness of the EAS algorithm. In other words, if the efficient solution is adjusted in accordance with the EAS algorithm, the obtained result is always more optimal than the original scheme.

4. Experimental Analysis

In this section, simulation experiment is conducted to illustrate the effectiveness of the proposed method. For VTD to be the precise algorithm, we only analyze the effectiveness of OSR and EAS.

4.1. Experimental Parameters. The proposed algorithms are implemented by Matlab2007 on a PC with Pentium IV 3.06 GHz CPU, 2 GB memory. As far as we know, there are no accepted benchmarks yet in scheduling problem of high-altitude airships, so the random models are used to construct the application scenario and simulate the battlefield area with $200 \times 300 \, \text{km}^2$.

The main parameters for high-altitude airships simulation are listed in Table 1.

We divide task number into ten levels from 30 to 300 for offering the flexibility to simulate the various workloads on high-altitude airships. The positions of tasks are generated randomly in battlefield, and their deadline is distributed in a uniform distribution spanning over the active period of airship.

Table 2 gives the configuration of environment parameters employed in our experiment.

Additionally, in order to reduce the calculation complexity of experiments, we let the constant α of (16) be equal to "1." Then the calculation results of total energy consumption are relation values instead of real values.

4.2. The Effectiveness of the OSR Algorithm. The heuristic algorithm OSR is employed to solve the task sorting problem. The effectiveness of this algorithm will directly affect the number of value tasks. In order to test the performance of the OSR algorithm, we have compared it with the EDF algorithm and Greedy algorithm.

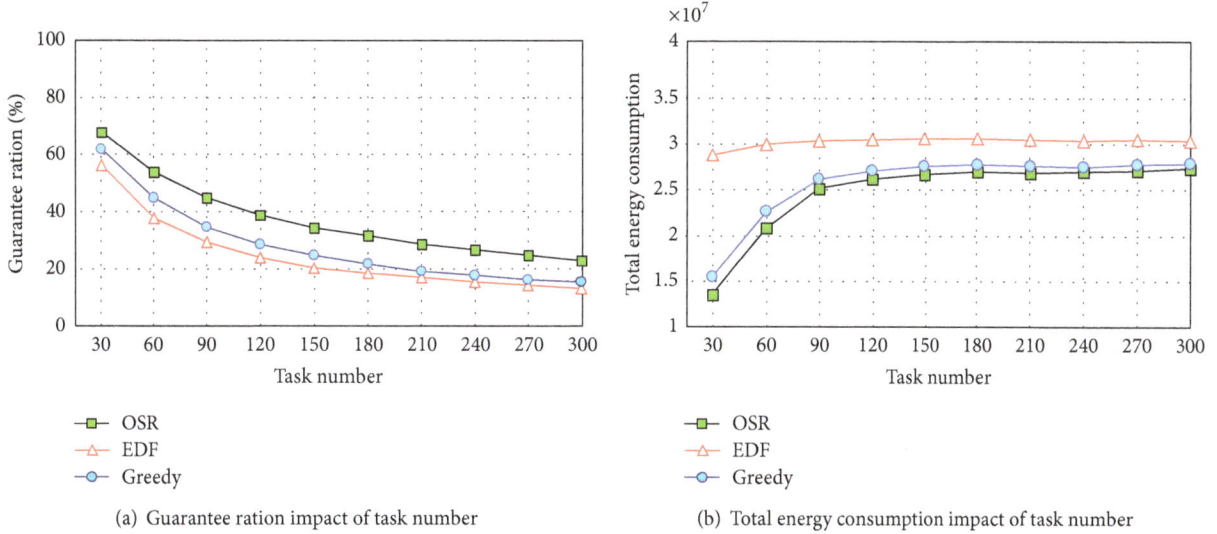

(a) Guarantee ration impact of task number

(b) Total energy consumption impact of task number

FIGURE 3: The scheduling results of OSR algorithm.

Figure 3 shows that OSR obtains a higher guarantee ration than EDF and Greedy. In various task scales, the guarantee ration obtained by OSR can be 12.84% higher than that of EDF and 8.89% higher than that of Greedy, which shows a very high scheduling performance. It can be seen in Figure 3(a) that as the quantity of tasks increases, the number of value tasks also increases gradually. However, the task guarantee ration shows a descending tendency. As the quantity of tasks increases, the number of tasks compatible with the airship observation also increases, so that more executable tasks appear. However, the observation capability of airship is limited. Thus, the increasing rate of value task will be much lower than the growth speed of task as the quantity of tasks increases to a certain degree. As for this, the task guarantee ration tends to decrease. As is shown in Figure 3(b), as the task quantity increases, the total energy consumption of airship also increases. However, when the task amount reaches the range of 30~120, the value task amount has a higher growth speed, leading to a sharp increase in energy consumption. By contrast, when task amount is around 120~300, the number of value tasks is nearly saturated, so that the growth of energy consumption slows down.

4.3. The Effectiveness of the EAS Algorithm. The EAS algorithm is used to realize energy-saving optimization of the scheduling scheme based on the deadline distribution of value tasks. In order to test the performance of this algorithm, its scheduling results will be compared with the performance before optimization (called HSA algorithm), that is, to compare the optimized solution with the efficient solution. According to Conclusion 1, the optimized solution has the same task guarantee ration as the efficient solution. As for this, the only parameter to be tested is the energy-saving performance of the EAS algorithm. The statistical indexes include the total energy consumption and energy consumption per tasks (ECPT).

It can be observed in Figure 4 that the total energy consumption of the EAS algorithm is always lower than that of the HSA algorithm in different task scales. This conclusion complies with Theorem 10. When keeping the same task guarantee ration, the total energy consumption of the EAS algorithm is 8.25% lower than that of the HSA algorithm, showing a higher energy saving performance. It can be seen in Figure 4(a) that the total energy consumption firstly increases sharply and then slows down as the task number increases. This phenomenon is in line with the analysis conclusion in Figure 3(b). The main reason for this trend is the variation tendency of the value task quantity. In Figure 4(b), we may find that ECPT reduces gradually as the task scale increases. As the area of battlefield is a constant, increase of the task scale will improve the density of task in the area. Therefore, the airship is able to execute more tasks within a short route, which in return reduces the energy consumption caused by cruising. In the meantime, we may find that the ECPT of the EAS algorithm is lower than that of the HSA algorithm in different task scales. This is because the EAS algorithm always adopts the most optimal cruising speed in each route, which reduces the energy consumption to the maximum degree.

5. Conclusions

The emergency scheduling problem on the high-altitude airship for imaging observation tasks is a multiobjective combination optimization issue. In research, this paper mainly made the following contributions

(1) The task execution process of airship is analyzed, and the method to detect the value tasks is provided. In this paper, a power-velocity model was also constructed by considering the influence of the wind field on airship's cruising. Moreover, the programming model of this problem is presented by proposing

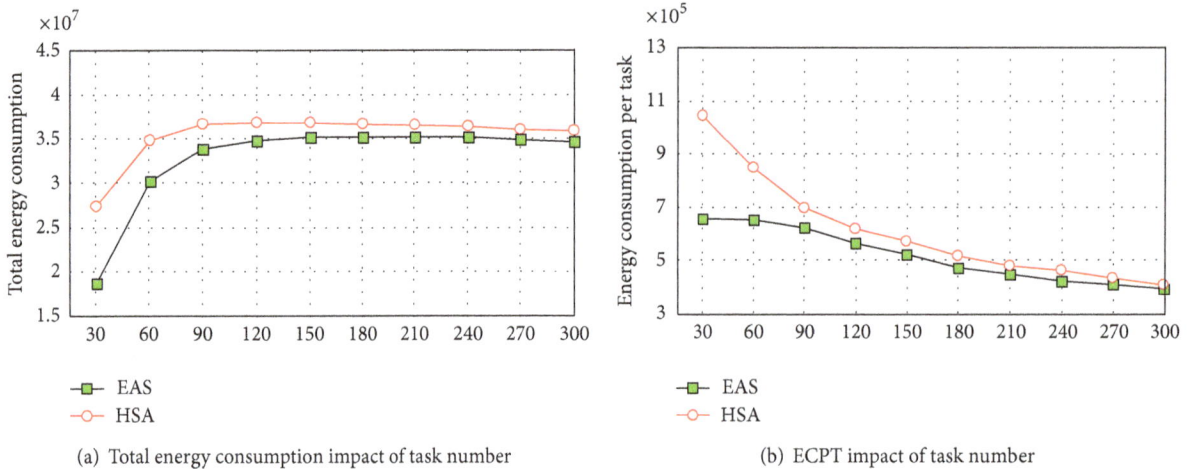

(a) Total energy consumption impact of task number

(b) ECPT impact of task number

FIGURE 4: The scheduling results of EAS algorithm.

the optimal objectives and listing corresponding constraints.

(2) In order to simplify the solution process, a hierarchy optimization framework which divides the original problem into three subproblems is provided in the paper. As is shown in the experiment, this method is valid in reducing the solution space of the original problem, which is beneficial to efficiently obtain the scheduling scheme.

(3) The OSR algorithm is proposed to rank the priority execution order of tasks. The EDF algorithm and Greedy algorithm only use the deadline and cruising distance as the basis of task sorting. By contrast, the OSR algorithm considers both factors at the same time, so as to produce more reasonable results.

(4) The EAS algorithm is employed to optimize the scheduling scheme with minimum energy consumption as the objective. This algorithm can adjust the cruising speed of airship in each leg according to the deadline distribution of value tasks. As for this, the total energy consumption of airship can be reduced with no influence on the task guarantee ration. It is of great importance to extend the duration of airship in the observation activity.

Also for our future work, we plan to develop the cooperation scheduling problem of multiple airship in observation activity. This research will be performed based on the research in this paper, with the aim to integrate observation resources and to improve the overall observation efficiency.

References

[1] Y. Liang and Y. Jin, "Heuristic algorithm for resource leveling problem in emergency scheduling," *Computer Integrated Manufacturing Systems*, vol. 15, no. 6, pp. 1165–1171, 2009.

[2] J. Wang, J. Li, H.-Z. Chen, and N. Jing, "Multi-objective imaging scheduling approach of earth observation satellite for emergent

conditions," *Acta Electronica Sinica*, vol. 36, no. 9, pp. 1715–1722, 2008.

[3] C. He, X. M. Zhu, and D. S. Qiu, "Cooperative scheduling method of multi-satellites for imaging reconnaissance in emergency condition," *Systems Engineering and Electronics*, vol. 34, no. 4, pp. 726–731, 2012.

[4] C. L. Nickol, L. L. Kohout, M. D. Guynn, and T. A. Ozoroski, "High altitude long endurance air vehicle analysis of alternatives and technology requirements development," in *Proceedings of the 45th AIAA Aerospace Sciences Meeting*, pp. 12653–12669, January 2007.

[5] Y. F. Yin and S. C. Huang, "Optimization deployment of multi-sensor platforms in near-space based on adaptive genetic algorithm," in *Proceedings of the International Conference on Information Engineering and Computer Science (ICIECS '09)*, pp. 1–5, December 2009.

[6] A. Moutinho and J. R. Azinheira, "A gain-scheduling approach for airship path-tracking," in *Proceedings of the 3rd International Conference on Informatics in Control, Automation and Robotics (ICINCO '06)*, pp. 82–88, August 2006.

[7] D. K. Schmidt, J. Stevens, and J. Roney, "Near-space station-keeping performance of a large high-altitude notional airship," *Journal of Aircraft*, vol. 44, no. 2, pp. 611–615, 2007.

[8] T. Tozer, P. Hendrick, and B. Sträter, "Developing a European research strategy in the high altitude aircraft and airship sector," in *Proceedings of the 7th AIAA Aviation Technology, Integration, and Operations Conference*, pp. 554–568, September 2007.

[9] J.-M. Park, B.-J. Ku, Y.-S. Kim, and D.-S. Ahn, "Technology development for wireless communications system using stratospheric platform in Korea," in *Proceedings of the 13th IEEE International Symposium on Personal, Indoor and Mobile Radio Communications (PIMRC '02)*, pp. 1577–1581, September 2002.

[10] T. C. Hong, B. J. Ku, J. M. Park, D.-S. Ahn, and Y.-S. Jang, "Capacity of the WCDMA system using high altitude platform stations," *International Journal of Wireless Information Networks*, vol. 13, no. 1, pp. 5–17, 2006.

[11] C. He, D. S. Qiu, and J. Liu, "Cooperative scheduling of imaging observation tasks for high-altitude airships based on propagation algorithm," *The Scientific World Journal*, vol. 2012, Article ID 548250, 13 pages, 2012.

[12] L. F. Bertuccelli and M. L. Cummings, "Operator choice modeling for collaborative UAV visual search tasks," *IEEE Transactions on Systems, Man, and Cybernetics A*, vol. 42, no. 5, pp. 1088–1099, 2012.

[13] B. Lee, P. Park, C. Kim et al., "Power managements of a hybrid electric propulsion system for UAVs," *Journal of Mechanical Science and Technology*, vol. 26, no. 8, pp. 2291–2299, 2012.

[14] F. Marinelli, S. Nocella, F. Rossi, and S. Smriglio, "A Lagrangian heuristic for satellite range scheduling with resource constraints," *Computers & Operations Research*, vol. 38, no. 11, pp. 1572–1583, 2011.

[15] P. Wang, G. Reinelt, P. Gao, and Y. Tan, "A model, a heuristic and a decision support system to solve the scheduling problem of an earth observing satellite constellation," *Computers and Industrial Engineering*, vol. 61, no. 2, pp. 322–335, 2011.

[16] L. F. Tan, Q. P. Tan, J. J. Xu et al., "Formal verification of signature-monitoring mechanisms by model checking," *Computer Science and Information Systems*, vol. 9, no. 4, pp. 1431–1451, 2012.

[17] J. J. Rao, Z. B. Gong, J. Luo, Z. Jiang, S. Xie, and W. Liu, "Robotic airship mission path-following control based on ANN and human operator's skill," *Transactions of the Institute of Measurement and Control*, vol. 29, no. 1, pp. 5–15, 2007.

[18] H. F. Tan, C. Wang, and C. G. Wang, "Progress of new type stratospheric airships for realization of lightweight," *Acta Aeronautica et Astronautica Sinica*, vol. 31, no. 2, pp. 257–264, 2010.

[19] N. Bessert and O. Frederich, "Nonlinear airship aeroelasticity," *Journal of Fluids and Structures*, vol. 21, no. 8, pp. 731–742, 2005.

[20] Y. P. Ren, Z. W. Tian, and Z. N. Wu, "Some aerodynamics problems of airship," *Acta Aeronautica et Astronautica Sinica*, vol. 31, no. 3, pp. 431–443, 2010.

[21] H. F. Wang, B. F. Song, B. Liu, and W. An, "Exploring configuration design of high altitude airship," *Journal of Northwestern Polytechnical University*, vol. 25, no. 1, pp. 56–60, 2007.

[22] Y. P. Ma and K. W. Sun, "Research on multi-power management system of high-altitude airship," in *Proceedings of the Asia-Pacific Power and Energy Engineering Conference (APPEEC '10)*, pp. 1–4, March 2010.

[23] H. F. Wang, B. F. Song, and L. K. Zuo, "Effect of high-altitude Airship's attitude on performance of its energy system," *Journal of Aircraft*, vol. 44, no. 6, pp. 2077–2080, 2007.

[24] S. Q. Chen, H. F. Wang, and B. F. Song, "Modeling and dynamic simulation study of big inertia propulsion system of high altitude airship," in *Proceedings of the 2nd International Conference on Artificial Intelligence, Management Science and Electronic Commerce (AIMSEC '11)*, pp. 4065–4068, August 2011.

[25] C. Jordi, S. Michel, and E. Fink, "Fish-like propulsion of an airship with planar membrane dielectric elastomer actuators," *Bioinspiration & Biomimetics*, vol. 5, no. 2, Article ID 026007, 2010.

[26] D. K. Schmidt, "Modeling and near-space stationkeeping control of a large high-altitude airship," *Journal of Guidance, Control, and Dynamics*, vol. 30, no. 2, pp. 540–547, 2007.

[27] S. Hong, S. Baoyin, and Y. Qiupingl, "Study of the solar power system of stratospheric airships," *Chinese Space Science and Technology*, vol. 29, no. 1, pp. 26–31, 2009.

[28] X.-L. Fang, X.-X. Liu, F. Wang, and K.-F. Liu, "Research on modeling technology for a high altitude airship," in *Proceedings of the International Conference on Advanced in Control Engineering and Information Science (CEIS '11)*, pp. 747–751, August 2011.

[29] Y. W. Li, M. Nahon, and I. Sharf, "Airship dynamics modeling: a literature review," *Progress in Aerospace Sciences*, vol. 47, no. 3, pp. 217–239, 2011.

[30] R. J. Szczerba, P. Galkowski, I. S. Glickstein, and N. Ternullo, "Robust algorithm for real-time route planning," *IEEE Transactions on Aerospace and Electronic Systems*, vol. 36, no. 3, pp. 869–878, 2000.

5

A Dynamic Scheduling Method of Earth-Observing Satellites by Employing Rolling Horizon Strategy

Qiu Dishan, He Chuan, Liu Jin, and Ma Manhao

Science and Technology on Information Systems Engineering Laboratory, National University of Defense Technology, Changsha 410073, China

Correspondence should be addressed to He Chuan; chuanhe@nudt.edu.cn

Academic Editors: A. De Felice, K. Kudela, and M. Ragulskis

Focused on the dynamic scheduling problem for earth-observing satellites (EOS), an integer programming model is constructed after analyzing the main constraints. The rolling horizon (RH) strategy is proposed according to the independent arriving time and deadline of the imaging tasks. This strategy is designed with a mixed triggering mode composed of periodical triggering and event triggering, and the scheduling horizon is decomposed into a series of static scheduling intervals. By optimizing the scheduling schemes in each interval, the dynamic scheduling of EOS is realized. We also propose three dynamic scheduling algorithms by the combination of the RH strategy and various heuristic algorithms. Finally, the scheduling results of different algorithms are compared and the presented methods in this paper are demonstrated to be efficient by extensive experiments.

1. Introduction

The mission of an earth-observing satellite (EOS) is to scout targets with a certain range of ground to produce high-resolution photographs [1–4]. According to the working mechanism, the imaging sensors load in EOS can be divided into visible light imaging, microwave imaging, infrared imaging, and so forth. Since they have numerous merits such as rapider response, broad coverage range, longer duration, and freedom from airspace boundaries, EOS supports many important services, such as military surveillance, geodesy and navigation, remote sensing, and monitoring.

Nowadays, EOS is attracting more and more interests worldwide be accompanied with the dramatic increase of the demand for imaging service. One major research trend is that the single satellite used in early reconnaissance is replaced by cooperation of large satellites, yielding the socalled multi-satellite application. Unfortunately, as the number of satellites grows large, the traditional manual coordination will no longer be feasible because multisatellite scheduling (MSS) is an NP-hard combinatorial optimization problem [5]. Therefore, the researches on MSS refer to assigning observation resource to match various imaging requirements as indispensable.

There exist numerous studies on scheduling algorithm for multi-satellite to realize automated resource planning. Wang et al. [6] proposed a hybrid ant colony optimization (HACO) algorithm to overcome the disadvantage of current ant colony optimization (ACO) algorithm which is more easily plunged into local optimal solution in solving MSS problem. Jian and Cheng [7] constructed an integer programming model of MSS based on the analysis of the resource constraints and task characteristic. To solve this combinatorial optimization problem, they provided a genetic particle swarm optimization (PSO) algorithm which searches only in the appointed integer space. He et al. [8] presented a cooperative scheduling architecture of multiple satellites by converting this scheduling problem into a main problem and a subproblem. In addition, an improved PSO algorithm was used to solve MSS problem by taking task benefits as an optimization objective.

However, those above researches have been primarily focused on static scheduling problem of EOSs. It is usually assumed that the imaging tasks have been submitted before scheduling and their information is acquired. In practice, the requests from customers are continuously delivered, which lead to the imaging tasks arriving one by one. The most significant feature of dynamic scheduling is time urgency; that is, the task must be completed within a specified time limit, or it will lose its execution value caused by failure. The satellite imaging reconnaissance mission generally has a deadline, which can reflect its execution urgency. Execution of the task must be completed within the specified deadline; otherwise, the expected benefits will not be obtained. The traditional static scheduling methods tend to overlook the timeliness feature of imaging task, which makes them inapplicable to the dynamic scheduling problem of imaging satellite.

At present, there are a few works on the dynamic scheduling problem of EOSs. Baolin et al. [9] described a new satellite mission scheduling algorithm based on constraint satisfaction problem (CSP). Zhu et al. [10] considered the problems of satellite scheduling for realizing optimal disaster rescue and proposed a hybrid algorithm to solve this multiple objects optimization problem. Wang et al. [11] considered the characteristic of EOS in dynamic scheduling and proposed a rule-based heuristic algorithm to solve this problem. The study mentioned above still has the following shortages.

(1) It is difficult for these proposed algorithms to generate a task planning within a short time. A dynamic planning process consists of repeated scheduling events, and the traditional intelligence algorithm (IA) has high timing complexity, which cannot rapidly generate the planning scheme. Therefore, the high-efficiency heuristic algorithm should be used to address the dynamic scheduling problem of EOSs.

(2) The impacts of scheduling time on the available tasks were not considered. Since tasks are dynamic arrivals, the planning system collects dissimilar task sets at different scheduling times. Thus, the task set should be determined based on current scheduling time before the scheduling.

(3) The constraints during dynamic scheduling have not been adequately considered. Many constraints (i.e., the storage capacity, maximum swing angle, and continuous observation time) which have been simplified in static scheduling should be considered in dynamic scheduling.

In this paper, we tackle the above challenges imposed on the dynamic scheduling problem of EOSs by handling the impacts of deadline constraint and scheduling time on planning scheme. The integer programming model is constructed based on various constraints in actual reconnaissance activities, and the rolling horizon (RH) strategy and heuristic algorithms are employed to solve this model.

The remainder of this paper is organized as follows. Section 2 describes the dynamic scheduling problem of EOSs, and establishes the integer programming model. Section 3 designs the scheduling architecture and algorithms based on the RH optimization. Simulation results and performance analysis are given in Section 4. Finally, Section 5 concludes the paper with some future research directions.

2. Problem Description and Modeling

The EOS operates in the space in a certain orbit as shown in Figure 1. The view coverage of EOS can be formed on the ground by the subsatellite point of satellite platform as well as the view angle, swing angle, and tilting angle of satellite payload [12–14].

The purpose in addressing the dynamic scheduling problem of EOSs is to appoint observation resources and execution time for the dynamical submitted tasks with various constraints, so as to maximize the task benefits of reconnaissance activity and minimize the resource consumption as far as possible.

The imaging tasks in dynamic scheduling problem are submitted to the planning system in independent times compared to the static scheduling problem which can obtain all the tasks in advance. The dynamic scheduling system only acquires the information of arrived tasks but can not gain the situation of following tasks. Hence it needs to trigger multiple scheduling in order to cope with the new tasks which are successively submitted to the scheduling system. Therefore, the dynamic scheduling algorithms designed in this paper should have the overall coordination capacity; that is, the algorithm should be able to timely adjust the execution scheme of planned tasks for executing the emergency tasks submitted later, so as to maximize task benefits. For future reference, we summarize main notations used in this paper as the following:

$T_p = [t_{start}, t_{end}]$ is the reconnaissance activity period of satellites, where t_{start} is the starting time and t_{end} is the ending time of the observation activity;

$T_S = \{t_S^0, t_S^1, \ldots, t_S^p\}$ is the scheduling time set, where t_S^i $(i \geq 1)$ is the ith scheduling time, $t_R^0 = t_{start}$ is the virtual initial scheduling time, $T_{period} = (t_S^p - t_S^0)/p$ is the average scheduling period, and p is the quantity of scheduling activities during the reconnaissance activity period;

Task = $\{task_1, task_2, \ldots, task_m\}$ is the imaging task set, where $task_i$ is the ith task, and m is the task number;

Sat = $\{sat_1, sat_2, \ldots, sat_n\}$ is the satellite resource set, where sat_j is the jth satellite, and n is the satellite number;

$TW_i = \bigcup_{j=1}^n \bigcup_{k=1}^{q_{i,j}} tw_{i,j}^k$ is the observation opportunity window set of $task_i$ between its arrival time a_i and deadline d_i, where $q_{i,j}$ is the opportunity window number of $task_i$ on sat_j, $tw_{i,j}^k = [ws_{i,j}^k, we_{i,j}^k]$ is the kth opportunity window of $task_i$ on sat_j, and $ws_{i,j}^k$ and $we_{i,j}^k$ are the starting time and ending time of $tw_{i,j}^k$, respectively.

In addition, $task_i$ has an execution benefit p_i and it requires a continuous observation time d_i. If $task_i$ can be executed before its deadline, assume ts_i is the starting time, te_i is the ending time, and h_i is the required storage capacity. Furthermore, the satellite has a swing angle gh_i and a tilting angle gw_i to execute $task_i$. Assume a_j is the view angle of sat_j, β_j is the maximum swing angle, γ_j is the maximum tilting angle, ω_j is the swing speed, ρ_j is the tilting speed, tc_j is the position transfer stabilization time, r_j is the maximum position transfer frequency, and M_j is the memory capacity.

The decision of scheduling times is affected by many factors, for example, the quantity and density of task, the upload period of satellite instructions, and the communication capacity of control center. The decision variable of the dynamic scheduling problem provided in this paper is as follows:

$$x_{i,j}^k = \begin{cases} 1, & \text{if } task_i \text{ is executed in } tw_{i,j}^k, \\ 0, & \text{otherwise.} \end{cases} \tag{1}$$

Assume t_S^k is the current scheduling time and the maximum task benefits are the optimization object, then we can build the integer programming model of the dynamic scheduling problem as follows [15]:

$$\max \quad Z = \sum_{i=1}^{m}\sum_{j=1}^{n}\sum_{k=1}^{q_{i,j}} x_{i,j}^k p_i$$

$$\text{s.t.} \begin{cases} \sum_{j}^{n}\sum_{k}^{q_{i,j}} x_{i,j}^k \leq 1, \\ [ts_i, te_i] \bigcap [ts_{i'}, te_{i'}] = \emptyset, & (1) \\ a_i \leq t_S^k \leq ts_i, \\ \quad \text{if } \sum_{j=1}^{n}\sum_{k=1}^{q_{i,j}} x_{i,j}^k = 1, & (2) \\ [ts_i, te_i] \subset \bigcup_{k=1}^{q_{i,j}} tw_{i,j}^k \subset [t_S^k, e_i], \\ \quad \text{if } \sum_{k=1}^{q_{i,j}} x_{i,j}^k = 1, & (3) \\ te_i - ts_i \geq d_i, \\ \quad \text{if } \sum_{j=1}^{n}\sum_{k=1}^{q_{i,j}} x_{i,j}^k = 1, & (4) \\ te_i + \max\left(\dfrac{|gh_i - gh_{i'}|}{\omega_j}, \dfrac{|gw_i - gw_{i'}|}{\rho_j}\right) + tc_j \leq ts_{i'}, \\ \quad \text{if } (te_i \leq ts_{i'}) \wedge \left(x_{i,j}^k = x_{i',j}^{k'}\right), & (5) \\ \sum_{i}^{m}\sum_{k}^{q_{i,j}} x_{i,j}^k h_i \leq M_j, & (6) \\ \sum_{i}^{m}\sum_{k}^{q_{i,j}} x_{i,j}^k \leq r_j, & (7) \\ i, i' \in [1, 2, \ldots, m], j \in [1, 2, \ldots, n] \\ k \in [1, 2, \ldots, q_{i,j}], k' \in [1, 2, \ldots, q_{i',j}], \end{cases}$$

$$\tag{2}$$

where Z is the optimization objective; constraint (1) means each task only need to be executed once, and the execution process does not involve preemptive service; constraint (2) means the scheduling system only assigns the arrived tasks and their deadline must later than current scheduling time;

constraint (3) means that if any task is executed, the execution time should between current scheduling time and its deadline; constraint (4) means that if the task can be executed, its execution time should not be shorter than the required continuous observation time; constraint (5) means any two tasks assigned to the same satellite should have an adequate time interval between their execution times to ensure the sensor of satellite can adjust its gesture; constraint (6) means that the memory store any task should not exceed its maximum capacity; constraint (7) means that the gesture transfer number of satellite in one orbit should not be higher than the allowable frequency.

3. Dynamic Scheduling Approach

3.1. Architecture of Algorithm. In this paper, the dynamic scheduling algorithm based on the RH strategy [16–18] is proposed to allocate tasks which have special characteristic such as the independent arrival time and deadline. This method can conduct scheduling (or rescheduling) forward for a task set through rolling advance, so as to effectively handle the uncertain factors brought by the dynamic tasks. The heuristic algorithms are incorporated with RH strategy to reduce the timing complexity of scheduling while each scheduling requires one replanning. In other words, the task set that needs to be scheduled is determined by the RH strategy firstly, and then the heuristic algorithms are used to assign satellites and execution periods to each task.

3.2. Rolling Horizon Strategy. The basic method of RH strategy is to divide the tasks into multiple task sets with certain overlaps based on the arrival sequence, and the division can be continuously updated along with the scheduling time. Each scheduling will decide and only assign its task set, which is called as rolling horizon. The new tasks are continuously added to the rolling horizon, and the finished tasks are gradually deleted with the advancement of the scheduling time, so as to realize the update of rolling horizon. The advantage of RH strategy is that it can decompose the complicated dynamic scheduling problem into multiple simple static scheduling sub-problems, and the optimization solution of previous problem is replaced with the optimized solutions of sub-problems, so that the complexity of the original problem will be reduced.

3.2.1. Task States. In general, tasks will go through four states based on current scheduling time: a new task, waiting task, running task, and finished task. One task may be scheduled in different time, thus the state of task is dynamic; that is, a task might be in two states in different scheduling.

In the example shown in Figure 2, t_S^k is current scheduling time. If $task_i$ has been planned during the $(k-1)$th scheduling (i.e., $a_i \leq t_S^{k-1}$), then $task_i$ is a finished task while $te_i \leq t_S^k$ (such as $task_1$); $task_i$ is a running task if $tb_i < t_S^k \leq te_i$; $task_i$ is a waiting task if $tb_i > t_S^k$. If $task_i$ has not been planned in the $(k-1)$th scheduling (i.e., $t_S^{k-1} \leq a_i \leq t_S^k$), then $task_i$ is a new task.

3.2.2. Rolling Horizon. The rolling horizon is used to store the tasks that need be scheduled currently. There are two key elements about rolling-horizon: the quantity and state of tasks in rolling-horizon. From the perspective of task quantity, the more the tasks fall into the rolling horizon, the stronger the capacity of scheduling system to obtain comprehensive task information is, which is important to acquire the better solution. But the timing complexity of scheduling algorithm will also be aggravated. From the perspective of task state, the rolling-horizon consists of the running tasks, waiting tasks, and new tasks generally. Among them, the processing method for the running task is an important criterion to distinguish the preemptive and nonpreemptive scheduling. The later scheduling mode is out of the interest of this paper; that is, the rolling-horizon only includes the waiting tasks and new tasks. It should be noticed that during actual scheduling, the rescheduling of waiting task will not consume any additional resources because waiting task has not been executed yet.

3.2.3. Trigger Mode. The arrangement of scheduling time is the key factor which affects the application efficiency of RH strategy, and it is mainly determined by the trigger mode of scheduling. The general trigger modes include the following types.

(i) *Event Trigger Mode.* The scheduling is triggered while the scheduling environment is changed or a manual intervention occurred, such as a new task arrived, satellite number changed, or a scheduling requirement was sent from decision-making section. The event-trigger mode is sensitive to the scheduling environment, and it can assign the emergency tasks in time. However, this mode might cause the scheduling algorithm to have a high timing complexity due to frequent scheduling, and result in hard to generate a planning scheme rapidly.

(ii) *Period Trigger Mode.* In this mode, the scheduling event will be trigged after a certain time interval, and this time interval can be a uniform constant or dynamic variable. The period-trigger mode has capability to ensure the stable frequency of scheduling, but it can not provide timely scheme for emergency tasks with high timeliness, and also has the fault that it can not adjust scheme in dealing with the change of the satellite number.

(iii) *Mixed Trigger Mode.* Mixed-trigger mode is the combination of the aforementioned two modes. It can timely allocate the emergency tasks with less time consumption, so it is an adaptive mode to be employed in the scheduling of real-time system and dynamic system.

The mixed-trigger mode is adopted in this paper, and the scheduling time for period factors and event factors is embodied in the elements belonging to T_s. The RH strategy based on the mixed-trigger mode is depicted as follows.

In Algorithm 1, each scheduling time $t_S^k \in T_S$ will trigger one rescheduling (see line 1), and the scheduling frequency during T_p is $O(P)$. The criterion described in Figure 1 is used to divide the arrived tasks into the finished task, running tasks, waiting tasks, and new tasks based on current scheduling time (see lines 3~12); the timing complexity is $O(m)$. The tasks in rolling-horizon are sorted and assigned to the satellite resources and execution time (see line 13). The timing complexity of this step is mainly depends on the assigning algorithm; thus we assume it is $O(A)$ temporarily. It only takes $O(1)$ to execute other lines. Hence, the timing complexity of Algorithm 1 is calculated as $O(P)[O(m) + O(A)]$.

The above algorithm needs to assign satellite resources and execution time for each task, so $O(A)$ is usually much larger than $O(m)$. The timing complexity of Algorithm 1 will be dominated by $O(A)$ if the scheduling frequency (or scheduling interval) is a constant. Then, the additional timing complexity caused by RH strategy is mainly related to the scheduling frequency but not affected by other factors such as the satellite quantity. At present, the intelligence algorithms (IA) are usually used to assign satellites and execution time to tasks in the static scheduling problem. These algorithms generally require multiple iterations to realize optimization of the scheduling scheme, and have a high timing complexity. The imaging tasks in dynamic scheduling have the timeliness feature, which requires the planning scheme to satisfy the deadline possibly, and, furthermore, the planning scheme should be rapidly generated and can be fast adjusted to adapt to the change of task set by scheduling algorithm. In Algorithm 2, the heuristic algorithm is employed to match the satellites and the tasks in order to shorten the generation time of the planning scheme. Meanwhile, the RH strategy can well adjust the previous planning scheme of planned tasks based on the change of rolling-horizon. The timing complexity of the heuristic algorithm embedded in RH strategy will be analyzed in detail in the next section.

3.3. Heuristic Algorithms. This paper has proposed the heuristic algorithms AIS, DIS, and WIS based on the arrival time priority, the deadline priority, and the waiting time priority inspired by the earlier arrived time first (EAT) algorithm [19] and earlier deadline first (EDF) algorithm [20, 21] and considering the task importance.

Let $RH = \{task_{k1}, task_{k2}, \ldots, task_{km}\}$ denote the rolling-horizon in the kth scheduling. The basic parameters (including the arrival time, deadline, waiting time, and importance) of all tasks are standardized as follows:

$$f\left(x_{k_i}\right) = \frac{x_{k_i} - \min_{task_{k_i} \in RH}\left(x_{k_i}\right) + 1}{\max_{task_{k_i} \in RH}\left(x_{k_i}\right) - \min_{task_{k_i} \in RH}\left(x_{k_i}\right) + 1}. \quad (3)$$

After the standardization, we record the basic parameters of $task_{k_i}$ as AS_{k_i}, DS_{k_i}, CS_{k_i}, and WS_{k_i} respectively. The arrival time priority degree of $task_{k_i}$ is given as follows:

$$AI_{k_i} = \frac{AS_{k_i}}{WS_{k_i}}. \quad (4)$$

The deadline priority degree of $task_{k_i}$ is as follows:

$$DI_{k_i} = \frac{DS_{k_i}}{WS_{k_i}}. \quad (5)$$

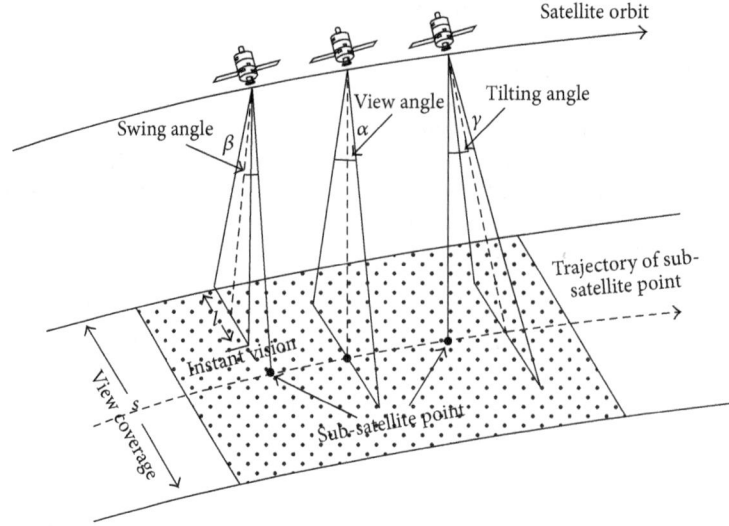

FIGURE 1: The observation field of EOS.

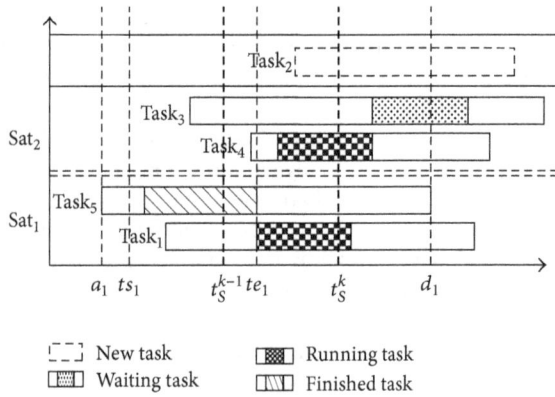

FIGURE 2: Task states based on current scheduling time.

The waiting time priority degree of task$_{k_i}$ is as follows:

$$CI_{k_i} = \frac{CS_{k_i}}{WS_{k_i}}. \quad (6)$$

The main steps of the heuristic algorithm adopted in this paper are described as follows.

Step 1. The three priority degrees of each task in rolling-horizon are calculated.

Step 2. The tasks in the rolling-horizon are sorted by their different priority-degrees, and the ranking results can be obtained corresponding to the AIS, DIS, and WIS algorithms, respectively.

Step 3. The assignment strategy based on windows conflict index (WCI) used to assign satellites and execution time for each task based on ranking results, and the scheduling scheme is generated after all tasks have been assigned.

In the aforementioned steps, WCI denotes the total impact on the unassigned tasks when a task is allocated to an available opportunity window. If task$_i$ is appointed to the kth opportunity window of sat$_j$, the WCI of task$_i$ can be calculated as follows:

$$WCI_{i,j}^k = \sum_{task_{i'} \in Waiting_i} \sum_{k'=1}^{q_{i',j}} \frac{g\left(tw_{i,j}^k, tw_{i',j}^{k'}\right)}{we_{i',j}^{k'} - ws_{i',j}^{k'}} p_{i'}, \quad (7)$$

where Waiting$_i$ is used to store tasks scheduled after task$_i$ according to the ranking result and $g(tw_{i,j}^k, tw_{i',j}^{k'})$ can be calculated as follows:

$$g\left(tw_{i,j}^k, tw_{i',j}^{k'}\right)$$

$$= \begin{cases} 0, & \text{if } w_{i,j}^k \bigcap tw_{i',j}^{k'} = \emptyset, \\ \min\left(we_{i,j}^k, we_{i',j}^{k'}\right) \\ \quad - \max\left(ws_{i,j}^k, ws_{i',j}^{k'}\right), & \text{otherwise.} \end{cases} \quad (8)$$

Let Queue = $\{task_{b1}, task_{b2}, \ldots, task_{bf}\}$ denote the task ranking result and EQueue the finished task set delivered by Algorithm 1. The pseudocode of heuristic algorithm is described in Algorithm 2.

In Algorithm 2, the tasks in the rolling horizon are scheduled in order (see line 1), and the timing complexity is $O(m)$. The valid opportunity windows of task$_i$ are checked to analyze the confliction with the running tasks (see lines 3~9); the timing complexity is $O(Tnm)$. It takes $O(Tn)$ to calculate the WCI of the valid opportunity windows (see lines 11~13). The timing complexity of other lines is $O(1)$, so the timing complexity of Algorithm 2 is $O(m)[O(Tnm) + O(Tn)] = O(Tnm^2)$. Furthermore, the timing complexity of Algorithm 1 is added to Algorithm 2; then the dynamic scheduling algorithm proposed in this paper has a timing complexity $O(P)[O(m) + O(Tnm^2)] = O(PTnm^2)$. Specially, the scheduling number and the reconnaissance activity

(1) **for** each timing instant t_S^k in set T_S **do**
(2)   RH ← NULL; Finished ← NULL; /∗Initialization∗/
(3)   **for** each task task$_i$ arrived before t_S^k **do**
(4)   **if** $\exists x_{i,j}^k \leftarrow 0, tb_i \leftarrow$ NULL, $te_i \leftarrow$ NULL **then**/∗Delete the planning scheme of task$_i$∗/
(5)   Add task$_i$ to set RH; /∗RH is the rolling-horizon∗/
(6)   **else if** $\sum_{j=1}^{n} \sum_{k=1}^{q_{i,j}} x_{i,j}^k = 0$ and $a_i > t_S^{k-1}$ **then**/∗task$_i$ is a waiting task∗/
(7)   Add *task$_i$* to set RH;
(8)   **end if**
(9)   **end for**
(10)   Sort all task in set RH, and schedule each task by **Heuristic**
       Algorithm in order; /∗The pseudocode of this algorithm is given in the following Algorithm 2∗/
(11) Add task$_i$ to set Finished;
(12) Update the scheduling decisions;
(13) **end for**

ALGORITHM 1: The pseudocode of RHO strategy.

(1) **for** each task$_{b_i}$ in RH **do**
(2)   Validwindow ← w_{b_i}; $ts_{b_i} \leftarrow$ NULL; $te_{b_i} \leftarrow$ NULL;
(3)   **for** each $tw_{b_i,j}^k$ in time window set Validwindow **do**
(4)   **for** each task$_{b_{i'}}$ in task set Finished **do**
(5)   **if** $[ws_{i,j}^k, we_{i,j}^k] \bigcap [ts_i, te_i] \neq \emptyset$ **then**
(6)   Delete $tw_{i,j}^k$ from time window set Validwindow;
(7)   **end if**
(8)   **end for**
(9)   **end for**
(10)   Remove task$_{b_i}$ from RH to Finished;
(11)   **if** time window set Validwindow $\neq \emptyset$ **then**
(12)   **for** each $tw_{i,j}^k$ in time window set Validwindow **do**
(13)   Calculate $WCI_{i,j}^k$ of $tw_{i,j}^k$;
(14)   **end for**
(15)   Assign task$_{b_i}$ by WCI strategy, and calculate ts_i, te_i;
(16)   **end if**
(17) **end for**

ALGORITHM 2: The pseudocode of heuristic algorithm.

period have a proportional relation if the rescheduling time interval is constant, then the timing complexity of dynamic scheduling algorithm is $O(T^2 m^2 n)$.

We incorporate RH strategy with AIS, DIS, and WIS to yield three new algorithms named RH-AIS, RH-DIS, and RH-WIS, respectively. Meanwhile, AIS, DIS, and WIS can also be used to solve the dynamic scheduling problem separately; that is, only new tasks are scheduled in each scheduling by those heuristic algorithms. The six algorithms mentioned before are compared in the following experiment to evaluate the efficiency of the RH strategy.

4. Evaluation

The proposed algorithms are implemented by Matlab2007 on a laptop with Pentium IV 3.06 GHz CPU, 2 GB memory, and Windows XP operating system. The experimental scenarios are generated randomly for there has been no benchmark in

the field of satellite scheduling yield. The operating points of simulated experiment are given as follows [22–24].

(1) The reconnaissance activity period is from March 21, 2010, to March 22, 2010, and the scheduling period is two hours; that is, $T_R = \{0\,\text{h}, 2\,\text{h}, \ldots, 24\,\text{h}\}$.

(2) The imaging tasks are generated in the area with a longitude $0° \sim 150°$ and latitude $-30° \sim 60°$ randomly. The task quantity varied from 100 to 400. The arrival time gap between two adjacent tasks is subject to the negative exponential distribution, with a density of 0.1. Set the execution value of task from 1 to 10, the required continuous time $3 \sim 5$ minute, and the occupied storage $2 \sim 4$ G, and the deadline is a random variable generated between the arrival time of the task and the ending time of the observation activity, which abides by the uniform distribution.

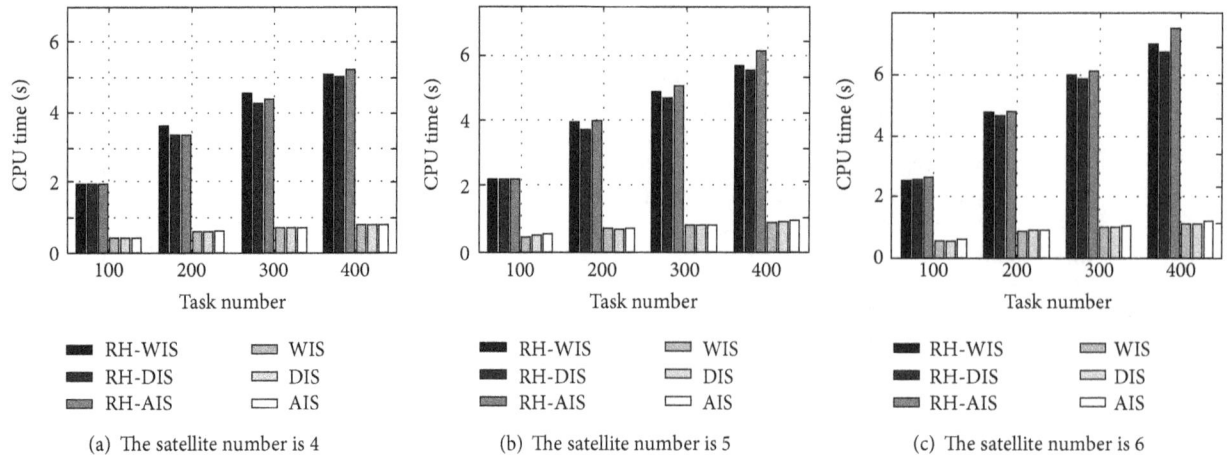

FIGURE 3: Search time of algorithm in different satellite quantities.

(3) The satellite quantity varied from 4 to 6, the memory storage is 240 G, the field angle is 3°, the maximum sway angle is 35°, the maximum tilting angle is 40°, and the maximum number of position transfers within a single orbit is no more than 5.

For the convenience of description, the dynamic scheduling problem of n satellites and m tasks is recorded as $m \times n$. The twelve experiments are designed in different problem scales, and the performance metrics include the task benefits and guarantee ration. The scheduling results of various algorithms are displayed in Table 1.

From Table 1, we can observe that the RH strategy can effectively improve the performance of heuristic algorithms. The task benefits of the three heuristic algorithms embedding RH strategy increase by 16.19%, 15.82%, and 8.29%, respectively. This is because the algorithms without embedded RH strategy only operate the new tasks, while the embedded one can adjust the scheduling scheme of waiting tasks for executing the new tasks which can not be executed previously; hence a better scheduling solution can be achieved. Among the rolling-horizon scheduling algorithms (RH-AIS, RH-DIS, and RH-WIS), the task benefits obtained by RH-WIS are slightly higher than those of RH-DIS and RH-AIS while the satellites have adequate observation capacity to execute imaging tasks (for the problem scales 100×4, 100×5, and 100×6), and the resolution of RH-DIS is better than that of the other algorithms while the satellites have inadequate observation capacity. This is because both RH-DIS and RH-WIS have considered the deadline feature of tasks in scheduling scheme; it is the precondition to execute the valuable task. A large amount of high-timeliness tasks cannot be executed in their deadlines if the satellites have inadequate observation capacity, then RH-DIS gives priority to the tasks with earlier deadlines, so that the task guarantee ration is increased, and higher task benefits can be obtained. Most tasks can be effectively executed if the satellites have adequate observation capacity, then RH-AIS schedules tasks based on the deadline distribution but also considers the impact of executing current task on follow-up tasks, so that higher task

benefits can be obtained on the aspect of overall planning effects.

From Figure 3, the CPU time of six algorithms is compared in different satellite numbers. The timing complexity of scheduling algorithms is aggravated after embedding the RH strategy; thus the generation speed of the planning scheme is reduced. In the scenarios with three different satellite numbers, the average CPU times of algorithms incorporating RH strategy are increased 3.76, 3.73, and 3.74 times, respectively. The increased CPU time of rolling-horizon scheduling does not have special change with raising satellite number, which means the impact caused by RH strategy on the timing complexity of scheduling algorithm is weakly related to the satellite number, and this is consistent with the previous conclusion in the timing complexity analysis of Algorithm 2. Although the RH strategy will delay the generation of planning scheme, it also satisfies the timelessness of dynamic scheduling because the longest CPU time of the rolling-horizon scheduling is only 7.5 s, which is a low level.

Set the scheduling period from one to twelve hours in order to analyze the impact of scheduling time interval on the overall performance of planning algorithm. The six algorithms are tested in different problem scales and shown in Figure 4.

From Figure 4, the task benefits are reduced overall as the scheduling period extends gradually. This is because the extensive scheduling period will increase the rejected tasks for their deadline can not be satisfied; hence the low task benefits are obtained for satellites unable to execute the tasks in time. The heuristic algorithms which have not been embedded into RH strategy (including RH-AIS, RH-DIS, and RH-WIS) still have high task benefits in different scheduling period. Let the ratio between the task number n and the satellite number m as the approximate workload of the satellite resources; then a scheduling period between 1~8 h does not have a significant impact on the task benefits if the satellite resource has a light workload. However, the task benefits will be sharply reduced if the scheduling period is extended to 8~12 h. There are high task benefits if we maintain the scheduling period within 1~6 h to adapt the increased

TABLE 1: Simulation results in different scenarios.

Problem scale	RH-CIS		RH-DIS		RH-AIS		CIS		DIS		AIS	
	Task benefits	Guarantee ration	Task benefits	Guarantee ration	Task benefits	Guarantee ration	Task benefits	Guarantee ration	Task benefits	Guarantee ration	Task benefits	Guarantee ration
100 × 4	396	0.672	401	0.683	397	0.674	382	0.642	397	0.677	396	0.673
100 × 5	412	0.705	417	0.716	416	0.713	401	0.681	416	0.713	409	0.698
100 × 6	425	0.733	430	0.745	424	0.729	408	0.696	426	0.726	424	0.729
200 × 4	679	0.622	672	0.613	633	0.574	559	0.497	559	0.497	554	0.492
200 × 5	746	0.691	740	0.682	699	0.638	611	0.553	611	0.553	610	0.550
200 × 6	804	0.741	813	0.754	763	0.707	688	0.625	688	0.625	693	0.637
300 × 4	808	0.492	813	0.506	725	0.442	665	0.409	663	0.405	658	0.392
300 × 5	890	0.551	900	0.562	812	0.506	745	0.458	744	0.453	736	0.450
300 × 6	1027	0.649	1021	0.642	933	0.580	873	0.541	877	0.549	860	0.531
400 × 4	897	0.423	895	0.413	801	0.366	729	0.339	722	0.332	717	0.326
400 × 5	992	0.468	990	0.461	876	0.403	815	0.377	802	0.370	800	0.364
400 × 6	1152	0.549	1143	0.541	1025	0.487	959	0.448	953	0.441	946	0.435

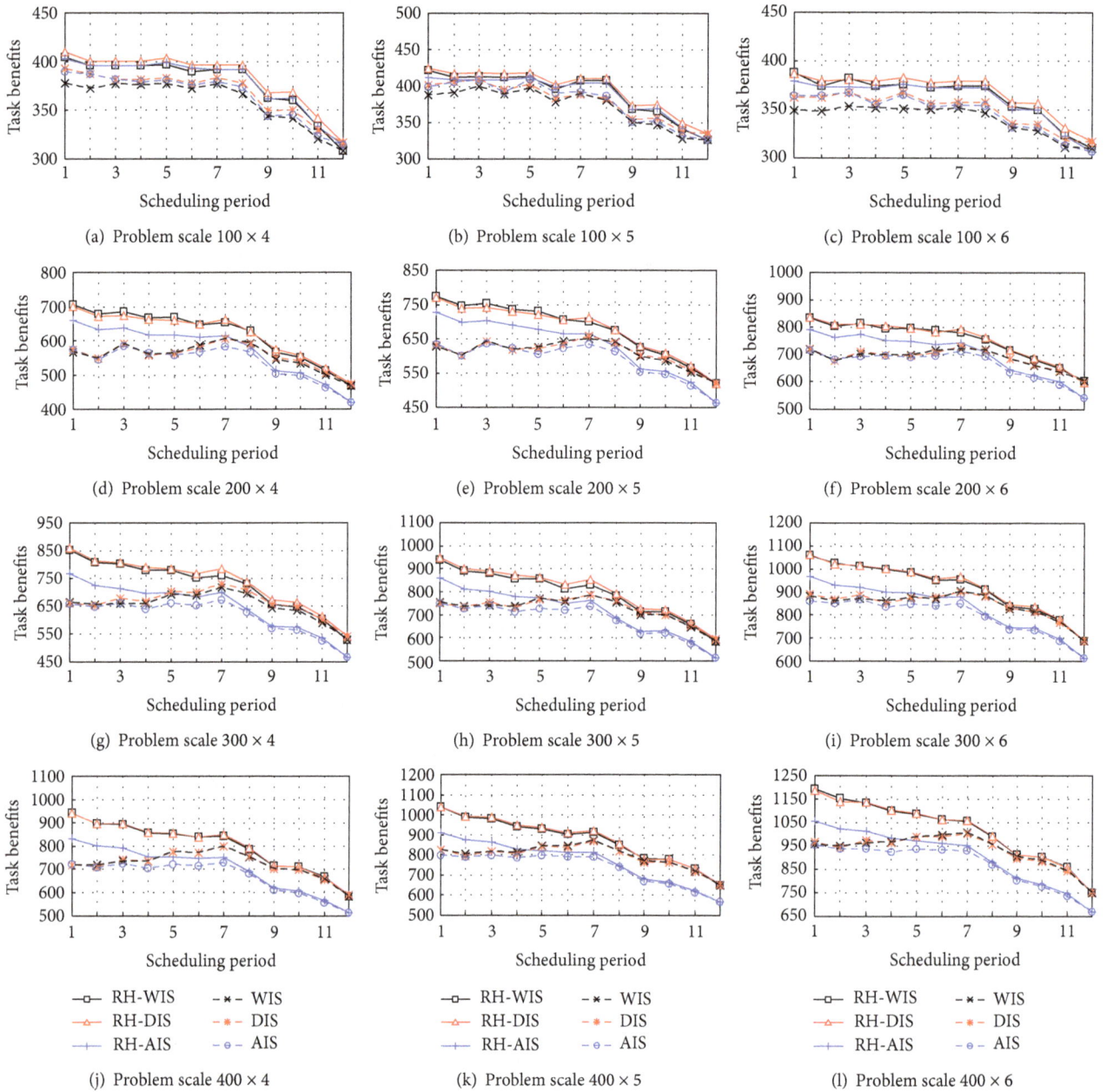

(a) Problem scale 100 × 4

(b) Problem scale 100 × 5

(c) Problem scale 100 × 6

(d) Problem scale 200 × 4

(e) Problem scale 200 × 5

(f) Problem scale 200 × 6

(g) Problem scale 300 × 4

(h) Problem scale 300 × 5

(i) Problem scale 300 × 6

(j) Problem scale 400 × 4

(k) Problem scale 400 × 5

(l) Problem scale 400 × 6

FIGURE 4: Performance of algorithms in different problem scales.

workload of the satellite resource. It is also appropriate to maintain the scheduling period within 1~3 h in accordance with the heavier workload. Furthermore, it should be ensured the scheduling period is no more than 2 h while the satellite resource has the heaviest workload. So far we conclude that the scheduling time is a key factor to affect the task benefits. In addition, the task benefits do not present monotonic change in accompany with the scheduling period as shown in Figure 4. For example, Figure 4(a) presents that the task benefits gradually increase with the extension of scheduling period within 1~4 h and fluctuate within 4~8 h but begin to reduce within 8~12 h. This is because there is a conflict

between the performance scheduling requirement of the planning system and the timeliness scheduling requirement of the tasks. The scheduling system only acquires the local task information if the scheduling period is short and yields inferior scheme which can not handle the following tasks with high timelessness. On the contrary, a large amount of high-timeliness tasks may not be finished before their deadline if the scheduling period is long, and the task benefits are reduced due to a low guarantee ration. Therefore, the scheduling period should be reasonably selected based on the satellite quantity and task density in actual application, so as to ensure the performance of scheduling system.

5. Conclusions and Future Work

This paper has studied the dynamic scheduling problem of EOSs. An integer programming model has been constructed by considering the independent arrival time and deadline of the imaging tasks. A dynamic scheduling algorithm based on the RH strategy which can be combined with multiple heuristic algorithms proposed in many researches and its timing complexity has been analyzed.

The scheduling algorithms adapted to RH strategy can effectively adjust to the planning scheme based on the satellite workload to execute the emergency tasks. The effectiveness of this strategy has been verified by comparing the scheduling results of six algorithms in the experiment. It is worth to restate that the RH strategy might cause a high time consumption to yield an optimized scheme if a large amount of tasks are involved in the scheduling system. This problem can be solved by limiting the size of the rolling horizon.

Also for our future work, we plan to research the detection method of scheduling time, which is a significant factor to impact the performance of scheduling algorithm and system. With the method in place, we will extend our algorithm to cooperative scheduling of EOSs; we will consider Qos requirements in our RH-WIS; we are going to combine the dynamic resources management into our scheme.

References

[1] F. Marinelli, S. Nocella, F. Rossi, and S. Smriglio, "A Lagrangian heuristic for satellite range scheduling with resource constraints," *Computers and Operations Research*, vol. 38, no. 11, pp. 1572–1583, 2011.

[2] P. Wang and Y. Tan, "A heuristic method for selecting and scheduling observations of satellites with limited agility," in *Proceedings of the 7th World Congress on Intelligent Control and Automation (WCICA'08)*, pp. 5292–5297, June 2008.

[3] T. Y. Mao, Z. Q. Xu, R. Hou et al., "Efficient satellite scheduling based on improved vector evaluated genetic algorithm," *Journal of Networks*, vol. 7, no. 3, pp. 517–523, 2012.

[4] W. J. Wolfe and S. E. Sorensen, "Three scheduling algorithms applied to the earth observing systems domain," *Management Science*, vol. 46, no. 1, pp. 148–168, 2000.

[5] S. M. Han, S. W. Baek, K. R. Cho, D. W. Lee, and H. D. Kim, "Satellite mission scheduling using genetic algorithm," in *Proceedings of the International Conference on Instrumentation, Control and Information Technology (SICE '08)*, pp. 1226–1230, August 2008.

[6] H. Wang, M. Xu, R. Wang, and Y. Li, "Scheduling earth observing satellites with hybrid ant colony optimization algorithm," in *Proceedings of the International Conference on Artificial Intelligence and Computational Intelligence (AICI '09)*, vol. 2, pp. 245–249, November 2009.

[7] L. Jian and W. Cheng, "Resource planning and scheduling of payload for satellite with genetic particles swarm optimization," in *Proceedings of the IEEE Congress on Evolutionary Computation (CEC '08)*, pp. 199–203, June 2008.

[8] C. He, X. M. Zhu, and D. S. Qiu, "Cooperative scheduling method of multi-satellites for imaging reconnaissance in emergency condition," *Systems Engineering and Electronics*, vol. 34, no. 4, pp. 721–731, 2012.

[9] S. Baolin, W. Wenxiang, and Q. Qianqing, "Satellites scheduling algorithm based on dynamic constraint satisfaction problem," in *Proceedings of the International Conference on Computer Science and Software Engineering (CSSE '08)*, vol. 4, pp. 167–170, December 2008.

[10] K. J. Zhu, J. F. Li, and H. X. Baoyin, "Satellite scheduling considering maximum observation coverage time and minimum orbital transfer fuel cost," *Acta Astronautica*, vol. 66, no. 1-2, pp. 220–229, 2010.

[11] J. M. Wang, J. F. Li, and Y. J. Tan, "Study on heuristic algorithm for dynamic scheduling problem of earth observing satellites," in *Proceedings of the 8th ACIS International Conference on Software Engineering, Artificial Intelligence, Networking, and Parallel/Distributed Computing (SNPD '07)*, vol. 1, pp. 9–14, August 2007.

[12] J. Wang, N. Jing, J. Li, and Z. H. Chen, "A multi-objective imaging scheduling approach for earth observing satellites," in *Proceedings of the 9th Annual Genetic and Evolutionary Computation Conference (GECCO '07)*, pp. 2211–2218, July 2007.

[13] D. T. Chi and Y. T. Su, "On a satellite coverage problem," *IEEE Transactions on Aerospace and Electronic Systems*, vol. 31, no. 3, pp. 891–896, 1995.

[14] S. K. Chronopoulos, C. T. Angelis, A. Koumasis, and P. Drakou, "Satellite coverage analysis for the investigation of real-time communication in selected areas," *WSEAS Transactions on Communications*, vol. 5, no. 10, pp. 1965–1972, 2006.

[15] G. Verfaillie, C. Pralet, and M. Lemaître, "Constraint-based modeling of discrete event dynamic systems," *Journal of Intelligent Manufacturing*, vol. 21, no. 1, pp. 31–47, 2010.

[16] C. He, X. M. Zhu, H. Guo et al., "Rolling-horizon scheduling for energy constrained distributed real-time embedded systems," *Journal of Systems and Software*, vol. 85, no. 4, pp. 780–794, 2012.

[17] I. M. Ovacik and R. Uzsoy, "Rolling horizon algorithms for a single-machine dynamic scheduling problem with sequence-dependent setup times," *International Journal of Production Research*, vol. 32, no. 6, pp. 1243–1263, 1994.

[18] L. Tang, S. Jiang, and J. Liu, "Rolling horizon approach for dynamic parallel machine scheduling problem with release times," *Industrial and Engineering Chemistry Research*, vol. 49, no. 1, pp. 381–389, 2010.

[19] U. Schmid and J. Blieberger, "Some investigations on FCFS scheduling in hard real time applications," *Journal of Computer and System Sciences*, vol. 45, no. 3, pp. 493–512, 1992.

[20] M. Bertogna and S. Baruah, "Tests for global EDF schedulability analysis," *Journal of Systems Architecture*, vol. 57, no. 5, pp. 487–497, 2011.

[21] F. Zhang and A. Burns, "Schedulability analysis for real-time systems with EDF scheduling," *IEEE Transactions on Computers*, vol. 58, no. 9, pp. 1250–1258, 2009.

[22] Y. Chen, M. Q. Zhou, and H. Zou, "Multi-satellite observation scheduling algorithm based on hybrid genetic particle swarm optimization," *Lecture Notes in Electrical Engineering*, vol. 136, pp. 441–448, 2012.

[23] C. He, D. S. Qiu, and J. Liu, "Cooperative scheduling of imaging observation tasks for high-altitude airships based on propagation algorithm," *The Scientific World Journal*, vol. 2012, Article ID 548250, 13 pages, 2012.

[24] W. C. Lin and D. Y. Liao, "A tabu search algorithm for satellite imaging scheduling," in *Proceedings of IEEE International Conference on Systems, Man and Cybernetics (SMC '04)*, vol. 2, pp. 1601–1606, October 2004.

6

Turbine Rotor Disk Health Monitoring Assessment Based on Sensor Technology and Spin Tests Data

Ali Abdul-Aziz and Mark Woike

NASA Glenn Research Center, Cleveland, OH 44135, USA

Correspondence should be addressed to Ali Abdul-Aziz; ali.abdul-aziz-1@nasa.gov

Academic Editors: P.-C. Chen and M. Curioni

The paper focuses on presenting data obtained from spin test experiments of a turbine engine like rotor disk and assessing their correlation to the development of a structural health monitoring and fault detection system. The data were obtained under various operating conditions such as the rotor disk being artificially induced with and without a notch and rotated at a rotational speed of up to 10,000 rpm under balanced and imbalanced state. The data collected included blade tip clearance, blade tip timing measurements, and shaft displacements. Two different sensor technologies were employed in the testing: microwave and capacitive sensors, respectively. The experimental tests were conducted at the NASA Glenn Research Center's Rotordynamics Laboratory using a high precision spin system. Disk flaw observations and related assessments from the collected data for both sensors are reported and discussed.

1. Introduction

The strive to develop a robust health monitoring system to detect rotating engine component malfunctions is among the key areas of interest for engine companies and the associated aviation industry. Typically, health monitoring is performed using sensor systems and other similar means that are capable of functioning under harsh and severe environmental operating conditions. Such systems are to operate without interference with the overall operation of the engine. However, implementation of such technology is highly dependent on many factors and among them setting up specific types of experiments to simulate representative turbine engine conditions and frequent mishaps that the engine encounters during operation. Supportive studies like analytical verification and modeling are equally important in order to verify and complement the experimental findings. Testing under high temperature and wireless technology using durable and effective sensor technologies is also highly desirable.

Health monitoring is not only confined to sensor technology since there are many other ways of conducting such inspection which are mostly nondestructive evaluation-based approaches. These approaches are widely used in the aviation industry to track engine component performance and durability. They are further used to locate cracks and other anomalies before they become a risk factor that leads to catastrophic failure. Nevertheless, some if not most of these techniques can be both costly and impractical, in particular, when it comes to inspecting complex geometries and large structures [1]. Therefore, the urge for developing systematic, reliable and realistic diagnostic tools to detect damage and monitor the health of key components in the engine, such as rotor disks and turbine blades, is highly in need. It is greatly fundamental to maintaining engine safety, dependability, and life [2].

The NASA Aviation Safety Program under the Vehicle Systems Safety Technology (VSST) project is taking the lead in partnership with the Federal Aviation Administration, Aviation Industry and the Department of Defense [3], to promote the development of these technologies to improve and reduce the fatal aviation accidents and assist safety as a whole. This effort is being carried out at NASA Glenn Research Center through the Optical Instrumentation and NDE branch by conducting controlled spin experiments of turbine engine rotor like test articles to explore various sensing advancements for local and global detection of rotor damage. Comparison of test data for baseline disks without any damage with that of a disk with artificially induced damage, a small crack or a notch, is performed to appraise the findings. Hence, this paper presents experimental results obtained from spin tests of a rotor disk and their association to the development of a structural health monitoring and fault detection system.

2. Technical Approach

The experimental work in this study considered a conceptual design of a disk with machined teeth to imitate compressor or turbine blades and provide a cost effective test article to simulate crack initiation and propagation; see Figure 1. The central region of the disk is counter-bored on both sides to create the rim, web, and bore regions of a typical turbine disk. The machined teeth on the rim simulate tip passing, but they trim down the blade mass loading on the web and bore usually experienced in most rotors. The goal is to induce changes in radial tip displacement without disk yielding in order to test the instrumentation and to then initiate and grow cracks by machining and/or increasing rotational speed. Two sensor types (capacitive and microwave) are employed to capture the blade tip clearance both for health monitoring and comparison purposes. An eddy current sensor is also included in the system to measure the shaft displacements.

Figure 1 illustrates a description of the disk specimen and the induced crack notch along with the tip clearance probes. The test specimen disk has an outside diameter of 23.495 cm (9.25 in), a bore and an outside rim thickness of 2.54 cm (1 in) and 3.175 cm (1.25 in), the thickness of the web is 0.254 cm (0.10 in), and the cross section and height of the blades are 3.175 cm × 0.330 cm (1.25 in × 0.13 in) and 0.838 cm (0.33 in), respectively. It has rotor-like blades, a total of 32, evenly spaced around the circumference. Eight holes, 0.508 cm (0.20 in) diameter each, were drilled through the disk halfway in the rim. The holes were spaced every 45°, and they were designed for future studies as possible mass add-on points or notch initiation sites. The disk specimen is made out of nickel base material alloy Haynes X-750 and it weighs approximately 4.88 Kg (10.75 Lb).

The notch had a width of 0.381 mm (0.015 in) as per wire thickness and burn area of the electric discharge machining (EDM) process. The notch region was intentionally selected to be in the web area since finite element analysis results revealed that this section encounters the highest stress level in the disk during the spin operation [1, 2]. Technical considerations were emphasized to preserve system consistency of all the operating parameters and other experimental conditions during the removal-reinstallation process of the disk specimen in both situations, baseline no-notch and notch states.

3. Sensor Technology

3.1. Capacitive Probe Sensor.

For blade tip clearance measurements, a capacitive sensor system was installed; see Figure 1(a). These types of sensors are based on a direct current (DC) offset, an offsetting of a signal from zero where it refers to a direct current voltage, rather than a modulation technique which is a method used to digitally represent sampled analog signals. The capacitive sensors are designed to monitor the electrical property of "capacitance" to initiate and take measurements. The capacitance is a function of the physical dimensions (geometry) of the conductors and the permittivity of the dielectric. It is defined as a field that exists between two conductive surfaces within some rational proximity. Capacitance is directly proportional to the surface area of the conductor plates and inversely proportional to the separation distance between the plates. Variations in the distance between the surfaces lead to changes in the capacitance rate. This rate change is used by the sensors to indicate the difference in position of a target. High-performance displacement sensors use small sensing surfaces and as a result are positioned close to the targets (0.25–2 mm). The DC voltage, in conjunction with the motion of the rotor, allowed the current system to record three channels at a rate of 1 MHz each.

3.2. Microwave Sensor Background and Theory.

The microwave tip clearance sensor system works on principles that are similar to a short-range radar system. The tip clearance probe is both a transmitting and receiving antenna [3–10]. The sensor emits a continuous microwave signal and measures the signal that is reflected off a passing blade. The motion of the blade modulates the reflected signal. The reflected signal is then compared to an internal reference signal and the phase difference directly corresponds to the distance to the blade. The system consists of two major components. The first component is the probe, (Figure 2(a)). The second component is the sensor electronics, (Figure 2(b)). The probe contains the transmitting and receiving antenna and is designed to be installed in the casing of the engine where it can measure the radial clearance between the face of the sensor and the turbine blade tips. The probes are made of high temperature material and are designed to operate in temperatures of 900°C uncooled, 1200°C with cooling air. Two generations of probes are in operation. The first-generation probes operate at 5.8 GHz and can measure clearance distances up to ~ 25 mm (i.e., one-half the radiating wavelength). The second-generation probes operate at 24 GHz and in theory can measure clearance distances up to ~6 mm. In regards to physical size, the first generation probes are approximately 14 mm in diameter and 26 mm long. The second-generation probes are approximately 9 mm in diameter and 19 mm long. This technology has an ultimate goal of obtaining clearance accuracies approaching 25 μm. Accuracies in this order were observed in the laboratory during testing [4]. A frequency response of up to 5 MHz is typical, with up to 25 MHz being possible with this technology.

The sensor electronics consist of the radio frequency (RF) generator, RF detector, and all of the associated hardware required to generate, measure, and convert the microwave signals into a displacement reading [4].

The sensor electronics are designed to be located off-board of the engine in an environmentally benign area. The probes are connected to the sensor electronics using a microwave rated coaxial cable. A rack-mounted PC is used to interface to the sensor electronics and run the data acquisition and display software. The data acquisition computer is connected to the sensor electronics through a network switch. The data acquisition computer is intended to be remotely

(a) Capacitive sensor probe

(b) 3D model with the Notch

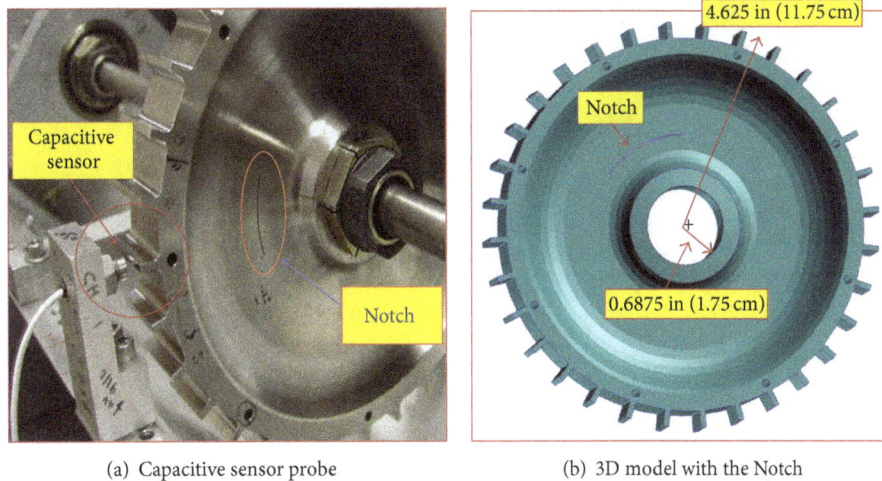

FIGURE 1: Test disk capacitive sensor assembly.

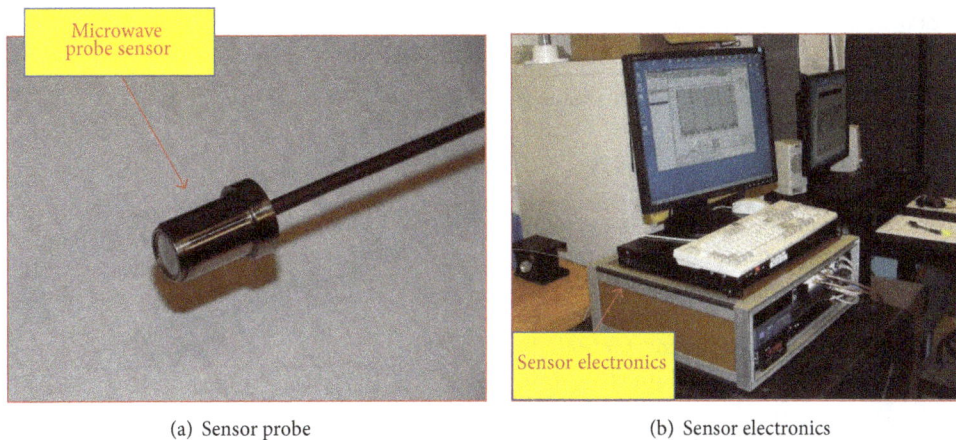

(a) Sensor probe

(b) Sensor electronics

FIGURE 2: Microwave sensor probe and its electronics setup.

located away from the sensor electronics in an area such as a control room using a CAT5E connection.

3.3. Capacitive and Microwave Sensors Performance. Figure 3 demonstrates a comparison between the two primary blade tip clearance sensors, microwave and capacitive, attached to the data system. The data collected for both sensors compared relatively well. However, the similarity is not as uniform, the similarity is not as uniform as anticipated which implies that certain calibration is needed to fine-tune the sensor system further and bring the data closer. Perhaps, adding an additional filtering or average process for the microwave sensor may result in improving the agreement between the two measurements. Nevertheless, each sensor system is operating as expected. Their role in the experiment configuration is to serve the same functionality in a different fashion and to test their performance.

4. Experimental Results

Spin tests were performed on the rotor disk and covered baseline runs with both undamaged and damaged disks via

the artificially induced notch, shown in Figure 1. The tests included spinning the rotor under various simulated engine mission profiles starting from a minimum rotational speed of 3000 up to a maximum of 10000 rpm. The controlled speed applied during the current testing was made with an acceleration-deceleration rate of 60 rpm/second. This insured passing the critical speed of 2,610 rpm and leading to post-critical state [11]. Figure 4 shows samples of two mission profiles that were used to test the rotors. These profiles were derived on the basis of revolutions per minute data obtained on different flights comprising different flight maneuvers [12].

Figure 4(a) is referred to as the constant engine power cycle profile; however the graph shown does not illustrate the constant behavior due to an input offset. The engine speed reaches 10,000 rpm in two steps, a take off with a brief hold up at 5000 rpm and 40 seconds hold at 10,000 rpm with a rapid decrease to 5000 rpm and a ramp up to 10,000 rpm for one repetitive cycle. Figure 4(b) shows another mission profile (engine cyclic power cycle) that allows the rotor disk to go through somewhat analogous series of events starting at a speed beyond the critical value [11] and up to 8500 rpm. These profiles are being used to imitate unusual engine conditions

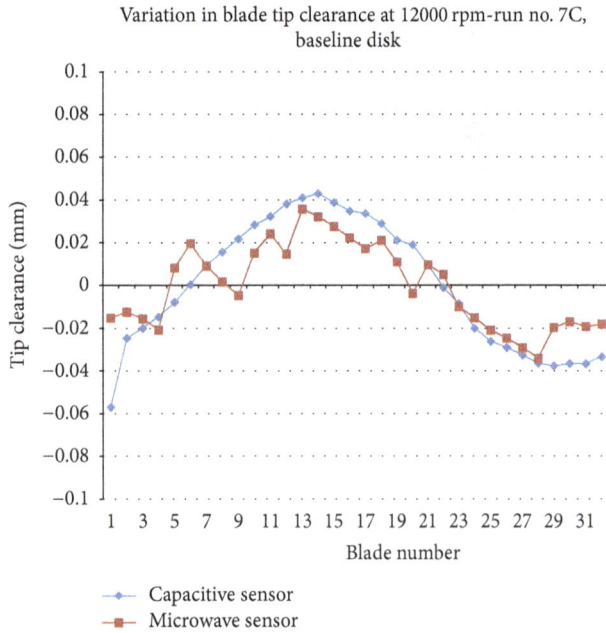

Figure 3: Sample of capacitive and microwave sensor data readings for a baseline disk at 12000 rpm.

and to help in evaluating the rotor performance under harsh and complex loading events in an attempt to fatigue the disk with the expectation that all the existing anomalies would appear in the test data. Additionally, under these conditions, these experiments supplied valuable assessments for both the crack detection scheme and the structural durability of the disk materials.

4.1. Spin Test Results. Experimental data under both mission profiles are represented in Figures 5 and 6. The data in Figure 5 are produced under the constant power cycle mission for both the baseline no-notch and the notched disks. The mission history is shown along with the trace vibration vector and Bode plots for the phase and amplitude response. Bode plots are a very useful way to represent the gain and phase of a system as a function of frequency. This is referred to as the frequency domain behavior of a system. The magnitude and phase plots determine the phasor representation of the transfer function at any frequency. It is typically used for transient analysis in both run-up and run-down tests. It can help identify the resonance speed of a rotor or examine the rotor dynamics on an order basis. The x-axis in a Bode plot is speed or frequency, which enables seeing the changes in magnitude and phase over speed or frequency.

A clear observation of data variation is noted in Figures 5 and 6 between the two plots. For instance, a circular loop representation for the trace of vibration vector (disk vibration response) is seen for the baseline disk, while a gap in the loop is present for the notched disk. This behavior hints that a difference in the vibration response for the two structures is present signifying the existence of some type of irregularity. Such observation has been reported in [13–17], where a crack in the rotor disk is documented via the presence of distorted

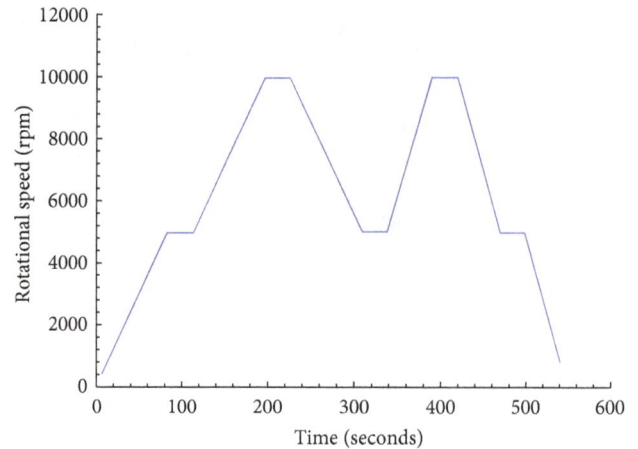

(a) Constant engine power cycle profile

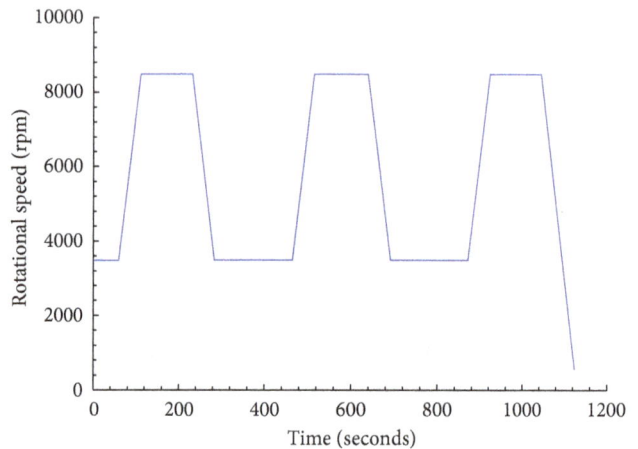

(b) Engine cyclic power cycle

Figure 4: Simulated engine mission history test profiles.

trace of the vibration vector distribution and a rise in the phase and amplitude response upon surpassing the 1st critical speed; *see Figure 6 for additional clarifications and captions notation, axes labels and units.*

This conduct is certainly noticeable in Figure 5(a) for the notched disk. A rise in the phase magnitude response is noted. Also, it is noted that the peak is at a critical frequency (5000–9000 rpm), and then it begins to settle out at maximum speed close to critical frequency. So the damping ratio keeps the curve from flattening compared to no crack, (Figure 5(b)). Note how the magnitude phase graphs no longer represent a complete circle in the notched disk case, (Figure 7(a)).

This is a sign of a crack growing and is detected from the plot of a cycle worth of data. At the same time the magnitude graph (Figure 5(a)) has started a ω^2 (ω is the rotational speed) rise after settling past the critical frequency. A rise in amplitude and constant phase is typically an indication of the growth of a crack, assuming that some internal movement

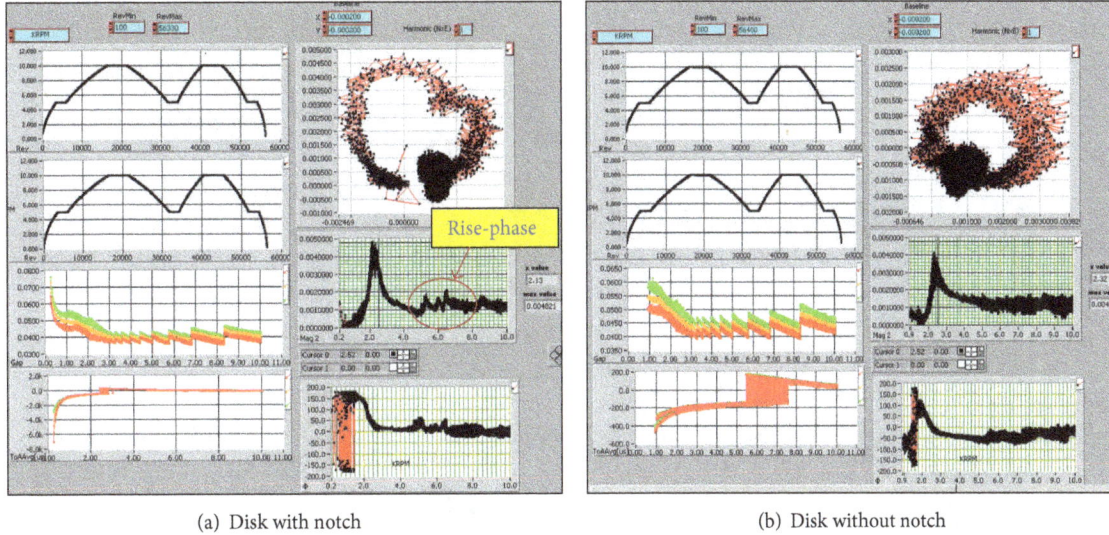

(a) Disk with notch

(b) Disk without notch

FIGURE 5: Bode Plots 9 minutes comparison test of the disk with and without notch, capacitive sensor data, Engine cyclic power profile at 5–10 Krpm rotational speed.

of the rotating structure does not cause the unbalance [13]. In this case the system is tracking a crack growing in the disk. Therefore, examination of the above data has verified that the detection scheme based on the blade tip clearance response allows identifying the presence of some sort of fault in the rotor disk. However, further confirmation is warranted throughconducting more tests for different rotors under similar operating conditions to authenticate that this type of behavior in the vibration response is accurately due to some existing structural defects in the rotor rather than a system-related unbalance [18–20].

Figure 6 shows the results obtained under the cyclic run time mission profile. The trace of vibrations vector in the Bode plot contains rather distorted data with an incomplete circular shape. Also, the amplitude shows a mild rise at the 5000–6000 rpm range which substantiates the theory of the presence of an unbalance condition or an anomaly state as noted in the data presented in Figure 5. However, the manifestation of such observations remains not as straightforward but it underlines or it confirms the presence of some type of defect. Still, further work to confirm this scrutiny is needed.

The data reported in Figure 7 shows the test output of the microwave sensor for a constant amplitude engine history profile. Only data for the takeoff portion of the profile up to the first 300 revolutions is shown, (Figure 7(a)). This is for a 1500 seconds long test at 100 rpm acceleration/deceleration rate for a notched disk. The response of the microwave sensor is very similar to that of the capacitive sensor; it has been introduced into the testing scheme to investigate its applicability and performance for engine health monitoring applications. And as mentioned earlier, the microwave tip clearance sensor system works on principles that are similar to a short-range radar system and are different than those for the capacitive sensor. The probe is both a transmitting and a receiving antenna; it emits a continuous microwave signal and measures the signal that is reflected off a rotating blade.

Figure 7(b) shows a magnified phase amplitude output produced by the data obtained from the microwave sensor for a spin test of a notched disk at 10,000 rpm rotational speed. It is obviously noted that a rise in the phase exhibiting a second-degree order (ω^2) is recognized at a speed range of 7000 to 10000 rpm. This supports the observations made earlier for the capacitive sensor data concerning the crack detection phenomena in the rotor and the similarity of the microwave sensor performance; see Figure 5.

4.2. *Unbalance Test Results.* An unbalance test was performed to institute a baseline database for the rotor at various operating conditions and to support investigating and evaluating the vibration response under nonordinary service environment such as imbalance situations. The test was conducted at 10,000 rpm at 100 seconds acceleration/deceleration rate and lasted 4.33 minutes long. The test covered a standard mission profile under transient ramp up, cruise, and ramp down conditions. Test conditions were kept the same as those applied during the non-unbalance state to enable precise assessments of the imbalance factors and their impact on the rotor vibrations response under consistent and refined test margins for the same rotor. Figure 8(a) shows a photo of the disk used for the unbalance tests with labels indicating the sites of the locations of the weight (0.5 gram) during the tests.

As noted, two cases were considered: one case with the extra weight being along the notch side (position A) and another one with weight being across from the notch (position B). Additionally, the purpose of the unbalance test is to check the capability of the sensors technology not only in crack detection, but also in predicting other major malfunctions in the rotor system such as unbalance provision. This further supports the theory which relates to rotor design; whereas the rotational velocity of any rotating object

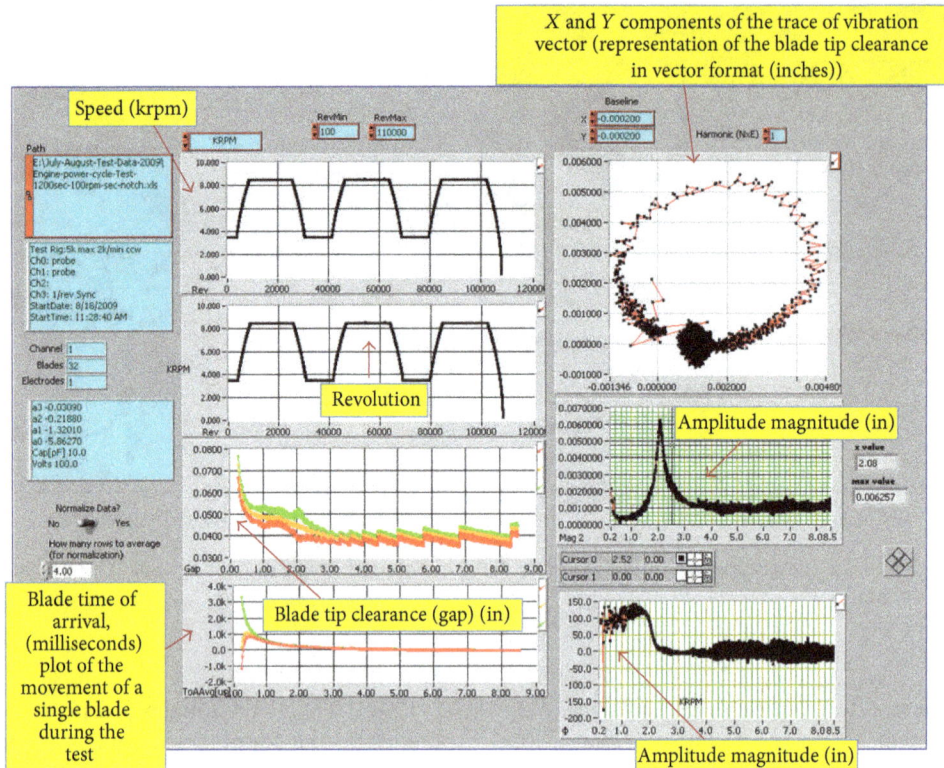

FIGURE 6: Engine constant power cycle test: capacitive sensor data at 8500 rpm, 1200 sec long at 100 rpm acceleration/deceleration rate for the Notched disk.

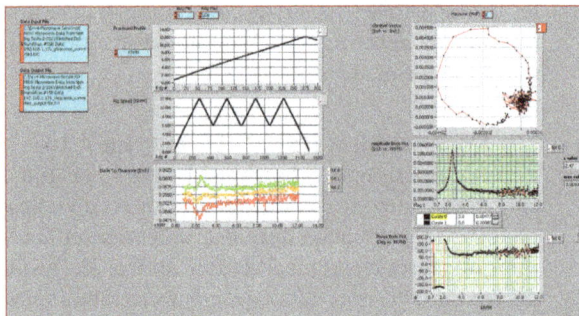

(a) Bode plot output

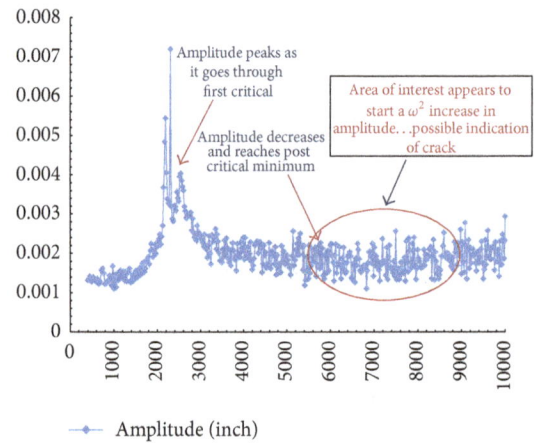

(b) Vibration amplitude as a function of rotational speed

FIGURE 7: Simulated engine mission history profile (constant amplitude) test, microwave sensor data.

increases, its level of vibration often passes through a maximum at what is called a critical speed [11]. This is commonly excited by unbalance of the rotating structure. If the amplitude of vibration at these critical speeds is excessive, catastrophic failure can occur. For this reason, it is typically recommended that in large rotors design the appropriate approach is to use physical prototypes and tests in order to ensure safe rotating machinery design and balance it well.

Figure 8(b) shows a test that runs for the notched rotor disk under unbalance test with the mass being at position B as indicated in Figure 8(a). Position B is on the opposite side of the notch. This location was intentionally chosen to investigate imbalance effects not only at positions within the region surrounding the notch, but also at areas outside the notch section and further out. Test results at "position A" are not shown due to space limitation and since similar results were observed as well.

It is also noted that the trace of vibration vector shown in Figure 8(b) shows a very irregular distribution signifying that the imbalance had eliminated the typical complete circular

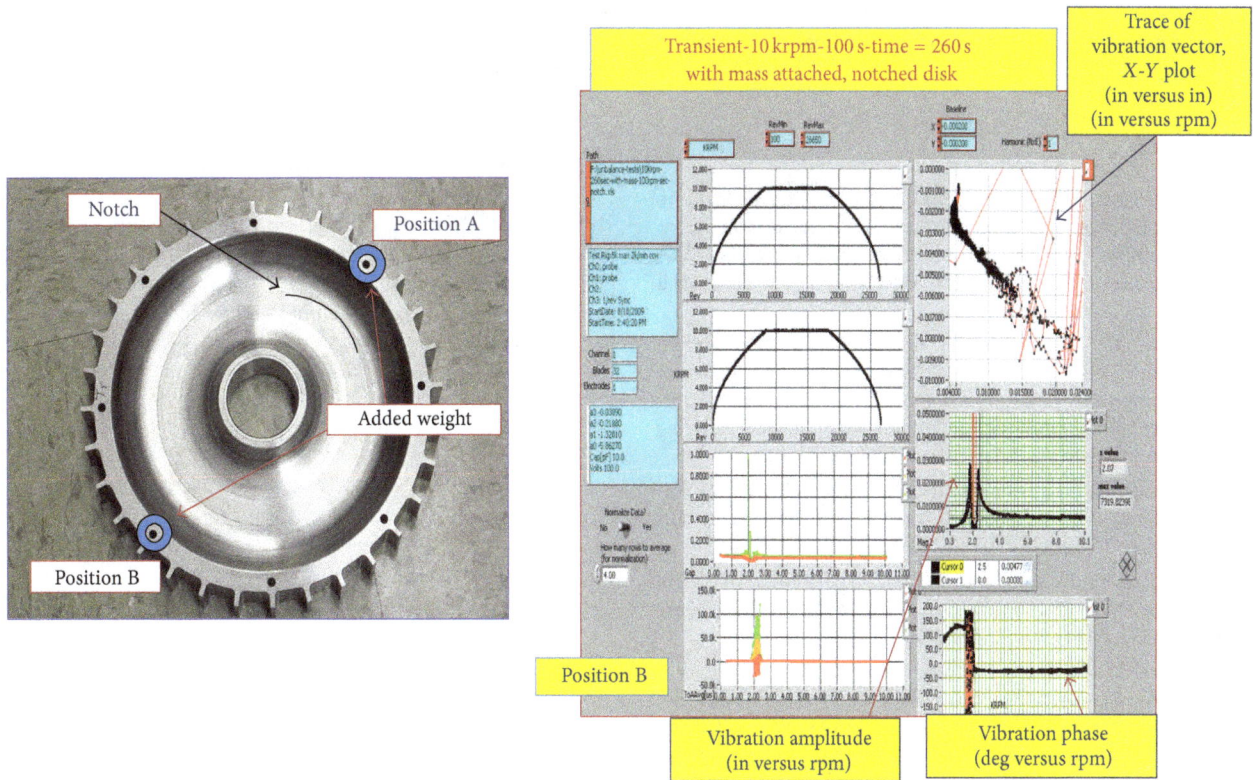

(a) Unbalance layout with added weight and locations

(b) Unbalance vibration response

FIGURE 8: Tip clearance variation with blade numbers and simulated rotor unbalance layout.

shape usually encountered when a clean baseline rotor disk without any balance weight or damage is in operation. This further designates that having an imbalance state whether it is due to the existence of a notch or other factors within the system will lead to nonordinary shape of the trace of the vibration vector curve. Additionally, these data confirm that upon altering the disk weight by adding or removing mass, the state of imbalance is imminent and the tip clearance magnitude will depend on the added mass size and location. Further, the findings of these results substantiated the purpose of these tests to help determine that the blade tip clearance sensors have recognized the significance of these critical parameters and their influence on the disk performance.

5. Conclusions

The work performed in this paper involved conducting spin experiments on the rotor disk with and without an artificially induced notch at different rotational speed levels. Bode plots of data from these tests provided indications of differences between the undamaged rotor with that of the damaged one. Crack growing signs have been detected where the amplitude magnitude in the Bode plot has started a ω^2 rise after settling past the critical frequency. This growth of the magnitude is a noteworthy indicator of a crack growth or another fault which confirms similar results cited by prior studies. Microwave and capacitive sensors delivered closely similar data for most of

the blades except for some where the measurements were slightly dissimilar which can be attributed to factors such as equipment calibration and apparatus instrumentation fine tuning.

The trace of vibration vector which represents the imbalance mass showed very inconsistent distribution compared to data obtained from a balanced disk. This showed that the imbalance had eliminated the complete circular shape typically encountered when a clean baseline rotor disk without any damage is in operation. Additionally, this indicated that having an imbalance state whether it is due to the existence of a notch or other factors within the system, the trace of the vibration vector curve will have an asymmetrical non-circular and highly distorted shape. This leads to the conclusion that the data obtained from spin testing of the rotor to some extent showed that the detection scheme based on the blade tip clearance response is capable of identifying the presence of defects in the rotor.

Lastly, the experimental data enabled exploring the difference in the vibration response between a baseline and a damaged rotor suggesting when the existence of some type of anomaly is present. Also, the combined sensor technology, which included the capacitive, microwave and eddy current probes, supplemented the tests with ample evidence and allowed exploring the changes in the disk vibration response at different operating conditions. However, further work and more testing must be continued to develop, improve and link this experimental investigation to put forward a more precise and accurate appraisal of monitoring the health of rotating components.

References

[1] A. Abdul-Aziz, M. R. Woike, J. D. Lekki, and G. Y. Baaklini, "Development of a flaw detection/health monitoring scheme for turbine engine rotating components," in *presented at the AIAA Infotech at Aerospace*, AIAA 2010-3329, Atlanta, Ga, USA, April 2010.

[2] A. Abdul-Aziz, M. R. Woike, G. Abumeri, J. D. Lekki, and G. Y. Baakilini, "NDE using sensor based approach to propulsion health monitoring of a turbine engine disk," in *Health Monitoring of Structural and Biological Systems*, vol. 7295 of *Proceeding SPIE*, San Diego, Calif, USA, March 2009.

[3] U. S. Department of Transportation and Federal Aviation Administration, *Engine Damage-Related Propulsion System Malfunctions*, DOT/FAA/AR-08/24, 2008.

[4] M. Woike, A. Abdul-Aziz, and T. Bencic, *A Microwave Blade Tip Clearance Sensor for Propulsion Health Monitoring*, NASA/TM-2010-216736, AIAA-2010-3308, 2010.

[5] M. Woike, J. Roeder, C. Hughes, and T. Bencic, *Testing of a Microwave Blade Tip Clearance Sensor at the NASA Glenn Research Center*, NASA/TM-2009-215589, AIAA-2009-1452, 2009.

[6] A. Abdul-Aziz, M. Woike, J. Lekki, and G. Y. Baaklini, *Health Monitoring of a Rotating Disk Using A Combined Analytical-Experimental Approach*, NASA Technical Memorandum, NASA/TM-2009-215675, 2009.

[7] T. A. Holst, T. R. Kurfess, S. A. Billington, J. L. Geisheimer, and J. L. Littles, *Development of an Optical-Electromagnetic Model of a Microwave Blade Tip Sensor*, AIAA-2005-4377, 2005.

[8] J. L. Geisheimer, S. A. Billington, T. Holst, and D. W. Burgess, *Performance Testing of a Microwave Tip Clearance Sensor*, AIAA-2005-3987, 2005.

[9] J. L. Geisheimer, S. A. Billington, and D. W. Burgess, *A Microwave Blade Tip Clearance Sensor for Active Clearance Control Applications*, AIAA-2004-3720, 2004.

[10] T. A. Holst, *Analysis of spatial filtering in phase-based microwave measurements of turbine blade tips [M.S. thesis]*, Georgia Institute of Technology, Atlanta, Ga, USA, 2005.

[11] A. L. Gyekenyesi, J. T. Sawicki, and W. C. Haase, "Modeling disk cracks in rotors by utilizing speed dependent eccentricity," *Journal of Materials Engineering and Performance*, vol. 19, no. 2, pp. 207–212, 2010.

[12] U. S. Department of Transportation, Federal Aviation Administration, and Office of Aviation Research, *Turbine Engine Fan Disk Crack Detection Test*, DOT/FAA-AIAR-04/28, Washington, DC, USA, 2004.

[13] W. C. Hass and M. J. Drumm, "Detection, discrimination and real-time tracking of cracks in rotating disks," in *IEEE Aerospace Conference*, Big sky, March 2002.

[14] M. Drumm and C. W. Haase, *High Performance Rotor Health Monitoring*, Exsell Instruments, Acton, Mass, USA, 2000.

[15] W. C. Hasse, "High performance rotor health monitoring," in *Proceedings the SAE Conference on Engine Condition Monitoring*, Montreal, Canada, October 1999.

[16] W. C. Hasse and M. J. Drumm, "Detection, discrimination and real-time tracking of cracks in rotating disks," in *Proceedings of the 9th International Symposium on Transport Phenomena and Dynamics of Rotating Machinery (ISROMAC-9 '02)*, Honolulu, Hawaii, USA, February 2002.

[17] A. S. Sekhar and B. S. Prabhu, "Condition monitoring of cracked rotors through transient response," *Mechanism and Machine Theory*, vol. 33, no. 8, pp. 1167–1175, 1998.

[18] D. E. Bentlty, "Detecting cracked shafts at earlier levels," *Orbit Magazine, Bently Nevada*, vol. 3, no. 2, 1982.

[19] J. Wauer, "On the dynamics of cracked rotors: a literature survey," *Applied Mechanics Review*, vol. 43, no. 1, pp. 13–17, 1990.

[20] A. K. Koul and R. V. Dainty, "Fatigue fracture of aircraft engine compressor disks," in *Rotating Equipment, Handbook of Case Histories in Failure Analysis*, vol. 1, pp. 241–250, 2002.

7

A Bat Algorithm with Mutation for UCAV Path Planning

Gaige Wang,[1,2] Lihong Guo,[1] Hong Duan,[3] Luo Liu,[1,2] and Heqi Wang[1]

[1] *Changchun Institute of Optics, Fine Mechanics and Physics, Chinese Academy of Sciences, Changchun 130033, China*
[2] *Graduate School of Chinese Academy of Sciences, Beijing 100039, China*
[3] *School of Computer Science and Information Technology, Northeast Normal University, Changchun 130117, China*

Correspondence should be addressed to Lihong Guo, guolh@ciomp.ac.cn

Academic Editors: E. Acar and I.-S. Jeung

Path planning for uninhabited combat air vehicle (UCAV) is a complicated high dimension optimization problem, which mainly centralizes on optimizing the flight route considering the different kinds of constrains under complicated battle field environments. Original bat algorithm (BA) is used to solve the UCAV path planning problem. Furthermore, a new bat algorithm with mutation (BAM) is proposed to solve the UCAV path planning problem, and a modification is applied to mutate between bats during the process of the new solutions updating. Then, the UCAV can find the safe path by connecting the chosen nodes of the coordinates while avoiding the threat areas and costing minimum fuel. This new approach can accelerate the global convergence speed while preserving the strong robustness of the basic BA. The realization procedure for original BA and this improved metaheuristic approach BAM is also presented. To prove the performance of this proposed metaheuristic method, BAM is compared with BA and other population-based optimization methods, such as ACO, BBO, DE, ES, GA, PBIL, PSO, and SGA. The experiment shows that the proposed approach is more effective and feasible in UCAV path planning than the other models.

1. Introduction

Uninhabited combat aerial vehicle (UCAV) is one of inevitable trends of the modern aerial weapon equipment which develop in the direction of unmanned attendance and intelligence. Research on UCAV directly affects battle effectiveness of the air force and is fatal and fundamental research related to safeness of a nation. Path planning and trajectory generation is one of the key technologies in coordinated UCAV combatting. The flight path planning in a large mission area is a typical large scale optimization problem; a series of algorithms have been proposed to solve this complicated multiconstrained optimization problem, such as differential evolution [1], biogeography-based optimization [2, 3], genetic algorithm [4], ant colony algorithm [5] and its variant [6, 7], cuckoo search [8, 9], chaotic artificial bee colony [10], firefly algorithm [11, 12], and intelligent water drops optimization [13]. However, those methods can hardly solve the contradiction between the global optimization and excessive information.

In 1995, Storn and Price firstly proposed a novel evolutionary algorithm (EA): differential evolution (DE) [14], which is a new heuristic approach for minimizing possibly nonlinear and nondifferentiable continuous space functions. It converges faster and with more certainty than many other acclaimed global population-based optimization methods. This new method requires few control variables, which makes DE more robust and easy to use and lend itself very well to parallel computation.

First presented in [15], the bat-inspired algorithm or bat algorithm (BA) is a metaheuristic search algorithm, inspired by the echolocation behavior of bats with varying pulse rates of emission and loudness. The primary purpose of a bat's echolocation is to act as a signal system to sense distance.

However, in the field of path planning for UCAV, no application of BA algorithm exists yet. In this paper, we use an original BA and an improved modified BA algorithm to solve UCAV path planning problem. Here, we add mutation operation in DE between bats to propose a new metaheuristic algorithm according to the principle of BA, and then an improved BA algorithm is used to search the optimal or suboptimal route with complicated multiconstraints. To investigate the feasibility and effectiveness of our proposed approach, it is compared with BA and other population-based optimization methods, such as ACO, BBO, DE, ES, GA, PBIL, PSO, and SGA under complicated combating environments. The simulation experiments indicate that our hybrid metaheuristic method can generate a feasible optimal route for UCAV more effectively than other population-based optimization methods.

The remainder of this paper is structured as follows. Section 2 describes the mathematical model in UCAV path planning problem. Subsequently, the principle of the basic BA is explained in Section 3, and then an improved BA with mutation for UCAV path planning is presented in Section 4 and the detailed implementation procedure is also described in this section. The simulation experiment is conducted in Section 5. Finally, Section 6 concludes the paper and discusses the future path of our work.

2. Mathematical Model in UCAV Path Planning

Path planning for UCAV is a new low altitude penetration technology to achieve the purpose of terrain following and terrain avoidance and flight with evading threat, which is a key component of mission planning system [16]. The goal for path planning is to calculate the optimal or suboptimal flight route for UCAV within the appropriate time, which enables the UCAV to break through the enemy threat environments, and self-survive with the perfect completion of mission. In our work, we use the mathematical model in UCAV path planning in [1], which is described as follows.

2.1. Problem Description. Path planning for UCAV is the design of optimal flight route to meet certain performance requirements according to the special mission objective and is modeled by the constraints of the terrain, data, threat information, fuel, and time. In this paper, firstly the route planning problem is transformed into a D-dimensional function optimization problem (Figure 1).

In Figure 1, we transform the original coordinate system into new coordinate whose horizontal axis is the connection line from starting point to target point according to transform expressions shown as (1), where the point (x, y) is coordinate in the original ground coordinate system O_{XY}; the point (x', y') is coordinate in the new rotating coordinate system $O_{X'Y'}$; θ is the rotation angle of the coordinate system. One has

$$\theta = \arcsin \frac{y_2 - y_1}{\left| \vec{AB} \right|},$$

$$\begin{pmatrix} x \\ y \end{pmatrix} = \begin{pmatrix} \cos\theta & \sin\theta \\ -\sin\theta & \cos\theta \end{pmatrix} \cdot \begin{pmatrix} x' \\ y' \end{pmatrix} + \begin{pmatrix} x_1 \\ y_1 \end{pmatrix}. \tag{1}$$

Then, we divide the horizontal axis X' into D equal partitions and then optimize vertical coordinate Y' on the vertical line for each node to get a group of points composed by vertical coordinate of D points. Obviously, it is easy

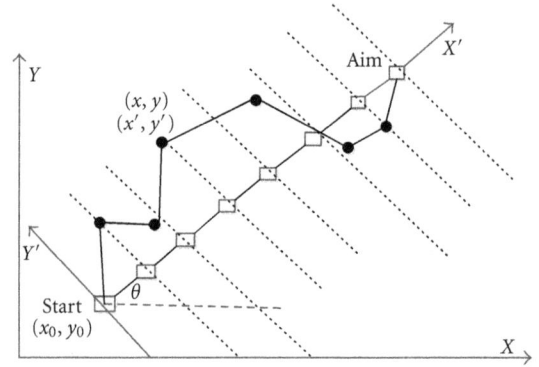

FIGURE 1: Coordinates transformation relation.

to get the horizontal abscissas of these points. We can get a path from start point to end point through connecting these points together, so that the route planning problem is transformed into a D-dimensional function optimization problem.

2.2. Performance Indicator. A performance indicator of path planning for UCAV mainly contains the completion of the mandate of the safety performance indicator and fuel performance indicator, that is, indicators with the least threat and the least fuel.

Minimum of performance indicator for threat

$$\min J_f = \int_0^L w_t \, dl, \quad L \text{ is the length of the path.} \tag{2}$$

Minimum of performance indicator for fuel

$$\min J_f = \int_0^L w_f \, dl, \quad L \text{ is the length of the path.} \tag{3}$$

Then the total performance indicators for UCAV route

$$\min J = kJ_t + (1 - k)J_f, \tag{4}$$

where w_t is the threat cost for each point on the route; w_f is fuel cost for each point on the path which depends on path length (in this paper, $w_f \equiv 1$); $k \in [0,1]$ is balanced coefficient between safety performance and fuel performance, whose value is determined by the special task UCAV performing; that is, if flight safety is of highly vital importance to the task, then we choose a larger k, while if the speed is critical to the aircraft task, then we select a smaller k.

2.3. Threat Cost. When the UCAV is flying along the path L_{ij}, the total threat cost generated by N_t threats is calculated as follows:

$$w_{t,L_{ij}} = \int_0^{L_{ij}} \sum_{k=1}^{N_t} \frac{t_k}{\left[(x - x_k)^2 + (y - y_k)^2 \right]^2} \, dl. \tag{5}$$

To simplify the calculations (as shown in Figure 2), each path segment is discretized into five subsegments and the threat cost is calculated on the end of each subsegment. If the

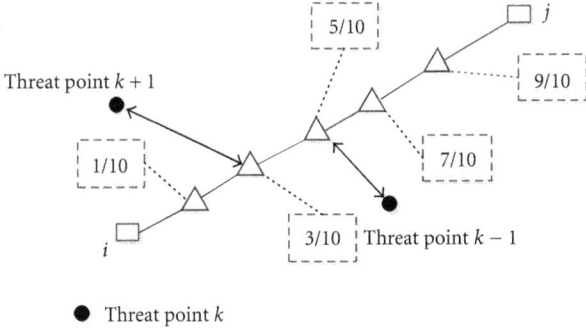

Figure 2: Modeling of the UCAV threat cost [6].

distance from the threat point to the end of each subsegment is within threat radius, we can calculate the responding threat cost according to

$$
w_{t,L_{ij}} = \frac{L_{ij}^5}{5} \sum_{k=1}^{N_t} t_k \left(\frac{1}{d_{0.1,k}^4} + \frac{1}{d_{0.3,k}^4} + \frac{1}{d_{0.5,k}^4} + \frac{1}{d_{0.7,k}^4} + \frac{1}{d_{0.9,k}^4} \right),
$$

(6)

where L_{ij} is the length of the subsegment connecting node i and node j; $d_{0.1,k}$ is the distance from the 1/10 point on the subsegment L_{ij} to the kth threat; t_k is threat level of the kth threat.

As fuel cost related to flight length, we can consider $w_f = L$, for simplicity, and fuel cost of each edge can be expressed by $w_{f,L_{ij}} = L_{ij}$.

3. Bat Algorithm (BA)

The bat algorithm is a new swarm intelligence optimization method, in which the search algorithm is inspired by social behavior of bats and the phenomenon of echolocation to sense distance.

3.1. Mainframe of BA. In [17], for simplicity, bat algorithm is based on idealizing some of the echolocation characteristics of bats, which are following approximate or idealized rules.

(1) All bats apply echolocation to sense distance, and they always "know" the surroundings in some magical way.

(2) Bats fly randomly with velocity v_i and a fixed frequency f_{min} at position x_i, varying wavelength λ, and loudness A_0 to hunt for prey. They can spontaneously accommodate the wavelength (or frequency) of their emitted pulses and adjust the rate of pulse emission $r \in [0, 1]$, depending on the proximity of their target.

(3) Although the loudness can change in different ways, it is supposed that the loudness varies from a minimum constant (positive) A_{min} to a large A_0.

Based on these approximations and idealization, the basic steps of the bat algorithm (BA) can be described as shown in Algorithm 1. In BA, each bat is defined by its position x_i^t, velocity v_i^t, frequency f_i, loudness A_i^t, and the emission pulse rate r_i^t in a d-dimensional search space. The new solutions x_i^t and velocities v_i^t at time step t are given by

$$
f_i = f_{min} + (f_{max} - f_{min})\beta,
$$

$$
\mathbf{v}_i^t = \mathbf{v}_i^{t-1} + (\mathbf{x}_i^t - \mathbf{x}_*)f_i,
$$

(7)

$$
\mathbf{x}_i^t = \mathbf{x}_i^{t-1} + \mathbf{v}_i^t,
$$

where $\beta \in [0, 1]$ is a random vector drawn from a uniform distribution. Here $\mathbf{x}_*$ is the current global best location (solution) which is located after comparing all the solutions among all the n bats. Generally speaking, depending on the domain size of the problem of interest, the frequency f is assigned to $f_{min} = 0$ and $f_{max} = 100$ in practical implementation. Initially, each bat is randomly given a frequency which is drawn uniformly from $[f_{min}, f_{max}]$.

For the local search part, once a solution is selected among the current best solutions, a new solution for each bat is generated locally using random walk

$$
\mathbf{x}_{new} = \mathbf{x}_{old} + \varepsilon A^t,
$$

(8)

where $\varepsilon \in [-1, 1]$ is a scaling factor which is a random number, while $A_t = \langle A_i^t \rangle$ is the average loudness of all the bats at time step t.

The updates of the velocities and positions of bats have some similarity to the procedure in the standard particle swarm optimization [18] as f_i in essence controls the pace and range of the movement of the swarming particles. To some degree, BA can be considered as a balanced combination of the standard particle swarm optimization and the intensive local search controlled by the loudness and pulse rate.

Furthermore, the loudness A_i and the rate r_i of pulse emission update accordingly as the iterations proceed as shown in

$$
A_i^{t+1} = \alpha A_i^t, \qquad r_i^{t+1} = r_i^0[1 - \exp(-\gamma t)],
$$

(9)

where α and γ are constants. In essence, α is similar to the cooling factor of a cooling schedule in the simulated annealing [19]. For simplicity, we set $\alpha = \gamma = 0.9$ in this work.

3.2. Algorithm BA for UCAV Path Planning. In BA, the standard ordinates are inconvenient to solve UCAV path planning directly. In order to apply BA to UCAV path planning, one of the key issues is to transform the original ordinate into rotation ordinate by (1).

Fitness of bat i at position x_i is determined by the threat cost by (4), and the smaller the threat cost, the smaller the fitness of bat i at position x_i. Each bat is encoded by D-dimensional deciding variables. And then, we use BA to optimize the path planning to get the best solution that is optimal flight route for UCAV. At last, the best solution is inversely converted to the original ordinates and output. The algorithm BA for UCAV path planning is shown as Algorithm 2.

Begin
  Step 1: Initialization. Set the generation counter $t = 1$; Initialize the population of NP bats P
              randomly and each bat corresponding to a potential solution to the given problem;
              define loudness A_i, pulse frequency Q_i and the initial velocities v_i ($i = 1, 2, \ldots, NP$); set
              pulse rate r_i.
  Step 2: While the termination criteria is not satisfied **or** $t <$ MaxGeneration **do**
              Generate new solutions by adjusting frequency, and updating velocities
              and locations/solutions [(7)]
              if (rand $> r_i$) **then**
                  Select a solution among the best solutions;
                  Generate a local solution around the selected best solution
              end if
              Generate a new solution by flying randomly
              Generate a new solution by flying randomly
              if (rand $< A_i$ and $f(x_i) < f(x_*)$) **then**
                  Accept the new solutions
                  Increase r_i and reduce A_i
              end if
              Rank the bats and find the current best x_*
              $t = t + 1$;
  Step 3: end while
  Step 4: Post-processing the results and visualization.
End.

ALGORITHM 1: Bat Algorithm.

Begin
  Step 1: Initialization. Set the generation counter $t = 1$; Initialize the population of NP bats
              P randomly and each bat corresponding to a potential solution to the given
              problem; define loudness A_i, pulse rate r_i, pulse frequency Q_i and the initial
              velocities v_i ($i = 1, 2, \ldots, NP$).
  Step 2: Generating rotation coordinate system. Transform the original coordinate system
              into new rotation coordinate whose horizontal axis is the connection line from
              starting point to target point according to (1); convert
              battlefield threat information to the rotation coordinate system and divide the axis
              X' into D equal partitions. Each feasible solution, denoted by
              $P = \{p_1, p_2, \ldots, p_D\}$, is an array indicated by the composition of D coordinates
              which are the floating-point numbers
  Step 3: Evaluate the threat cost J for each bat in P by (4)
  Step 4: while The halting criteria is not satisfied or $t <$ MaxGeneration **do**
              Generate new solutions by adjusting frequency, and updating velocities
              and locations/solutions [(7)]
              if (rand $> r_i$) **then**
                  Select a solution among the best solutions;
                  Generate a local solution around the selected best solution
              end if
              Generate a new solution by flying randomly
              if (rand $< A_i$ and $J_i < J_*$) **then**
                  Accept the new solutions
                  Increase r_i and reduce A_i
              end if
              Rank the bats and find the current best x_*
              $t = t + 1$
  Step 5: end while
  Step 6: Inversely transform the coordinates in final optimal path into the original
              coordinate, and output
End.

ALGORITHM 2: Algorithm of BA for UCAV path planning.

4. Bat Algorithm with Mutation (BAM)

The differential evolution (DE) algorithm, proposed by Storn and Price [14], is a simple evolutionary algorithm (EA) which generates new candidate solutions by combining the parent individual and a few other individuals of the same population. A candidate substitutes the parent only if it has better fitness. This is a rather greedy selection scheme which often overtakes traditional EAs. Advantages of DE are easy implementation, simple structure, speed, and robustness.

In general, the standard DE algorithm is adept at exploring the search space and locating the region of global optimal value, but it is not relatively good at exploiting solution. On the other hand, standard BA algorithm is usually quick at the exploitation of the solution though its exploration ability is relatively poor. Therefore, in this paper, a hybrid metaheuristic algorithm by inducing mutation in differential evolution into bat algorithm, so-called bat algorithm with mutation (BAM), is used to solve the path planning for UCAV. The difference between BAM and DE is that the mutation operator is used to improve the original BA generating new solution for each bat with a probability $1 - r$ originally using random walk. In this way, this method can explore the new search space by the mutation of the DE algorithm and exploit the population information with BA and therefore can overcome the lack of the exploitation of the DE algorithm. In the following, we will show the algorithm BAM which is a variety of DE and BA. Firstly, we describe a mainframe of BAM, and then an algorithm BAM for UCAV path planning is shown.

4.1. Mainframe of BAM. The critical operator of BAM is the hybrid differential evolution mutation operator, which composes the mutation operation in differential evolution with the BA. The core idea of the proposed hybrid mutation operator is based on two considerations. First, poor solutions can take in many new used features from good solutions. Second, the mutation operator of DE can improve the exploration of the new search space. In this way, we composed mutation operation into BAM which modifies the solutions with poor fitness in order to add diversity of the population to improve the search efficiency.

For bat algorithm, as the search relies entirely on random walks, a fast convergence cannot be guaranteed. Described here for the first time, a main modification of adding mutation operator is made to the BA, including two minor modifications, which are made with the aim of speeding up convergence, thus making the method more practical for a wider range of applications but without losing the attractive features of the original method.

The first modification is that we use fixed frequency f and loudness A instead of various frequency f_i and A_i^t. Similar to BA, in BAM, each bat is defined by its position x_i^t, velocity v_i^t, the emission pulse rate r_i^t, the fixed frequency f, and loudness A in a d-dimensional search space. The new solutions x_i^t and velocities v_i^t at time step t are given by

$$\mathbf{v}_i^t = \mathbf{v}_i^{t-1} + \left(\mathbf{x}_i^t - \mathbf{x}_*\right) f,$$
$$\mathbf{x}_i^t = \mathbf{x}_i^{t-1} + \mathbf{v}_i^t, \tag{10}$$

where $\mathbf{x}_*$ is the current global best location (solution) which is located after comparing all the solutions among all the n bats. In our experiments, we make $f = 0.5$. Through a series of simulation experiments on path planning for UCAV in Section 5.2, it was found that setting the parameter of pulse rate r to 0.6 and the loudness A to 0.95 produced the best results.

The second modification is to add mutation operator in an attempt to increase diversity of the population to improve the search efficiency and speed up the convergence to optima. For the local search part, once a solution is selected among the current best solutions, a new solution for each bat is generated locally using random walk by (8) when ξ is larger than pulse rate r, that is, $\xi > r$, where $\xi \in [0, 1]$ is a random real number drawn from a uniform distribution; while when $\xi \leq r$, we use mutation operator in DE updating the new solution to increase diversity of the population to improve the search efficiency by

$$\mathbf{x}_{\text{new}} = x_{r_1}^t + F\left(x_{r_2}^t - x_{r_3}^t\right), \tag{11}$$

where F is the mutation weighting factor, while r_1, r_2, and r_3 are uniformly distributed random integer numbers between 1 and NP. Through testing on path planning for UCAV in Section 5.2, it was found that setting the parameter of mutation weighting factor F to 0.5 in (11) and scaling factor ε to 0.1 in (4) produced the best results.

Based on above-mentioned analyses, the mainframe of the bat algorithm with mutation (BAM) can be described as shown in Algorithm 3.

4.2. Algorithm BAM for UCAV Path Planning. BAM can adapt to the needs of UCAV path planning, while optimization algorithms can improve the BA fast search capabilities and increase the search to the global possible optimum solution. Fitness for bat i at position x_i is represented by the objective function shown as (4) in UCAV path planning model, the smaller the threat value, the lower the fitness for bat i at position x_i.

Based on the above analysis, the pseudo code of improved BA-BAM for UCAV path planning is described as shown in Algorithm 4.

5. Simulation Experiments

In this section, we look at the performance of BAM as compared with other population-based optimization methods, such as ACO, BBO, DE, ES, GA, PBIL, PSO, and SGA. Firstly, we compare performances between BAM and other population-based optimization methods on the different parameters the maximum generation *Maxgen* and the dimension of converted optimization function D, and then we compare performances between BAM and BA on the different parameters loudness A, pulse rate r, weighting factor F, and scaling factor ε (where F and ε only for BAM).

To allow a fair comparison of running times, all the experiments were performed on a PC with an AMD Athlon(tm) 64 X2 Dual Core Processor 4200+ running at 2.20 GHz, 1024 MB of RAM, and a hard drive of 160 GB.

```
Begin
    Step 1: Initialization. Set the generation counter t = 1; Initialize the population of NP
            bats P randomly and each bat corresponding to a potential solution to the
            given problem; define loudness A; set frequency Q and the initial velocities v;
            set pulse rate r and weighting factor F.
    Step 2: Evaluate the quality f for each bat in P determined by f(x).
    Step 3: While the termination criteria is not satisfied or t < MaxGeneration do
            Sort the population of bats P from best to worst by order of quality f for each bat;
            for i = 1 : NP (all bats) do
                Select uniform randomly r₁ ≠ r₂ ≠ r₃ ≠ i
```
$r_4 = \lceil NP * \text{rand} \rceil$
$v_i^t = v_i^{t-1} + (v_i^t - x_*) \times Q$
$x_i^t = x_i^{t-1} + v_i^t$
```
                if (rand > r) then
```
$x_u^t = x_* + \alpha \varepsilon^t$
```
                else
```
$x_u^t = x_{r_1}^t + F(x_{r_2}^t - x_{r_3}^t)$
```
                end if
                Evaluate the fitness for the offspring xᵤᵗ, xᵢᵗ, x_{r₄}ᵗ
                Select the offspring x_kᵗ with the best fitness among the offsprings
                    xᵤᵗ, xᵢᵗ, x_{r₄}ᵗ
                if (rand < A) then
                    x_{r₄}ᵗ = x_kᵗ;
                end if
            end for i
            t = t + 1;
    Step 4: end while
    Step 5: Post-processing the results and visualization;
End.
```

ALGORITHM 3: Bat algorithm with mutation.

TABLE 1: Information about known threats.

No.	Location (km)	Threat radius (km)	Threat grade
1	(45,50)	10	2
2	(12,40)	10	10
3	(32,68)	8	1
4	(36,26)	12	2
5	(55,80)	9	3

Our implementation was compiled using MATLAB R2011b (7.13) running under Windows XP SP3. No commercial BBO tools or other population-based optimization tools were used in the following experiments.

5.1. General Performance of BAM. In this subsection, firstly we will present the supposed problem we use to test the performance of BAM. We use the parameters of battle field environments described as [1]. Supposed that there exists the following map information, UCAV flight from start point (10, 10) to end point (55, 100). In the flight course, there exist five threat areas. Their coordinates and corresponding threat radii are shown as in Table 1. Also, we set balanced coefficient between safety performance and fuel performance $k = 0.5$.

In order to explore the benefits of BAM, in this subsection we compared its performance on UCAV path

planning problem with BA and eight other population-based optimization methods, which are ACO, BBO, DE, ES, GA, PBIL, PSO, and SGA. ACO (ant colony optimization) [20] is a swarm intelligence algorithm for solving computational problems which is based on the pheromone deposition of ants. Biogeography-based optimization (BBO) [21–23] is a new evolutionary algorithm (EA) developed for global optimization which is a generalization of biogeography to EA. DE (differential evolution) [14] is a simple but excellent optimization method that uses the difference between two solutions to probabilistically adapt a third solution. An ES (evolutionary strategy) [24] is an algorithm that generally distributes equal importance to mutation and recombination, and that allows two or more parents to reproduce an offspring. A GA (genetic algorithm) [25] is a search heuristic that mimics the process of natural evolution. PBIL (probability-based incremental learning) [26] is a type of genetic algorithm where the genotype of an entire population (probability vector) is evolved rather than individual members. PSO (particle swarm optimization) [18, 27] is also a swarm intelligence algorithm which is based on the swarm behavior of fish, and bird schooling in nature. A stud genetic algorithm (SGA) [28] is a GA that uses the best individual at each generation for crossover.

Except an ad hoc explain, in the following experiments, we use the same MATLAB code and parameters settings for other population-based optimization methods in [21, 29].

Begin

Step 1: Initialization. Set the generation counter $t = 1$; Initialize the population of NP bats P randomly and each bat corresponding to a potential solution to the given problem; define pulse frequency Q; set loudness A_i, the initial velocities v_i and pulse rate r_i $(i = 1, 2, \ldots, NP)$; set weighting factor F.

Step 2: Generating rotation coordinate system. Transform the original coordinate system into new rotation coordinate whose horizontal axis is the connection line from starting point to target point according to (1);
convert battlefield threat information to the rotation coordinate system and divide the axis X' into D equal partitions. Each feasible solution, denoted by $P = \{p_1, p_2, \ldots, p_D\}$, is an array indicated by the composition of D coordinates which are the floating-point numbers

Step 3: Evaluate the threat cost J for each bat in P by (4)

Step 4: while The halting criteria is not satisfied or $t < \text{MaxGeneration}$ **do**

   Sort the population of bats P from best to worst by order of threat cost J

   for each bat;

   for $i = 1 : NP$ (all bats) **do**

      Select uniform randomly $r_1 \neq r_2 \neq r_3 \neq i$

      $r_4 = \lceil NP * \text{rand} \rceil$

      $v_i^t = v_i^{t-1} + (v_i^t - x_*) \times Q$

      $x_i^t = x_i^{t-1} + v_i^t$

      if (rand $> r$) **then**

         $x_u^t = x_* + \alpha \varepsilon^t$

      else

         $x_u^t = x_{r_1}^t + F(x_{r_2}^t - x_{r_3}^t)$

      end if

      Evaluate the fitness for the offsprings $x_u^t, x_i^t, x_{r_4}^t$

      Select the offspring x_k^t with the best fitness among the offsprings $x_u^t, x_i^t, x_{r_4}^t$

      if (rand $< A$) **then**

         $x_{r_4}^t = x_k^t$;

      end if

   end for i

   Evaluate the threat cost for each bat in P by (4).

   Sort the population of bats P from best to worst by order of threat cost J

   for each bat;

   $t = t + 1$;

Step 5: end while

Step 6: Inversely transform the coordinates in final optimal path into the original coordinate, and output

End.

ALGORITHM 4: Algorithm of BAM for UCAV path planning.

To compare the different effects among the parameters *Maxgen* and *D*, we ran 100 Monte Carlo simulations of each algorithm on the above UCAV path planning problem to get representative performances. For simplicity, we subtract 50 from the actual value; that is, if a value is 0.4419 in the following table, then its corresponding value 50.4419 is its true value. We must point out that we mark the best value with italic and bold font for each algorithm in Tables 2–5.

5.1.1. Effect of Maximum Generation: Maxgen. The choice of the best maximum generation of metaheuristic algorithm is always critical for specific problems. Increasing the maximum generation will increase the possibility of reaching optimal solution, promoting the exploitation of the search space. Moreover, the probability to find the correct search direction increases considerably. The influence of maximum generation is investigated in this sub-subsection. For all the population-based optimization methods, all the parameter settings are the same as above mentioned, only except for maximum generation *Maxgen* = 50, *Maxgen* = 100, *Maxgen* = 150, *Maxgen* = 200, and *Maxgen* = 250. The results are recorded in Tables 2, 3, 4, and 5 after 100 Monte Carlo runs. Table 2 shows the best minima found by each algorithm over 100 Monte Carlo runs. Table 3 shows the worst minima found by each algorithm over 100 Monte Carlo runs. Table 4 shows the average minima found by each algorithm, averaged over 100 Monte Carlo runs. Table 5 shows the average CPU time consumed by each algorithm, averaged over 100 Monte Carlo runs. In other words, Tables 2, 3, and 4 show the best, worst, and average performance of each algorithm,

TABLE 2: Best normalized optimization results on UCAV path planning problem on different *Maxgen*. The numbers shown are the best results found after 100 Monte Carlo simulations of each algorithm.

Parameter			Algorithm									
Popsize	*Maxgen*	D	ACO	BA	BAM	BBO	DE	ES	GA	PBIL	PSO	SGA
30	**50**	20	10.7202	4.0662	**0.6208**	7.0272	2.4179	9.6276	1.2604	100.0527	2.7827	1.7370
30	**100**	20	10.8912	4.7582	**0.4900**	4.7484	0.8503	10.6318	1.5073	98.3640	2.3469	1.3218
30	**150**	20	9.9096	4.1112	**0.4724**	4.2311	0.5319	11.1469	1.0991	71.2093	2.3738	1.1559
30	**200**	20	12.3080	3.1463	**0.4590**	2.7765	0.5047	11.2403	1.0792	72.8244	3.4276	0.7595
30	**250**	20	7.1358	4.4072	**0.4636**	2.6109	0.4792	12.3745	1.0640	74.9071	2.5221	1.0166

TABLE 3: Worst normalized optimization results on UCAV path planning problem on different *Maxgen*. The numbers shown are the worst results found after 100 Monte Carlo simulations of each algorithm.

Parameter			Algorithm									
Popsize	*Maxgen*	D	ACO	BA	BAM	BBO	DE	ES	GA	PBIL	PSO	SGA
30	**50**	20	18.7099	39.0832	**11.7494**	30.2785	25.3999	41.9676	10.2501	**464.2014**	28.6115	13.0102
30	**100**	20	18.4316	29.9962	**9.7666**	32.1868	18.6288	38.5875	8.2047	**339.5171**	25.7065	11.0529
30	**150**	20	17.4223	31.1293	**7.6952**	29.5695	13.8150	46.0828	10.5257	**457.5577**	29.6341	13.3517
30	**200**	20	17.2147	24.9732	**6.7334**	41.5292	10.4226	31.3944	6.7466	**308.6347**	33.0709	7.5385
30	**250**	20	16.9896	24.7175	**3.3564**	19.5894	8.9560	34.8908	8.9162	**201.3705**	27.3858	13.5830

respectively, while Table 5 shows the average CPU time consumed by each algorithm.

From Table 2, we see that BAM performed the best on all the groups, while DE performed the second best on the 5 groups especially when *Maxgen* = 150, 200, and 250. Table 3 shows that PBIL was the worst at finding objective function minima on all the five groups when multiple runs are made, while the BAM was the best on all the groups in the worst values. Table 4 shows that BAM was the most effective at finding objective function minima when multiple runs are made, while DE and SGA performed the second best on the 5 groups, and GA and SGA similarly performed the third best on the 5 groups. Table 5 shows that PBIL was the most effective at finding objective function minima when multiple runs are made, performing the best on all the 5 groups. By carefully looking at the results in Tables 2, 3, and 4, we can recognize that the values for each algorithm are obviously decreasing with the increasing *Maxgen*, while the performance of BAM increases little with the *Maxgen* increasing from 200 to 250, so we set *Maxgen* = 200 in other experiments. In sum, from Tables 2, 3, 4, and 5 we can draw the conclusion that the more the generations are, the smaller the objective function value we can reach, while the CPU time consumes more. Moreover, BAM performs better than other population-based optimization methods for the UCAV path planning problem with different maximum generation.

5.1.2. Effect of Dimensionality: D. In order to investigate the influence of the dimension on the performance of BAM, we carry out a scalability study comparing with other population-based optimization methods for the UCAV path planning problem with the dimensionality $D = 5$, $D = 10$, $D = 15$, $D = 20$, $D = 25$, $D = 30$, $D = 35$, and $D = 40$. The results are recorded in Tables 6, 7, 8, and 9 after 100

Monte Carlo runs. Table 6 shows the best minima found by each algorithm over 100 Monte Carlo runs. Table 7 shows the worst minima found by each algorithm over 100 Monte Carlo runs. Table 8 shows the average minima found by each algorithm, averaged over 100 Monte Carlo runs. Table 9 shows the average CPU time consumed by each algorithm, averaged over 100 Monte Carlo runs. In other words, Tables 6, 7, and 8 show the best, worst, and average performance of each algorithm, respectively, while Table 9 shows the average CPU time consumed by each algorithm.

From Table 6, we see that DE performed the best when $D = 10$, while BAM performed the best on the other groups when multiple runs are made. Table 7 shows that BA and ES were the worst when $D = 5$ and $D = 10$, respectively, and PBIL was the worst at finding objective function minima on all the other groups when multiple runs are made, while the DE, SGA, and GA were the best when $D = 5$, 10, and 15, respectively, and BAM was the best on the other groups in the worst values. Table 8 shows that DE and SGA were the most effective when $D = 5$ and 10, respectively, and BAM was the best on the other groups at finding objective function minima when multiple runs are made. Table 9 shows that PBIL was the most effective at finding objective function minima on all the groups. So, from the experimental results of this sub-subsection, we can conclude that the mutation operation between bats with a probability $1 - r$ during the process of generating new solutions has the ability to accelerate BA in general; especially the improvements are more significant at higher dimensionality. With the higher dimension, we are not always getting the better results with consuming more time; furthermore, the result is good enough when $D = 20$. In sum, in other experiments we should make $D = 20$ under the comprehensive consideration.

TABLE 4: Mean normalized optimization results on UCAV path planning problem on different *Maxgen*. The numbers shown are the minimum objective function values found by each algorithm, averaged over 100 Monte Carlo simulations.

Parameter			Algorithm									
Popsize	*Maxgen*	D	ACO	BA	BAM	BBO	DE	ES	GA	PBIL	PSO	SGA
30	**50**	20	16.3819	16.6782	**1.4842**	14.1072	12.3797	20.5653	4.2541	219.9368	10.0760	4.5491
30	**100**	20	16.2884	14.9048	**0.9337**	13.5705	6.0887	20.6706	3.5523	166.4567	9.1725	3.4353
30	**150**	20	16.1408	14.4874	**0.9123**	11.7978	3.7267	20.1996	3.4269	142.1862	9.5459	3.1636
30	**200**	20	16.3976	12.4323	**0.8000**	11.8224	2.6358	20.8610	3.0080	131.1306	8.9917	2.6434
30	**250**	20	16.1958	11.4213	**0.7422**	10.0553	1.9715	20.7600	2.9160	119.6745	7.8005	3.1409

TABLE 5: Average CPU time on UCAV path planning problem on different *Maxgen*. The numbers shown are the minimum average CPU time (sec) consumed by each algorithm.

Parameter			Algorithm									
Popsize	*Maxgen*	D	ACO	BA	BAM	BBO	DE	ES	GA	PBIL	PSO	SGA
30	**50**	20	1.1477	1.2389	2.5415	0.7540	1.0830	1.1045	1.0068	**0.5610**	0.9389	0.9733
30	**100**	20	2.2752	2.5180	5.0720	1.5041	2.1782	2.2028	1.9875	**1.0793**	1.8632	1.9253
30	**150**	20	3.4043	3.7411	7.3826	2.2564	3.2397	3.2778	2.9604	**1.5839**	2.7619	2.8612
30	**200**	20	4.5337	4.9930	9.7353	3.0201	4.3053	4.3581	3.9755	**2.0459**	3.6100	3.7132
30	**250**	20	5.6563	6.1668	12.2422	3.6952	5.4137	5.3999	4.9278	**2.6083**	4.6532	4.7580

5.2. Influence of Control Parameter. In [15], Yang concluded that if we adjust the parameters properly so that BA can outperform GA, HS (harmony search), and PSO. The choice of the control parameters is of vital importance for different problems. To compare the different effects among the parameters A, r, F, and ε (F and ε only for BAM), we ran 100 Monte Carlo simulations of BA and BAM algorithm on the above problem to get representative performances.

5.2.1. Loudness: A. To investigate the influence of the loudness on the performance of BAM, we carry out this experiment comparing BA for the UCAV path planning problem with the loudness $A = 0, 0.1, 0.2, \ldots, 0.9, 1.0$ and fixed pulse rate $r = 0.6$. All other parameter settings are kept unchanged. The results are recorded in Tables 10, 11, 12, and 13 after 100 Monte Carlo runs. Table 10 shows the best minima found by BA and BAM algorithms over 100 Monte Carlo runs. Table 11 shows the worst minima found by BA and BAM algorithms over 100 Monte Carlo runs. Table 12 shows the average minima found by BA and BAM algorithms averaged over 100 Monte Carlo runs. Table 13 shows the average CPU time consumed by BA and BAM algorithms, averaged over 100 Monte Carlo runs. In other words, Tables 10, 11, and 12 show the best, worst, and average performance of BA and BAM algorithm, respectively, while Table 13 shows the average CPU time consumed by BA and BAM algorithms.

From Table 10, we obviously see that BAM performed better (on average) than BA on all the groups, and BA and BAM reach the worst values 9.6473 and 11.9280 when $A = 0$, respectively, while BA and BAM reach the best values 4.0888 and 0.7774 when $A = 1.0$, respectively, among the optima when multiple runs are made. Table 11 shows evidently that BAM performed better (on average) than BA on all the groups (except $A = 0$), and BA as well as BAM reach the worst values 39.0584 and 38.6449 when $A = 0.1$ and $A = 0$,

respectively, while BA and BAM reach the best values 24.4673 and 8.8555 when $A = 0.1$ and $A = 1.0$, respectively, among the worst values when multiple runs are made. Table 12 shows that BAM performed better (on average) than BA on all the groups, and BA and BAM reach the worst values 20.3072 and 20.2230 when $A = 0$, respectively, while BA and BAM reach the best values 11.1174 and 2.7086 when $A = 1.0$, respectively, among the mean values when multiple runs are made. Table 13 shows that BA was more effective at finding objective function minima when multiple runs are made, performing the best on all the groups. By carefully looking at the results in Tables 10, 11, and 12, we can recognize that the threat value for BA and BAM is decreasing with the increasing A, and BA and BAM reach optima/minimum when A is equal or very close to 1.0, while BA and BAM reach maximum when A is equal or very close to 0. So, we set $A = 0.95$ which is very close to 1.0 in other experiments. In sum, from Tables 10, 11, 12, and 13, we can conclude that the mutation operation between bats during the process of the new solutions updating has the ability to accelerate BA in general.

5.2.2. Pulse Rate: r. To investigate the influence of the pulse rate on the performance of BAM, we carry out this experiment comparing with BA for the UCAV path planning problem with the pulse rate $r = 0, 0.1, 0.2, \ldots, 0.9, 1.0$ and fixed loudness $A = 0.95$. All other parameter settings are kept unchanged. The results are recorded in Tables 14, 15, 16, and 17 after 100 Monte Carlo runs. Table 14 shows the best minima found by BA and BAM algorithms over 100 Monte Carlo runs. Table 15 shows the worst minima found by BA and BAM algorithms over 100 Monte Carlo runs. Table 16 shows the average minima found by BA and BAM algorithms, averaged over 100 Monte Carlo runs. Table 17 shows the average CPU time consumed by BA and BAM

TABLE 6: Best normalized optimization results on UCAV path planning problem on different D. The numbers shown are the best results found after 100 Monte Carlo simulations of each algorithm.

Parameter			Algorithm									
Popsize	Maxgen	D	ACO	BA	BAM	BBO	DE	ES	GA	PBIL	PSO	SGA
30	200	5	10.1164	10.6909	**4.3575**	10.2341	4.3568	12.3746	5.2471	8.5576	5.6082	9.9596
30	200	10	7.4746	2.3600	1.3953	2.6157	**1.3952**	8.0656	1.5716	25.6821	2.1101	1.5498
30	200	15	9.8297	3.0757	**0.6094**	2.0896	0.6204	7.7408	0.8299	61.4656	3.2257	0.9700
30	200	20	10.0836	2.3950	**0.4679**	2.7765	0.4913	9.6276	0.8600	72.2897	2.3738	0.8426
30	200	25	11.5490	5.0173	**0.4484**	4.8474	0.6265	12.3169	1.5243	113.7537	2.3740	1.3743
30	200	30	13.8615	7.2470	**0.4671**	10.9403	1.1301	18.0090	1.7026	152.0173	3.6751	1.5147
30	200	35	16.9476	7.4484	**0.4795**	10.4147	1.2849	16.8613	2.1602	254.0060	5.4765	1.5319
30	200	40	17.6142	8.6500	**0.6028**	14.4997	3.9617	19.8244	2.4178	315.4459	5.5384	1.9406

TABLE 7: Worst normalized optimization results on UCAV path planning problem on different D. The numbers shown are the worst results found after 100 Monte Carlo simulations of each algorithm.

Parameter			Algorithm									
Popsize	Maxgen	D	ACO	BA	BAM	BBO	DE	ES	GA	PBIL	PSO	SGA
30	200	5	12.6928	**295.2557**	10.2403	119.9434	**9.7959**	62.1765	20.1888	22.9251	13.3267	22.6326
30	200	10	18.2565	58.7386	10.7242	42.1924	12.4821	**74.6665**	6.3799	64.0778	23.2604	**5.7899**
30	200	15	10.9917	35.7454	10.1928	34.8307	12.5250	50.3214	**8.1499**	**119.2700**	28.0228	9.9385
30	200	20	17.0266	33.7068	**3.7420**	32.1908	18.8897	38.7234	9.4820	**254.0913**	34.7133	11.6024
30	200	25	12.2373	24.9265	**3.5192**	30.9943	17.1415	33.4598	12.7971	**593.9572**	31.6741	16.0736
30	200	30	14.4647	30.0844	**10.2851**	61.7204	29.6529	37.4566	22.1291	**2011**	35.6656	14.0512
30	200	35	18.7271	32.7374	**8.8193**	37.9424	39.4435	46.6475	24.4790	**8424.28**	38.0578	15.6693
30	200	40	27.0641	33.2634	**8.4273**	49.5461	45.4130	44.3624	19.2098	**8856.06**	35.5090	22.5022

TABLE 8: Mean normalized optimization results on UCAV path planning problem on different D. The numbers shown are the minimum objective function values found by each algorithm, averaged over 100 Monte Carlo simulations.

Parameter			Algorithm									
Popsize	Maxgen	D	ACO	BA	BAM	BBO	DE	ES	GA	PBIL	PSO	SGA
30	200	5	11.4856	56.4830	9.0542	23.4238	**8.0557**	31.8202	10.5709	15.5053	10.0765	10.8836
30	200	10	12.5333	19.4251	2.7075	8.7776	3.1206	27.2252	2.3722	51.2935	7.2212	**2.2813**
30	200	15	10.2484	13.6018	**1.2318**	8.9120	2.3737	22.0792	2.1136	78.4948	7.7362	1.8973
30	200	20	16.3303	13.6305	**0.7609**	12.2883	3.0044	20.4717	2.9612	127.5765	9.9091	2.8621
30	200	25	11.5842	14.9017	**0.7093**	15.3698	4.6029	22.7244	3.7244	214.0821	10.3315	3.7238
30	200	30	13.9422	16.6162	**1.1067**	18.6997	11.4103	25.4016	5.3097	335.0904	12.7964	4.3798
30	200	35	18.3452	17.7033	**1.4617**	20.7753	19.1074	27.2172	6.0765	661.1281	13.8799	5.4943
30	200	40	24.7642	19.9737	**1.8769**	25.9148	28.7062	30.0177	7.6989	1174.90	15.1555	7.4237

TABLE 9: Average CPU time on UCAV path planning problem on different D. The numbers shown are the minimum average CPU time (sec) consumed by each algorithm.

Parameter			Algorithm									
Popsize	Maxgen	D	ACO	BA	BAM	BBO	DE	ES	GA	PBIL	PSO	SGA
30	200	5	1.93	1.87	3.58	1.26	2.03	2.13	2.21	**1.19**	2.27	2.12
30	200	10	2.86	3.08	5.28	1.86	2.64	2.88	2.83	**1.46**	2.79	2.81
30	200	15	3.61	3.92	7.61	2.31	3.47	3.60	3.43	**1.76**	3.31	3.34
30	200	20	4.50	4.95	9.83	3.02	4.24	4.33	3.93	**2.10**	3.73	3.84
30	200	25	5.57	5.83	12.06	3.37	5.00	4.96	4.43	**2.43**	4.22	4.39
30	200	30	6.44	7.30	14.40	3.86	5.58	5.84	4.91	**2.75**	4.70	4.90
30	200	35	7.30	8.39	16.84	4.46	6.23	6.63	5.65	**3.14**	5.16	5.39
30	200	40	8.34	9.60	19.34	4.97	6.71	7.34	6.06	**3.38**	5.68	5.96

TABLE 10: Best normalized optimization results on UCAV path planning problem on different A. The numbers shown are the best results found after 100 Monte Carlo simulations of BA and BAM algorithms.

Parameter		Algorithm	
A	r	BA	BAM
0	0.6	**9.6473**	**11.9280**
0.1	0.6	6.6352	2.3678
0.2	0.6	6.9462	1.9257
0.3	0.6	5.6720	1.2626
0.4	0.6	8.5845	1.0167
0.5	0.6	4.3351	1.1764
0.6	0.6	5.8639	0.9442
0.7	0.6	5.5337	0.9305
0.8	0.6	5.2613	0.9942
0.9	0.6	5.3534	1.0012
1.0	0.6	**4.0888**	**0.7774**

TABLE 12: Mean normalized optimization results on UCAV path planning problem on different A. The numbers shown are the minimum objective function values found by BA and BAM algorithms, averaged over 100 Monte Carlo simulations.

Parameter		Algorithm	
A	r	BA	BAM
0	0.6	**20.3072**	**20.2230**
0.1	0.6	18.3087	9.5999
0.2	0.6	16.7149	5.8455
0.3	0.6	14.5081	4.5573
0.4	0.6	15.2639	4.1528
0.5	0.6	13.3459	3.5631
0.6	0.6	12.6488	3.5585
0.7	0.6	12.6416	3.2400
0.8	0.6	11.9422	3.2585
0.9	0.6	12.5489	2.7930
1.0	0.6	**11.1174**	**2.7086**

TABLE 11: Worst normalized optimization results on UCAV planning problem on different A. The numbers shown are the worst results found after 100 Monte Carlo simulations of BA and BAM algorithms.

Parameter		Algorithm	
A	r	BA	BAM
0	0.6	33.4435	**38.6449**
0.1	0.6	**39.0584**	20.9711
0.2	0.6	38.7234	14.8265
0.3	0.6	29.4638	12.6085
0.4	0.6	36.3300	12.8931
0.5	0.6	26.0305	8.9874
0.6	0.6	28.7598	13.6922
0.7	0.6	26.2501	11.0757
0.8	0.6	**24.4673**	12.6789
0.9	0.6	29.4221	9.1986
1.0	0.6	25.7065	**8.8555**

TABLE 13: Average CPU time on UCAV path planning problem on different A. The numbers shown are the minimum average CPU time (Sec) consumed by BA and BAM algorithms.

Parameter		Algorithm	
A	r	BA	BAM
0	0.6	**4.81**	**9.77**
0.1	0.6	4.85	**8.75**
0.2	0.6	4.82	8.87
0.3	0.6	4.85	9.01
0.4	0.6	**4.88**	9.18
0.5	0.6	**4.88**	9.37
0.6	0.6	4.84	9.19
0.7	0.6	4.81	9.39
0.8	0.6	4.85	9.43
0.9	0.6	4.86	9.52
1.0	0.6	**4.88**	9.47

algorithm, averaged over 100 Monte Carlo runs. In other words, Tables 14, 15, and 16 shows the best, worst, and average performance of BA and BAM algorithm respectively, while Table 17 shows the average CPU time consumed by BA and BAM algorithms.

From Table 14, we obviously see that BAM performed better (on average) than BA on all the groups, and BA and BAM reach the worst values 5.1353 and 0.8536 when $r = 1.0$, respectively, while BA and BAM reach the best values 1.3626 and 0.4591 when $r = 0.1$ and $r = 0.2$, respectively, among the optima when multiple runs are made. Table 15 shows evidently that BAM performed better (on average) than BA on all the groups, and BA and BAM reach the worst value 30.9979 and 12.3230 when $r = 1.0$ and $r = 0.1$, respectively, while BA and BAM reach the best values 17.8310 and 6.4524 when $r = 0.2$ and $r = 0.7$, respectively, among the worst values when multiple runs are made. Table 16 shows that BAM performed better (on average) than BA on all

the groups, and BA and BAM reach the worst mean values 11.1155 and 2.9290 when $r = 1.0$, respectively, while BA and BAM reach the best mean values 6.8803 and 0.7729 when $r = 0.6$, respectively, among the mean values when multiple runs are made. Table 13 shows that BA was more effective at finding objective function minima when multiple runs are made, performing the best on all the groups. By carefully looking at the results in Tables 14, 15, and 16, we can recognize that the threat value for BA and BAM varies little with the increasing A, and BA and BAM reach mean optima/minima when r is equal or very close to 0.6, while BA and BAM reach maximum when r is equal or very close to 1.0. So, we set $r = 0.6$ in other experiments. In sum, from Tables 14, 15, 16, and 17 we can conclude that the mutation operation between bats during the process of generating new solutions has the ability to accelerate BA in general.

5.2.3. Weighting Factor: F. For the sake of investigating the influence of the weighting factor F on the performance

TABLE 14: Best normalized optimization results on UCAV path planning problem on different r. The numbers shown are the best results found after 100 Monte Carlo simulations of BA and BAM algorithms.

Parameter		Algorithm	
A	r	BA	BAM
0.5	**0**	1.9003	0.4709
0.5	**0.1**	**1.3626**	0.4598
0.5	**0.2**	2.4637	**0.4591**
0.5	**0.3**	1.7444	0.4669
0.5	**0.4**	1.7388	0.4633
0.5	**0.5**	2.6682	0.4600
0.5	**0.6**	2.4835	0.4614
0.5	**0.7**	1.8795	0.4702
0.5	**0.8**	2.2624	0.4711
0.5	**0.9**	3.5876	0.5019
0.5	**1.0**	**5.1353**	**0.8536**

TABLE 16: Mean normalized optimization results on UCAV path planning problem on different r. The numbers shown are the minimum objective function values found by BA and BAM algorithms, averaged over 100 Monte Carlo simulations.

Parameter		Algorithm	
A	r	BA	BAM
0.95	**0**	9.9001	0.8457
0.95	**0.1**	9.0247	1.0440
0.95	**0.2**	7.5435	1.0412
0.95	**0.3**	8.0903	1.0977
0.95	**0.4**	7.2892	0.8794
0.95	**0.5**	7.1919	0.9456
0.95	**0.6**	**6.8803**	**0.7729**
0.95	**0.7**	7.3384	0.8632
0.95	**0.8**	7.8609	0.8817
0.95	**0.9**	9.0251	1.1881
0.95	**1.0**	**11.1155**	**2.9290**

TABLE 15: Worst normalized optimization results on UCAV path planning problem on different r. The numbers shown are the worst results found after 100 Monte Carlo simulations of BA and BAM algorithms.

Parameter		Algorithm	
A	r	BA	BAM
0.95	**0**	29.4969	7.5532
0.95	**0.1**	25.5864	**12.3230**
0.95	**0.2**	**17.8310**	9.2830
0.95	**0.3**	30.4250	10.7164
0.95	**0.4**	21.2176	10.1811
0.95	**0.5**	20.8479	9.0701
0.95	**0.6**	17.8944	7.0315
0.95	**0.7**	28.7659	**6.4524**
0.95	**0.8**	28.7659	9.3845
0.95	**0.9**	21.9160	7.4771
0.95	**1.0**	**30.9979**	9.3678

TABLE 17: Average CPU time on UCAV path planning problem on different r. The numbers shown are the minimum average CPU time (sec) consumed by BA and BAM algorithms.

Parameter		Algorithm	
A	r	BA	BAM
0.95	**0**	5.02	**9.90**
0.95	**0.1**	4.96	9.88
0.95	**0.2**	5.03	9.83
0.95	**0.3**	4.95	9.76
0.95	**0.4**	4.99	**9.90**
0.95	**0.5**	**5.08**	9.83
0.95	**0.6**	5.06	9.86
0.95	**0.7**	5.11	9.82
0.95	**0.8**	4.97	9.79
0.95	**0.9**	5.00	9.66
0.95	**1.0**	**4.87**	**9.43**

of BAM, we carry out this experiment for the UCAV path planning problem with the weighting factor $F = 0, 0.1, 0.2, \ldots, 1.5$ and fixed scaling factor $\varepsilon = 0.1$. All other parameter settings are kept unchanged. The results are recorded in Table 18 after 100 Monte Carlo runs. Columns 1, 2, and 3 in Table 18 show the best, worst, and average performances of BAM algorithm, respectively, while Column 4 in Table 18 shows the average CPU time consumed by BAM algorithm.

From Table 18, we can recognize that the threat values for BAM varies little with the increasing F, and BAM reaches optimum/minimum on $F = 0.5$. So, we set $F = 0.5$ in other experiments. From Table 18 we can draw the conclusion that BAM is insensitive to the weighting factor F, so we do not have to fine-tune the parameter F to get the best performance for different problems.

5.2.4. Scaling Factor: ε. For the sake of investigating the influence of the scaling factor ε on the performance of BAM, we carry out this experiment for the UCAV path planning problem with the factor scaling factor $\varepsilon = 0, 0.1, 0.2, \ldots, 1.0$ and fixed weighting factor $F = 0.5$. All other parameter settings are kept unchanged. The results are recorded in Table 19 after 100 Monte Carlo runs. Columns 1, 2, and 3 in Table 19 shows the best, worst, and average performances of BAM algorithms respectively, while Column 4 in Table 19 shows the average CPU time consumed by BAM algorithm.

From Table 19, we can recognize that the values for BAM vary little with the increasing ε, and BAM reaches optimum/minimum and the worst/maximum on $\varepsilon = 0.1$ and $\varepsilon = 0$, respectively. So, we set $\varepsilon = 0.1$ in other experiments. From Table 19 we can draw the conclusion that BAM is insensitive to the scaling factor ε, so we do not have to fine-tune the parameter ε to get the best performance for different problems.

TABLE 18: Best normalized optimization results and average CPU time on UCAV path planning problem on different F. The numbers shown are the best results found after 100 Monte Carlo simulations of BAM algorithm.

Parameter		Algorithm			
		BAM			
F	ε	Best	Worst	Mean	CPU time (Sec)
0	0.1	0.7045	9.6041	**2.1675**	9.58
0.1	0.1	0.7806	11.3137	1.7250	9.69
0.2	0.1	**0.8138**	11.0360	1.8369	**9.75**
0.3	0.1	0.7421	9.1672	1.6696	9.72
0.4	0.1	0.6879	9.2307	1.6681	9.74
0.5	0.1	0.6843	9.1385	**1.5362**	9.78
0.6	0.1	0.6903	7.9582	1.9313	9.61
0.7	0.1	0.7193	7.8036	1.8229	9.65
0.8	0.1	0.6645	9.5210	2.0243	9.57
0.9	0.1	0.6964	9.3858	1.8923	9.58
1.0	0.1	0.7546	9.5856	1.8489	9.73
1.1	0.1	**0.6331**	13.0251	1.7814	9.72
1.2	0.1	0.6508	9.0815	1.6803	9.74
1.3	0.1	0.6694	9.9523	1.9246	**9.53**
1.4	0.1	0.6966	8.3557	1.6807	9.74
1.5	0.1	0.7112	**7.7598**	1.8652	9.62

TABLE 19: Best normalized optimization results and average CPU time on UCAV path planning problem on different ε. The numbers shown are the best results found after 100 Monte Carlo simulations of BAM algorithm.

Parameter		Algorithm			
		BAM			
F	ε	Best	Worst	Mean	CPU Time (sec)
0.5	**0**	**0.8817**	12.3484	2.9293	**9.08**
0.5	**0.1**	**0.4541**	3.4890	**0.7290**	**9.84**
1.0	**0.2**	0.4860	6.7479	0.9062	9.77
0.5	**0.3**	0.5158	7.8852	1.0775	9.76
0.5	**0.4**	0.5580	7.9747	1.0161	9.76
0.5	**0.5**	0.5570	8.6947	1.4135	9.67
0.5	**0.6**	0.6282	10.1393	1.3083	9.80
0.5	**0.7**	0.6363	10.8901	1.2981	9.82
0.5	**0.8**	0.6442	7.6385	1.5515	9.74
0.5	**0.9**	0.6779	9.9564	1.8094	9.68
0.5	**1.0**	0.7060	9.5250	1.9116	9.61

The simulation experiment performed in Sections 5.1 and 5.2 shows that the algorithm BAM we proposed performed the best but worst effectively when solving the UCAV path planning problem. From deep investigation, we can see that BAM cam reach minima when maximum generation *Maxgen* = 50 and population size Popsize = 30, while other population-based optimization methods cannot achieve satisfactory result under this condition; that is, BAM needs fewer maximum generation, less population size, and less time than other population-based optimization methods when arriving to the same performance. In sum,

the simulation implemented in Section 6 shows that the algorithm BAM we proposed performed the best and most absolutely effectively, and it can solve the UCAV path planning problem perfectly. Furthermore, comparing to other population-based optimization methods, the algorithm BAM is insensitive to the parameter loudness A, pulse rate r, weighting factor F, and scaling factor ε, so we do not have to fine-tune the parameters A, r, F, and ε to get the best performance for different problems.

5.3. Discussions. The BA algorithm is a simple, fast, and robust global optimization algorithm developed by X. S. Yang in 2010. However, it may lack the diversity of population between bats. Therefore, in this work, we add mutation operation between bats to the BA during the process of new solutions updating. And then, the BAM algorithm is proposed to solve the UCAV path planning. From the experimental results we can sum up the following:

(i) Our proposed BAM approach is effective and efficient. It can solve the UCAV path planning problem effectively.

(ii) The overall performance of BAM is superior to or highly competitive with BA and other compared state-of-the-art population-based optimization methods.

(iii) BAM and other population-based optimization methods were compared for different maximum generations and the dimension. Under majority conditions, BAM is significantly substantial better than other population-based optimization methods.

(iv) BAM and BA were compared for different loudness A and pulse rate r, weighting factor F, and scaling factor ε. Under almost all the conditions, BAM is far better than BA.

(v) The algorithm BAM is insensitive to the parameter loudness A and discovery rate r, weighting factor F, and scaling factor ε, so we do not have to fine-tune the parameters A, r, F, and ε to get the best performance for different problems.

6. Conclusion and Future Work

This paper presented a bat algorithm with mutation for UCAV path planning in complicated combat field environments. A novel type of BA model has been described for single UCAV path planning, and a modification is applied to mutate between bats during the process of the new generation generating. Then, the UCAV can find the safe path by connecting the chosen nodes while avoiding the threat areas and costing minimum fuel. This new approach can accelerate the global convergence speed while maintaining the strong robustness of the basic BA. The detailed implementation procedure for this improved metaheuristic approach is also described. Compared with other population-based optimization methods, the simulation experiments show that this improved method is a feasible and effective way in

UCAV path planning. It is also flexible, in complicated dynamic battle field environments and pop-up threats are easily incorporated.

In the algorithm of UCAV path planning, there are many issues worthy of further study, and efficient route planning method should be developed depending on the analysis of specific combat field environments. Currently, the hot issue contains self-adaptive route planning for a single UCAV and collaborative route planning for a fleet of UCAVs. As the important ways of improving aircraft survivability, adaptive route planning should analyze real-time data under the uncertain and dynamic threat condition; even it can re-modify preplanned flight path to improve the success rate of completing mission. The difficulty of the collaborative route planning for a fleet of UCAVs exists in coordination between the various UCAVs, including the fleet formation, target distribution, arrival time constraint, and avoidance conflict, each of which is a complicated question worthy of further study. Our future work will focus on the two hot issues and develop new methods to solve problem in UCAV path planning and replanning.

Acknowledgments

This work was supported by State Key Laboratory of Laser Interaction with Material Research Fund under Grant no. SKLLIM0902-01 and Key Research Technology of Electric-Discharge Nonchain Pulsed DF Laser under Grant no. LXJJ-11-Q80.

References

[1] H. B. Duan, X. Y. Zhang, and C. F. Xu, *Bio-Inspired Computing*, Science Press, Beijing, China, 2011.

[2] G. Wang, L. Guo, H. Duan, L. Liu, H. Wang, and M. Shao, "Path planning for uninhabited combat aerial vehicle using hybrid meta-heuristic DE/BBO algorithm," *Advanced Science, Engineering and Medicine*, vol. 4, no. 6, pp. 550–564, 2012.

[3] G. Wang, L. Guo, H. Duan, L. Liu, H. Wang, and M. Shao, "Hybridizing harmony search with biogeography based optimization for global numerical optimization," *Journal of Computational and Theoretical Nanoscience*. In press.

[4] Y. V. Pehlivanoglu, "A new vibrational genetic algorithm enhanced with a Voronoi diagram for path planning of autonomous UAV," *Aerospace Science and Technology*, vol. 16, pp. 47–55, 2012.

[5] W. Ye, D. W. Ma, and H. D. Fan, "Algorithm for low altitude penetration aircraft path planning with improved ant colony algorithm," *Chinese Journal of Aeronautics*, vol. 18, no. 4, pp. 304–309, 2005.

[6] H. Duan, Y. Yu, X. Zhang, and S. Shao, "Three-dimension path planning for UCAV using hybrid meta-heuristic ACO-DE algorithm," *Simulation Modelling Practice and Theory*, vol. 18, no. 8, pp. 1104–1115, 2010.

[7] H. B. Duan, X. Y. Zhang, J. Wu, and G. J. Ma, "Max-min adaptive ant colony optimization approach to multi-UAVs coordinated trajectory replanning in dynamic and uncertain environments," *Journal of Bionic Engineering*, vol. 6, no. 2, pp. 161–173, 2009.

[8] G. Wang, L. Guo, H. Duan, L. Liu, H. Wang, and B. Wang, "A hybrid meta-heuristic DE/CS algorithm for UCAV path

[9] G. Wang, L. Guo, H. Duan, H. Wang, L. Liu, and M. Shao, "A hybrid meta-heuristic DE/CS algorithm for UCAV three-dimension path planning," *The Scientific World Journal*, vol. 2012, Article ID 583973, 11 pages, 2012.

[10] C. Xu, H. Duan, and F. Liu, "Chaotic artificial bee colony approach to Uninhabited Combat Air Vehicle (UCAV) path planning," *Aerospace Science and Technology*, vol. 14, no. 8, pp. 535–541, 2010.

[11] G. Wang, L. Guo, H. Duan, L. Liu, and H. Wang, "A modified firefly algorithm for UCAV path planning," *International Journal of Hybrid Information Technology*, vol. 5, no. 3, pp. 123–144, 2012.

[12] A. H. Gandomi, X. S. Yang, S. Talatahari, and A. H. Alavi, "Firefly algorithm with chaos," *Communications in Nonlinear Science and Numerical Simulation*, vol. 18, no. 1, pp. 89–98, 2013.

[13] H. Duan, S. Liu, and J. Wu, "Novel intelligent water drops optimization approach to single UCAV smooth trajectory planning," *Aerospace Science and Technology*, vol. 13, no. 8, pp. 442–449, 2009.

[14] R. Storn and K. Price, "Differential evolution—a simple and efficient heuristic for global optimization over continuous spaces," *Journal of Global Optimization*, vol. 11, no. 4, pp. 341–359, 1997.

[15] X. S. Yang, "A new metaheuristic Bat-inspired Algorithm," *Studies in Computational Intelligence*, vol. 284, pp. 65–74, 2010.

[16] W. Ye and H. D. Fan, "Research on mission planning system key techniques of UCAV," *Journal of Naval Aeronautical Engineering Institute*, vol. 22, no. 2, pp. 201–207, 2007.

[17] X. S. Yang, *Nature-Inspired Metaheuristic Algorithms*, Luniver Press, Frome, UK, 2nd edition, 2011.

[18] J. Kennedy and R. Eberhart, "Particle swarm optimization," in *Proceedings of the IEEE International Conference on Neural Networks*, pp. 1942–1948, Perth, Australia, December 1995.

[19] S. Kirkpatrick, C. D. Gelatt Jr., and M. P. Vecchi, "Optimization by simulated annealing," *Science*, vol. 220, no. 4598, pp. 671–680, 1983.

[20] M. Dorigo and T. Stutzle, *Ant Colony Optimization*, MIT Press, Cambridge, UK, 2004.

[21] D. Simon, "Biogeography-based optimization," *IEEE Transactions on Evolutionary Computation*, vol. 12, no. 6, pp. 702–713, 2008.

[22] G. Wang, L. Guo, H. Duan, L. Liu, and H. Wang, "Dynamic deployment of wireless sensor networks by biogeography based optimization algorithm," *Journal of Sensor and Actuator Networks*, vol. 1, no. 2, pp. 86–96, 2012.

[23] H. Duan, W. Zhao, G. Wang, and X. Feng, "Test-sheet composition using analytic hierarchy process and hybrid metaheuristic algorithm TS/BBO," *Mathematical Problems in Engineering*, vol. 2012, Article ID 712752, 22 pages, 2012.

[24] H. Beyer, *The Theory of Evolution Strategies*, Springer, New York, NY, USA, 2001.

[25] D. E. Goldberg, *Genetic Algorithms in Search, Optimization and Machine learning*, Addison-Wesley, New York, NY, USA, 1998.

[26] B. Shumeet, "Population-based incremental learning: a method for integrating genetic search based function optimization and competitive learning," Tech. Rep. CMU-CS-94-163, Carnegie Mellon University, Pittsburgh, Pa, USA, 1994.

[27] A. H. Gandomi, G. J. Yun, X.-S. Yang, and S. Talatahari, "Chaos-enhanced accelerated particle swarm optimization,"

Communications in Nonlinear Science and Numerical Simulation, vol. 18, no. 2, pp. 327–340, 2013.

[28] W. Khatib and P. Fleming, "The stud GA: a mini revolution?" in *Proceedings of the 8th International Conference on Parallel Problem Solving from Nature*, vol. 1498, pp. 683–691, 1998.

[29] G. Wang, L. Guo, A. H. Gandomi et al., "Lévy-flight krill herd algorithm," *Mathematical Problems in Engineering*. In press.

8

Experimental Studies of Active and Passive Flow Control Techniques Applied in a Twin Air-Intake

Akshoy Ranjan Paul,[1] **Shrey Joshi,**[2] **Aman Jindal,**[2] **Shivam P. Maurya,**[3] **and Anuj Jain**[1]

[1] *Department of Applied Mechanics, Motilal Nehru National Institute of Technology Allahabad, Allahabad 211004, India*
[2] *Department of Mechanical Engineering, Motilal Nehru National Institute of Technology Allahabad, Allahabad 211004, India*
[3] *Department of Chemical Engineering, Motilal Nehru National Institute of Technology Allahabad, Allahabad 211004, India*

Correspondence should be addressed to Akshoy Ranjan Paul; arpaul2k@yahoo.co.in

Academic Editors: H. H. Funke and M. Yamamoto

The flow control in twin air-intakes is necessary to improve the performance characteristics, since the flow traveling through curved and diffused paths becomes complex, especially after merging. The paper presents a comparison between two well-known techniques of flow control: active and passive. It presents an effective design of a vortex generator jet (VGJ) and a vane-type passive vortex generator (VG) and uses them in twin air-intake duct in different combinations to establish their effectiveness in improving the performance characteristics. The VGJ is designed to insert flow from side wall at pitch angle of 90 degrees and 45 degrees. Corotating (parallel) and counterrotating (V-shape) are the configuration of vane type VG. It is observed that VGJ has the potential to change the flow pattern drastically as compared to vane-type VG. While the VGJ is directed perpendicular to the side walls of the air-intake at a pitch angle of 90 degree, static pressure recovery is increased by 7.8% and total pressure loss is reduced by 40.7%, which is the best among all other cases tested for VGJ. For bigger-sized VG attached to the side walls of the air-intake, static pressure recovery is increased by 5.3%, but total pressure loss is reduced by only 4.5% as compared to all other cases of VG.

1. Introduction

Twin air-intake ducts are widely used in aircrafts for the purpose of providing pressurized air to the air compressor of an aeroengine (i.e., gas turbine engine used in aircraft) in order to achieve sufficient thrust to accomplish necessary maneuvers. The importance of the air-intake can be understood from the fact that around 20% of the swept volume of the aircraft is required to be ingested by the air-intakes for the engine during normal cruise, while for climb and take-off, the proportions are even higher [1].

In order to increase the performance and to maintain the stability of an engine operation, air flow at the engine inlet face should be sufficiently decelerated, having low total pressure distortion, high uniformity of the flow with minimum cross-flow velocity components and swirl. Therefore, the task of air-intake is to maximize the static pressure recovery and the flow uniformity at the engine inlet face/compressor inlet called "aerodynamic inlet plane" (AIP). The static pressure is increased by making the air-intake long and diverging.

Twin-side air-intakes with Y-configuration are commonly used for ingesting atmospheric air to the engine of single-engine combat aircrafts. In such air-intakes, air is ingested from either side of the aircraft with its two individual S-shaped diverging limbs merging into a single diverging duct leading air to AIP. Curvature to the duct is provided to accommodate it in a smaller space. It is reported that inhomogeneous flow generated by the supersonic part of the air-intake causes flow separation and its subsonic part causes secondary flow due to centerline curvature of the air-intake. The S-bends forming the twin air-intake initiate the strong swirl which eventually manifests itself on formation of vortices and cross-stream pressure gradients.

The nonuniformity at the AIP causes an uneven impact loading at the downstream components, like compressor. Persistence of such condition may cause sudden failure of compressor parts (e.g., blades) during flight, which may lead to catastrophe. Therefore, these conditions are unacceptable from aerodynamic as well as structural viewpoints. Thus, good aerodynamic design of twin air-intake is a challenge to increase overall performance and stability of the aircraft by ensuring sufficient uniform air supply. Employing a passive flow control (in which no external energy or no additional mass is injected) or an active flow control techniques (in which external energy as well as additional mass is injected into the system) is the possible solutions to accomplish nearly uniform air supply. However, optimizing it for a wide range of speeds, altitudes, and maneuvers poses further challenges.

The surface-mounted vane-type submerged vortex generator (VG), which is an example of passive flow control device, is used on the internal surfaces of the twin air-intakes to mix the low-momentum boundary layer with a higher momentum core flow to help reduce or eliminate boundary layer separation. The microvortices generated by these VG arrays can also be used favourably to redirect secondary flows. In both cases, the goal is to improve the performance of the engine by increasing engine face pressure recovery and decreasing engine face pressure distortion.

Several researchers [2–6] have contributed towards the effective design of VG. This VG is thin plate of triangular or trapezoidal shapes and is placed normal to the surface and at a lateral angle to the flow (referred to as inclination angle or vortex generator angle, β). Reichert and Wendt [7] used a low-profile "wishbone" type vortex generator to improve the total pressure distortions and recovery performance of a diffusing duct. The configuration employing the largest vortex generator was most effective in reducing distortion but did not produce major total pressure recovery. In a recent study, Paul et al. [8] showed the usefulness of fin-type submerged VG in flow improvement of an S-shaped diffusing duct. In another study, Paul et al. [9, 10] used similar VGs in a twin air-intake duct for flow control, especially at the AIP. The computational study demonstrated the efficacy of co-rotating VG array in reducing the flow distortion at the engine face. Johnston and Nishi [11] used a spanwise array of small, skewed, and pitched type of active flow control device called "vortex generator jets" (VGJs) in a turbulent boundary layer and proved the existence of longitudinal vortices downstream of the jet holes similar to the vortices behind the solid vortices. Johnston et al. [12] in another study described the development of vortex from VGJ. They performed experiment on a low-speed free-surface water channel to investigate the effect of dominance of VGJ array for its various configurations.

Lin [6] experimentally studied different types of vortex generating devices for turbulent flow separation control at low speeds. They used submerged vortex generators (wheeler doublet and wishbone type), spanwise cylinders, large eddy breakup (LEBU) device at small angle of attack, and vortex generator jets (VGJs).

Sullerey and Pradeep [13] reported the effectiveness of VGJ in controlling secondary flows in rectangular S-shaped diffusing ducts (resemble to single-limb air-intake) having an area ratio (A_r) of 1.39 and a turning angle ($\Delta\beta$) of 21°/21°. The test was carried out for two inflow conditions: uniform

and distorted. The use of VGJ resulted in over a 30% decrease in total pressure loss coefficient (C_{TL}) and flow distortion coefficient (DC_{60}). But for distorted inflow, a combination of passive device (tapered fin VG) and active device (VGJ) was used to reduce C_{TL} by 25%.

Harrison et al. [14] conducted experiments on boundary-layer-ingesting serpentine air-intake located on the aft surface of a blended-wing-body aircraft. Both suction and blowing (circumferential and reverse pyramid types) were applied at various locations in the air-intake in order to simulate the use of fluidic VGJ. The objective of using the VGJ was to redistribute the ingested low-momentum fluid around the periphery of the diffuser in order to normalize the flow distortion at the engine face, and thereby decreasing the fatigue and increasing operational surge margin.

From the literature review, it is revealed that the active flow control technique is more robust and can be used in various flow fields as it gives an extra degree of freedom as compared to passive methods. This can be easily understood by the fact that in active flow control methodology, the parameter which controls the flow can be varied according to the flow field and is desirable. The passive flow control technique, like vane-type and solid VG, however, has other advantages, such as simplicity, ruggedness, and low cost. It has practical applications in stall control on airfoils and in diffusing ducts. But it has limitations such as it does not have the ability to provide a time-varying control action, whereas inserted VGJ could be time varying (rotating jets) and could be switched on/off or even an increase and decrease in magnitude of jet energy could be possible if desired. However, the comparative study of both techniques while applied in the flow control of twin air-intakes is missing in the current literature.

The objective of the present study is therefore to compare the two techniques of flow control (namely, passive and active) in a twin air-intake and to find out the efficacy of each of the techniques in terms of aerodynamic performance.

2. Experimental Methodology

The present study examines the effects of the two flow control techniques on the aerodynamic performance of a twin air-intake with turning angle ($\Delta\beta$) of 20° as shown in Figure 1 along with the coordinate system. Vane-type VG being passive flow control device is shown in Figure 3, whereas VGJ as shown in Figure 4 is used for active flow control in the study.

2.1. Twin Air-Intake Model. The schematic diagram of a twin air-intake (R_c = 420 mm, $\Delta\beta$ = 20°, A_r = 1.33) is shown in Figure 1. Planes A and F are rectangular inlets (75×75 mm^2) while planes B and E are inflexion planes. Both of the S-shaped individual limbs are merged at plane-C and AIP is located 10 mm prior to the exit of the air-intake having cross-section of 75×200 mm^2. A 75 mm long rectangular straight pipes are attached before inlets and beyond plane-D to reduce the atmospheric disturbances.

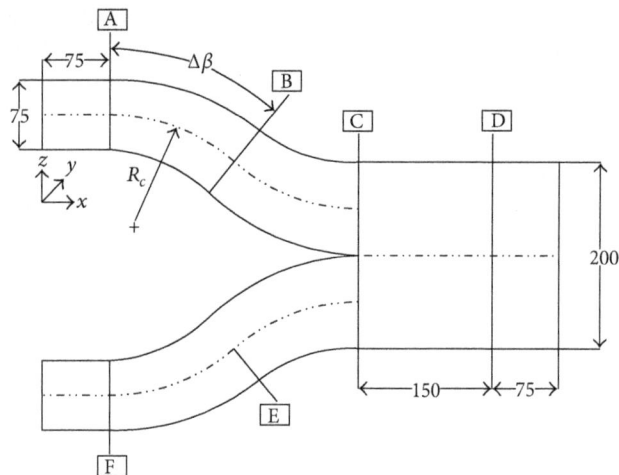

FIGURE 1: Schematic diagram of a twin air-intake.

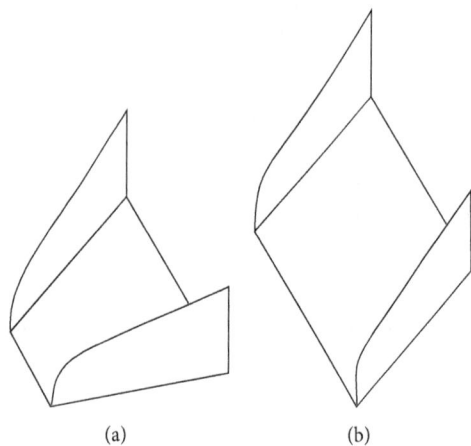

FIGURE 3: Nomenclature of a vane-type vortex generator (counter-rotating).

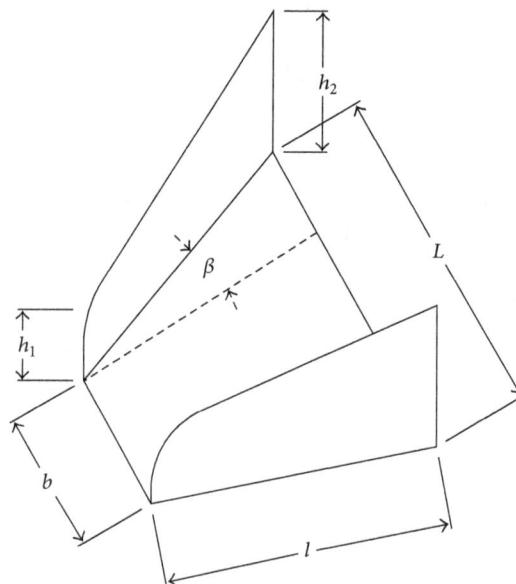

(a)   (b)

FIGURE 2: Vane-type vortex generators: (a) counter- and (b) corotating.

FIGURE 4: Vortex generator jet (VGJ).

2.2. Flow Control Devices. Two types of flow control devices—vane-type vortex generator (VG) and vortex generator jet (VGJ) are used in the study. VG is used for passive flow control and VGJ is used for active flow control in the air-intake.

2.2.1. Vane-Type Vortex Generator. The VG used in the study is designed as per Paul et al. [8], of trapezoidal shape, placed normal to the surface, and at various vortex generator angle (β). They are staggered as corotating (parallel) and counter-rotating (V-shape) configurations as shown in Figure 2. The dimensions and nomenclature of the vane-type VG are shown in Figure 3. Two different VG sizes used in the study are referred to as VG-1 (smaller) and VG-2 (larger). The design parameters of these VGs are given below.

VG-1: $\beta = 13.5°$, $b = 6.0$ mm, $h_1 = 2.0$ mm, $h_2 = 4.0$ mm, $L = 11.0$ mm, $\ell = 11.0$ mm

VG-2: $\beta = 27.0°$, $b = 6.0$ mm, $h_1 = 3.0$ mm, $h_2 = 6.55$ mm, $L = 18.0$ mm, $\ell = 13.2$ mm.

2.2.2. Vortex Generator Jet. Vortex generator jet (VGJ) is designed as given in the literature [11, 12] and is used for the study. Two stainless tubes of 2 mm diameter are provided at different pitch angles (0° and 45°) in the VGJ arrangement. The system is fitted in a 20 mm diameter rotating plug as shown in Figure 4, which enables to rotate it around 360° yaw angle. The jet was issued at a velocity ratio (i.e., jet velocity to free stream velocity of air) of 2.

2.3. Instrumentation. Digital micromanometer with a pressure scanner (make: Furness Controls, UK) is used to measure pressure, velocity when connected to a measuring instrument like pitot-static tube and wall-static pressure taps. A precalibrated five-hole static pressure probe with probe traverse system is used to carry out the steady-state measurements of three velocity components, inflow angles, static and total pressures simultaneously for a point in a flow field. A calibrated orifice meter (design as per ISO: 5167-2003) is used to provide the predetermined mass flow rate into the VGJ. Uncertainties associated in the experimentation are determined as per Kline [15] and are listed in Table 1.

TABLE 1: Least count and uncertainty in the measured parameters.

Parameter	Symbol	Instrument	Least count	Uncertainty used
Vertical traverse	y	Probe traversing mechanism	0.1 mm	± 0.1 mm
Angle	α, β	Probe orientation mechanism	$1°$	$\pm 1°$
Pressure	p	Digital micromanometer	0.001 N/m^2	$\pm 0.025\%$ of full scale deviation
Ambient temperature	T	Digital thermometer	$0.5°C$	$\pm 0.5°C$

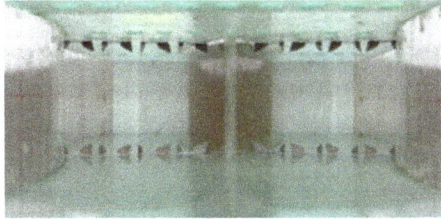

FIGURE 5: Location of VG (counterrotating) on the top-bottom walls of the air-intake.

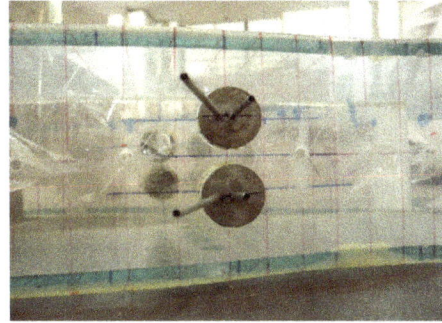

FIGURE 6: Location of VGJ on the side walls of the air-intake.

3. Experimental Procedure

Mass-averaged velocity of air is maintained at 20 m/s at both inlets (planes A and F) of twin air-intake. Additional air at a velocity of 40 m/s is issued through the VGJ connected at side walls of the air-intake. Both VG and VGJ are located at the inflexion planes (planes B and E) of individual limbs of the air-intake. VGs are attached to either top-bottom interior walls (eight on each of the two top and bottom walls) and side interior walls (three on each of the four side walls) of the air-intake as shown in Figure 5. The five different configurations of VG used in the study are as follows:

Case-1: no VG.

Case-2: counterrotating VG-1 array placed at top-bottom and side walls.

Case-3: corotating VG-1 array placed at top-bottom and side walls.

Case-4: counterrotating VG-2 array placed at top-bottom and side walls.

Case-5: corotating VG-2 array placed at top-bottom and side walls.

Likewise, two VGJs are affixed at each of the four interior side walls at the inflexion planes (B and E) of the air-intake as shown in Figure 6. Experiments were conducted for five various pitch and yaw combinations and are furnished below as well as depicted in Figure 7 for clarity. Study of the bare air-intake (i.e., without VG or VGJ) is referred to in the following sections as "Case-0."

Case-1: Pitch 90° (VGJs are directed perpendicular to the side walls).

Case-2: Pitch 45° and yaw 90° (Jets facing each other).

Case-3: Pitch 45° and yaw 180° both.

Case-4: Pitch 45° and yaw 0° both.

Case-5: Pitch 45° and yaw 45° (converging).

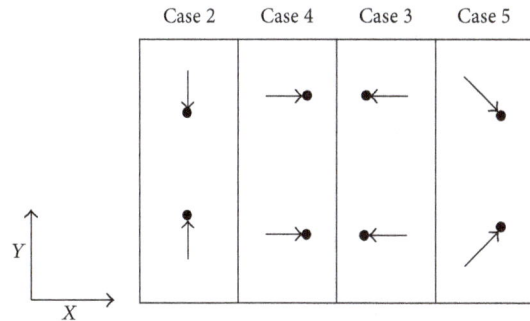

FIGURE 7: Different jet configurations of VGJ in the yaw (x-y) plane for pitch angle of 45°.

4. Results and Discussion

This section discusses the results obtained from experimentation using various flow control devices, namely, VG and VGJ.

4.1. Flow Control Using Vortex Generators (VGs). Different combinations of VG are tested and their results are tabulated in Table 2. Both geometry (VG height h_2 and VG angle) and locations of vane-type VG are varied to find out the optimum aerodynamics performance in twin air-intake. This type of VG arrays is used with the aim of manipulating the flow to reduce nonuniformity and total pressure distortion and possibly improve static pressure recovery by means of separation and secondary flow control. For this, two approaches are necessary. The first is to create a strong mixing between the boundary layer fluid and the main flow to produce a fairly uniform flow at the AIP. The second is to counter the effect of the secondary flows, which tend to accumulate the low energy boundary layer flow in one area, and redistribute

TABLE 2: Values of performance parameters for all the cases of VG.

VG sequence	Location	C_{PR}	C_{TL}	DC_{60}	S_{io}	σ_{xo}
No VG (Case-0)	—	0.452	0.113	0.259	0.010	5.420
VG-1 (Counter)	Side walls	0.473	0.120	0.253	0.011	5.359
	Top-bottom	0.456	0.112	0.248	0.011	5.270
VG-1 (Corot.)	Side walls	0.473	0.119	0.249	0.011	5.374
	Top-bottom	0.459	0.112	0.223	0.011	5.208
VG-2 (Counter)	Side walls	0.480	0.118	0.232	0.011	4.972
	Top-bottom	0.459	0.113	0.237	0.017	3.593
VG-2 (Corot.)	Side walls	0.476	0.108	0.228	0.013	5.370
	Top-bottom	0.461	0.112	0.217	0.017	3.438

the boundary layer evenly around the perimeter of the AIP, thus reducing the flow distortion [16].

It is found from the literatures [17, 18] that the vortex generators attached to the top and bottom walls of the air-intake effectively control the secondary flow instead of using it on side walls. This is evident from Table 2 that secondary flow non-uniformity (S_{io}) sometimes increases for the air-intakes with vortex generators attached to its side walls, whereas the same parameter only records a reduction if vortex generators are attached to its top-bottom walls irrespective of its heights.

Corotating VG array is useful in reducing flow separation if it is properly designed and located. The key advantage of co-rotating VG is their downstream effectiveness resulting in more effective usage of the vortex energy within the affected boundary layer. According to design wisdom, this type of VG has a few special advantages when used in twin air-intakes; namely: (a) the induced microvortices will remain close to the wall; consequently a "cleaner" core flow will result, and (b) the induced vortices will counteract the natural and often strong secondary flows, which can develop within the S-bend of each individual limb of such twin air-intakes. Counter-rotating VG array, on the other hand, has the disadvantages as compared to the co-rotating VG array, since the induced vortices tend to lift off the duct surface, thus reducing their effectiveness, causing higher loss in static pressure recovery and large total pressure distortion at AIP [19].

The use of co-rotating VG array is proved to be more successful as evident from Table 2. With effective use of such VG array, aerodynamic performance of nonfused air-intakes is improved in terms of C_{PR} (maximum rise by 6.2%), C_{TL} (maximum drop by 4.42%), DC_{60} (maximum reduction by 16.1%),and σ_{xo} (maximum reduction by 36.58%). From Table 2, co-rotating VG array when attached to the side walls of the air-intake offers the optimum performance.

4.2. Flow Control Using Vortex Generator Jets (VGJs).

The VGJ technique is a time-varying control action to optimize performance under a wide range of flow condition. For VGJ, the strength of longitudinal vortices is controllable by varying the jet speed. The values of different performance parameters are given in Table 3 for all the cases of VGJ tested in the study.

Jet dynamics is entirely controlled by vorticity. Generation of vortices at jet issuing source is completely dictated

TABLE 3: Values of performance parameters for all the cases of VGJs.

Parameter	Case-0	Case-1	Case-2	Case-3	Case-4	Case-5
ζ	0.68	0.74	0.67	0.72	0.69	0.74
C_{TL}	0.113	0.067	0.114	0.085	0.095	0.090
S_{io}	0.010	0.007	0.008	0.007	0.009	0.010
σ_{xo}	5.420	4.695	5.510	5.215	4.551	5.024
DC_{60}	0.218	0.236	0.209	0.272	0.168	0.251

with convection in axial direction and diffusion in crossways. At the onset of instability in jets, amplitude of fluctuating vortices shoots up and subsequently proper mixing of layers starts. The vortices roll up and finally, the merging of these vortices in the freestream controls the amount of momentum transferred between layers, which consequently affects the point of separation. The sole purpose of using a VG or a VGJ is to have proper mixing between layers such that the layer close to the walls gains energy to continue and not to separate out. In order to control flow separation, it is therefore worthwhile to consider certain factors affecting the onset of instability, vortex roll up, and vortex merging in a jet. Moreover, when two VGJs at each side wall simultaneously issue jet at 40 m/s, the effect of orientation of two jets and their interaction play an important role. The factors which affect the flow separation are therefore (a) disturbances in the freestream, (b) disturbances due to boundary layers, wakes and small recirculation zones, and (c) the interaction between the two jets.

As in case of a twin air-intake, the VGJs are applied at the inflexion plane from the side walls; hence considering the above listed factors, there are freestream disturbances. As the flow negotiates a curved path introducing secondary flow and turbulence due to centrifugal effects, small recirculation zones could also be noticed towards the inner side of the air-intake. Hence, there are three vortex interactions possible, one between the free stream disturbances and the jet vortices, second between recirculation vortices, and jet vortices and third between the two jet vortices itself. This incident together with turbulence superimposed becomes highly complicated flow phenomena. Prediction of the correct behavior of VGJ in a flow stream is only possible during an experiment in a laboratory. Study the effect of jets at different orientation is also important. Johnston and Nishi [11] presented a similar studying of jet behavior downstream when they interacted with each other on a flat plate. The present study is, however, a little different due to the wall curvature.

Case-1 observes the effect of jet inserted perpendicular to the main flow stream through air-intake (i.e., pitch angle = 90°). The case proves to be best as it exhibits low total pressure loss and high static pressure recovery as shown in Table 3. This clearly indicates that all possible interaction of vortices shedding from the jets is quite capable of mixing layers and transferring momentum, hence energizing the near wall flow. Johnston and Nishi [11] performed interaction of different orientations at 45° pitch and various yaw angles and observed similar trends.

When jets are oriented at a pitch angle of 45° and yaw angle of 90° (case-2), the jets converge the centerline. This case exhibits the least potential in terms of flow improvement and control and can be accounted to the fact that the vortex interaction produces momentum transfer but either insufficient or in reverse direction. The upper layer gets energized cascading energy from the near wall flow; hence, the separation point shifts upstream leading to higher pressure losses and lower static pressure recovery. This was exactly the behavior experienced by Johnston and Nishi [11] while using the converging jets.

Case-3 again includes jet oriented at a pitch angle of 45° and at a yaw angle of 180°; that is, the side view would show the jets aligned opposite to the flow. The case does not offer the best performance; however, reasonable improvement in performance parameters is observed. The results are well in agreement with the fact that when the jets are oriented at 180° to the main flow, two vortices streams are shed from each jet instead of one, though there are two jets; however, the performance parameters do not show the finest improvements for the reason that the two shed vortices are of relatively poorer strength and could not offer such an outcome that a single higher strength vortex could. The above is also true when jets are aligned with the flow at 0°; the case has been discussed under as case-4. Unlike case-3, the jets are now aligned to the direction of main flow keeping the jet orientation the same. This case shows slight improvement in the performance parameters as compared to the base value for the bare duct. Though being similar to case-3, nevertheless it does not show much improvement in performance as compared to case-3. The jets when aligned towards the flow shed two vortices of relatively low strength. But while the jets are aligned with the main flow direction, the interaction between the vortex and the free stream is not adequate since the vortices get carried away in the free stream relative to the amount of interaction occurred in case-1. It is, therefore, concluded that while the jets are aligned opposite to the main flow direction (case-3), the interaction between the flow and the vortices is adequately enough to provide an enhancement in performance.

In case-5, the jets are oriented at pitch and yaw angle both being 45°; that is, the jets look like diverging along the longitudinal mid-symmetry plane of the air-intake. The case shows improvement in flow parameters, the static pressure recovery

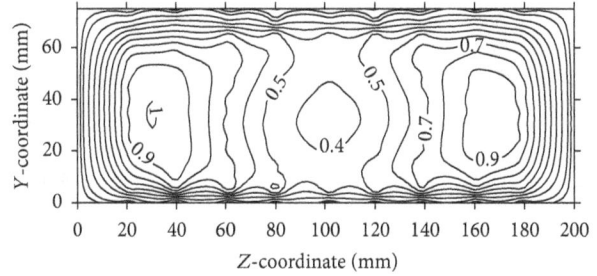

FIGURE 8: Normalized velocity contours at the outlet plane of bare air-intake.

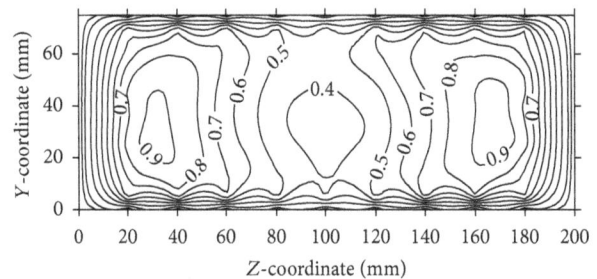

FIGURE 9: Normalized velocity contours at the outlet for air-intake (best VG case).

being highest but accompanied with relatively higher total pressure loss that makes the case less effective as compared to the best case (case-1).

It is, therefore, emerged from the above discussion that the VGJ when directed perpendicular to the air-intake side walls (case-1) offers the best aerodynamic performance.

4.3. Performance Comparison of Different Flow Control Techniques. A number of combinations are tested using flow control devices: VG and VGJ and their comparison in terms of normalized velocity contours are presented in Figures 8–10. Velocity components are calculated from the pressure data taken by five-hole pressure probe using calibration charts. For the sake of brevity, the details of calculation are not however included here. Velocity contours are drawn using graphic software (Surfer) and the spatial velocity data are interpolated using kriging technique to draw the velocity contours, which are detailed in Stein [20].

Figure 8 shows the velocity contours at the outlet plane (plane-D) of a bare air-intake (without any VG or VGJ). The values of local velocities are normalized by the free-stream velocity available at the inlets of the air-intake. The velocity contours at the outlet of the duct clearly indicate that the peak velocities occur near the side walls and low velocity core remains at the centre. From Figure 9, it is clear that the two velocity peaks on either side of the air-intake are reduced to an extent due to the use of VG, whereas the weak core flow still exists. However, a noticeable change in the flow pattern is observed in Figure 10, when VGJ is used in the air-intake. The weak core flow is disappeared as there is an increase in velocity in the middle of the outlet plane, whereas the velocity peaks near the two side walls are reduced to a large extent

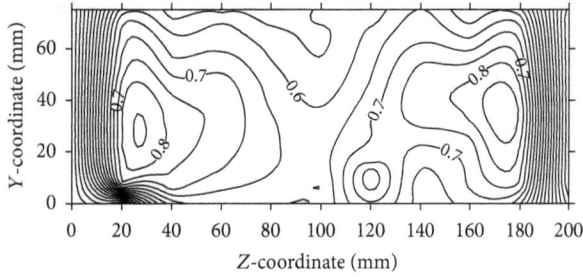

FIGURE 10: Normalized velocity contours at the outlet for air-intake (best VGJ case).

TABLE 4: Performance comparison of air-intake using flow control.

Flow control	C_{PR}	C_{TL}	DC_{60}	S_{io}	σ_{xo}
Bare air-intake	0.452	0.113	0.218	0.010	5.420
With VGJ	0.487	0.067	0.236	0.007	4.695
With VG	0.476	0.108	0.174	0.013	5.370

indicating the flow uniformity at the outlet plane of the air-intake. As a result, both the secondary and axial flow non-uniformities are reduced to a great extent (30% and 13%, resp.) as shown in Table 4.

Overall, the flow control of air-intake with VGJ promotes the highest static pressure recovery coefficient (C_{PR}), lowest total pressure loss coefficient (C_{TL}), maximum reduction of secondary flow non-uniformity (S_{io}), and axial flow non-uniformity (σ_{xo}). However, vane-type VG ensures minimum total pressure distortion coefficient (DC_{60}) as compared to the air-intake with VGJ. In cases where VGJ is used, the extra mass flow injected through it helps in decreasing the pressure losses.

5. Conclusions

Experimental studies on air-intake diffuser are conducted and various combinations of vane-type vortex generator (VG) and vortex generator jet (VGJ) are tested. The following conclusions are drawn from the current study.

(i) Comparison between the two flow control techniques shows that the use of VGJ is more effective as compared to VG. It is observed that VGJ has the potential to change the flow pattern drastically as compared to VG. Furthermore, for flow situations where stall control is not needed, parasitic drag can be avoided with the jet flow turned off. On the contrary, vane-type VG is always exposed in the flow and can increase drag. The VGJ technique accomplishes flow separation control only when it is necessary and therefore it is favored over vane-type VG for both design and off-design conditions.

(ii) In case of vortex generator jets, the best result is given when the jet is directed perpendicular to the side walls at a pitch angle of 90°. Two most important performance parameters—static pressure recovery is

increased by 7.8% and total pressure loss is reduced by 40.7% as compared to all other cases of VGJ.

(iii) In case of vane-type vortex generators, the best result is given when co-rotating VG-2 (big-sized VG) is attached to the side walls of the air-intake. In this case, static pressure recovery is increased by 5.3%, but total pressure loss is reduced by only 4.5% as compared to all other cases of VG.

Nomenclature

A: Cross-sectional area . . . mm
A_r: Area ratio, A_e/A_i
C_{PRi}:   Ideal static pressure recovery coefficient,
     $C_{PRi} = 1 - (1/A_r)^2$
C_{PR}:   Actual static pressure recovery coefficient,
     $C_{PR} = (p_s - p_{si})/((1/2)\rho U_{avi}^2)$
C_{TL}:   Total pressure loss coefficient,
     $C_{TL} = (p_{ti} - p_t)/((1/2)\rho U_{avi}^2)$
DC_{60}:   Distortion coefficient,
     $DC_{60} = (p_{te} - p_{t60})/((1/2)\rho U_{ave}^2)$
p_s: Static pressure
p_t: Total pressure at outlet
p_{t60}:   Total pressure at worst 60° sector
R_c: Radius of curvature
S_{io}:   Secondary flow non-uniformity,
     $S_{io} = \sum U_{yz}/n \times U_{avi}$
U_{av}:   Average velocity
V: Velocity component
β: Vortex generator angle
δ: Boundary layer thickness
$\Delta\beta$:   Turning angle of air-intake
σ_{xo}:   Axial flow non-uniformity index:
     $\sigma_{xo} = \sqrt{\sum (U_x - U_{xav})^2/n}$
ζ: Effectiveness of air-intake, $\zeta = C_{PR}/C_{PRi}$.

Subscript

av: Average
e: Exit
i: Inlet
x: Component in x-direction
yz: Cross-component.

Acknowledgments

The authors are grateful to the Department of Science & Technology, Government of India for providing the necessary financial support through DST-FIST Grant for the development of experimental facility necessary for conducting the research. The infrastructural support provided by MNNIT Allahabad is also acknowledged.

References

[1] J. Seddon, *Introduction To Intake Aerodynamics*, vol. 1 of *Lecture series 1988-04*, Von Karman Institute for Fluid Dynamics, Rhode Saint Genèse, Belgium, 1988.

[2] J. C. Lin, G. V. Selby, and F. G. Howrad, "Exploratory study of vortex-generating devices for turbulent flow separation control," AIAA Paper No. 91-004t2, 1991.

[3] J. C. Lin, "Control of turbulent boundary layer separation using micro-vortex generators," AIAA Paper No. 99-3404, 1999.

[4] A. Bernard, J. M. Foucaut, P. Dupont, and M. Stanislas, "Decelerating boundary layer: a new scaling and mixing length model," *AIAA Journal*, vol. 41, no. 2, pp. 248–255, 2003.

[5] L. Jenkins, S. A. Gorton, and S. Anders, "Flow control device evaluation for an internal flow with an adverse pressure gradient," AIAA Paper No. 2002-0266, 2002.

[6] J. C. Lin, "Review of research on low-profile vortex generators to control boundary-layer separation," *Progress in Aerospace Sciences*, vol. 38, no. 4-5, pp. 389–420, 2002.

[7] B. A. Reichert and B. J. Wendt, "An experimental investigation of S-duct flow control using arrays of low profile vortex generator," AIAA Paper No. 1993-0018, 1993.

[8] A. R. Paul, P. Ranjan, V. K. Patel, and A. Jain, "Comparative studies on flow control in rectangular S-duct diffuser using submerged-vortex generators," *Aerospace Science and Technology*, vol. 28, no. 1, pp. 332–343, 2013.

[9] A. R. Paul, P. Ranjan, R. R. Upadhyay, and A. Jain, "Passive flow controlin twin air-intakes," in *Proceedings of the WASET International Conference on Mathematical and Computational Methods in Science and Engineering (ICMCMSE '11)*, pp. 297–304, Paris, France, June 2011.

[10] A. R. Paul, P. Ranjan, R. R. Upadhyay, and A. Jain, "Passive flow control in twin air-intakes," *World Academy of Science, Engineering and Technology*, no. 77, article 56, pp. 297–304, 2011.

[11] J. P. Johnston and M. Nishi, "Vortex generator jets. Means for flow separation control," *AIAA Journal*, vol. 28, no. 6, pp. 989–994, 1990.

[12] J. P. Johnston, B. P. Mosier, and Z. U. Khan, "Vortex generating jets; effects of jet-hole inlet geometry," *International Journal of Heat and Fluid Flow*, vol. 23, no. 6, pp. 744–749, 2002.

[13] R. K. Sullerey and A. M. Pradeep, "Effectiveness of flow control devices on S-duct diffuser performance in the presence of inflow distortion," *International Journal of Turbo and Jet Engines*, vol. 19, no. 4, pp. 259–270, 2002.

[14] N. A. Harrison, J. Anderson, J. L. Fleming, and W. F. Ng, "Experimental investigation of active flow control of a boundary layer ingesting serpentine inlet diffuser," AIAA 2007-843, 2007.

[15] S. J. Kline, "The purposes of uncertainty analysis," *Journal of Fluids Engineering, Transactions of the ASME*, vol. 107, no. 2, pp. 153–160, 1985.

[16] A. J. Anabtawi, R. F. Blackwelder, P. B. S. Lissaman, and R. H. Liebeck, "An experimental study of vortex generators in boundary layer ingesting diffusers with a centerline inlet," AIAA Paper No. 1999-2110, 1999.

[17] R. K. Sullerey, S. Mishra, and A. M. Pradeep, "Application of boundary layer fences and vortex generators in improving performance of S-duct diffusers," *Journal of Fluids Engineering, Transactions of the ASME*, vol. 124, no. 1, pp. 136–142, 2002.

[18] R. K. Sullerey and A. M. Pradeep, "Secondary flow control using vortex generator jets," *Journal of Fluids Engineering, Transactions of the ASME*, vol. 126, no. 4, pp. 650–657, 2004.

[19] B. H. Anderson, P. S. Huang, W. A. Paschal, and E. Cavatorta, "A study on vortex flow control of inlet distortion in the re-engined 727-100 center inlet duct using computational fluid dynamics," AIAA Paper No. 1992-0152, 1992.

[20] M. L. Stein, *Statistical Interpolation of Spatial Data: Some Theory For Kriging*, Springer, New York, NY, USA, 1999.

9

Cooperative Scheduling of Imaging Observation Tasks for High-Altitude Airships Based on Propagation Algorithm

He Chuan, Qiu Dishan, and Liu Jin

Science and Technology on Information Systems Engineering Laboratory, National University of Defense Technology, Changsha 410073, China

Correspondence should be addressed to He Chuan, chuanhe@nudt.edu.cn

Academic Editors: E. Acar and A. F. B. A. Prado

The cooperative scheduling problem on high-altitude airships for imaging observation tasks is discussed. A constraint programming model is established by analyzing the main constraints, which takes the maximum task benefit and the minimum cruising distance as two optimization objectives. The cooperative scheduling problem of high-altitude airships is converted into a main problem and a subproblem by adopting hierarchy architecture. The solution to the main problem can construct the preliminary matching between tasks and observation resource in order to reduce the search space of the original problem. Furthermore, the solution to the sub-problem can detect the key nodes that each airship needs to fly through in sequence, so as to get the cruising path. Firstly, the task set is divided by using k-core neighborhood growth cluster algorithm (K-NGCA). Then, a novel swarm intelligence algorithm named propagation algorithm (PA) is combined with the key node search algorithm (KNSA) to optimize the cruising path of each airship and determine the execution time interval of each task. Meanwhile, this paper also provides the realization approach of the above algorithm and especially makes a detailed introduction on the encoding rules, search models, and propagation mechanism of the PA. Finally, the application results and comparison analysis show the proposed models and algorithms are effective and feasible.

1. Introduction

The high-altitude airship is a stay-in-air aerostat, which is lifted off by static buoyancy and moved forward by propulsion device. It is normally powered by regenerative energies, such as solar cells or regenerative fuel cells. The high-altitude airship can stay 20~100 km above the ground in the near space for a long time. This airspace belongs to neither the aviation field nor spaceflight field, but it has a broad prospect in many military applications. Once equipped with various payloads, it can be used to fulfill different missions, such as information acquisition, communication security, and battlefield transportation [1–4].

Nowadays, high-altitude airship located in the near space is attracting more and more interests worldwide. It is well known that some projects of high-altitude airship have been launched, for example, the Sky Tower [5] and Dark Sky Station [6] projects in USA, the Cargo Lifter [7] project in the European Union, the ETRI [8] project in Korea, and the Sky Net project in Japan [9]. China began its own high-altitude airship projects in 2002 and has already completed the design, manufacture and test flight of the low-altitude craft [10–12]. The verification airship has accomplished its low-altitude flight experiment in 2003. It is expected that the operational height of the final product is 20~22 km off ground, the payload can reach 1.8 t, and the duration would be more than one year.

As a new application platform, the high-altitude airship has many advantages compared with the imaging satellite [13–15] and conventional heavier-than-air (HTA) [16, 17] aircraft (e.g., unmanned aircraft vehicle). For instance, it has a longer endurance, a rapider response, a higher effectiveness-cost ratio, a more suitable operational altitude, and a higher viability [18–21].

Longer Endurance. Due to its specific flying mechanism, the high-altitude airship can successively work for several months or even longer. On the other hand, the working period of satellite is restrained by its orbit, and that of UAV is constrained by the fuel capacity.

Rapider Response. The lifting-off speed of the high-altitude airship can be up to 300 m/min. With such a fast lifting-off speed, it only requires 2 hours for the airship to get to the near space. In comparison, launching a new satellite requires about 40 days of preparations. Although UAV is a fast deployment aircraft, it also requires special auxiliary equipment for launching.

Higher Effectiveness-Cost Ratio. The high-altitude airship generally has a simpler structure and a lower cost in deployment, operation, and control. Furthermore, the high-altitude airship has a larger payload capacity than both satellite and UAV.

More Suitable Operational Altitude. The operational altitude of the high-altitude airship is higher than UAV but lower than the imaging satellite. Since stratosphere is below the ionosphere, the electromagnetic signal of airship would not be interfered by the ionized particles in the ionosphere.

Higher Viability. Airship has a naturally stealth ability for its nonmetal makeup. Both the electromagnetic and heat reflective areas of an airship are small. Moreover, the working altitude of the high-altitude airship is unlikely to be reached by normal antiaircraft weapons.

Those above features have turned high-altitude airship into an ideal imaging observation platform. In return, they also lead to many differences in control technologies (e.g., attitude control, driving control, pressure control, and position control) of high-altitude airships compared with imaging satellite and UAV. For example, altitude adjustments of airship require specific control technologies by capsule inflation and deflation.

Existing studies on high-altitude airship are scattered over a range of journals, conferences, books, and reports. Chen et al. [22] investigated the inertia propulsion system of high-altitude airship and established hydrodynamic models with three factors, that is, the velocity of the wind, the diameters, and the working altitude. Ma and Sun [23] studied the feasibility and technical difficulty of the Multipower system in high-altitude airship. Then, a new management method of the Multipower system is proposed based on the various flight modes. Rao et al. [24] focused on the mission path-following controller for the airship by applying artificial neural network (ANN). Bessert and Frederich [25] investigated the aerodynamic influence of high-altitude airship and presented a novel method to test the aerodynamics on the structural behavior of airship. The aforementioned researches mainly focused on the single high-altitude airship platform, which included cruising path control, energy system design, communication links optimization, and dynamics modeling. These researches were utilized to improve the performance of platform (e.g., reducing energy consumption, decreasing the cruising distance, and easing control). However, few studies have paid

attention to the task planning, which is one of the most demanding issues for airship applications.

There are also numerous distinctions in task scheduling between high-altitude airship and other platforms. Compared with imaging satellite, high-altitude airship can cruise independently above the target for fixed-point observation because of its slow speed and suspension ability. The task execution sequence of airship is completely determined by its path choosing. Thus, the task planning for high-altitude airship is free from orbit issues faced by imaging satellite. Moreover, the high moving speed of satellite enables it with multiple chances to observe the same target during the whole task period. In contrast, restrained by the low cruising speed, high-altitude airship's revisit cycle during the task period is longer. This is also a problem needing to be noticed during task planning for high-altitude airship. Different from UAV, high-altitude airship has high mobility as well as suspension observation ability; that is, high-altitude airship is able to continuously monitor a target by long time hovering. Although UAV is also able to conduct continuous monitoring over a fixed-point target, its continuous observations are actually comprised by several dispersed investigation activities restrained by the mobility and the cruising distance. Besides, UAV usually needs to take off and land in the base, which is the starting point and the ending point of the cruising path. On the contrary, the high-altitude airship is not restricted by this factor due to its long endurance.

In this paper, we drive into the cooperative scheduling problem of high-altitude airships toward the imaging observation tasks and discuss the optimization methods of task assigning and cruising path choosing. Multiairship scheduling is a combination optimization problem that is more complex than the single airship scheduling due to the fact that the former is required to optimize the task execution schemes in each airship simultaneously. Thus, a hierarchical scheduling framework is constructed to facilitate the problem solving process. The simulation results show the effectiveness of this strategy.

The remainder of this paper is organized as follows. Section 2 describes the collaborative programming problem of high-altitude airships and establishes the constraint programming model. Section 3 designs the collaborative scheduling frame and proposes the solution algorithms. Simulation results and performance analysis are given in Section 4. Finally, Section 5 concludes the paper with some discussion about future research.

2. Problem Description and Modeling

The application of the high-altitude airship into the earth imaging reconnaissance activity is an important supplement of other reconnaissance methods. The imaging payload of a high-altitude airship is usually installed in the task capsule and can swim or rotate within a certain angular range in order to observe the ground target. Figure 1 is the reconnaissance airship which is called RAID system and used by US troops in Iraq and Afghanistan. RAID began to be

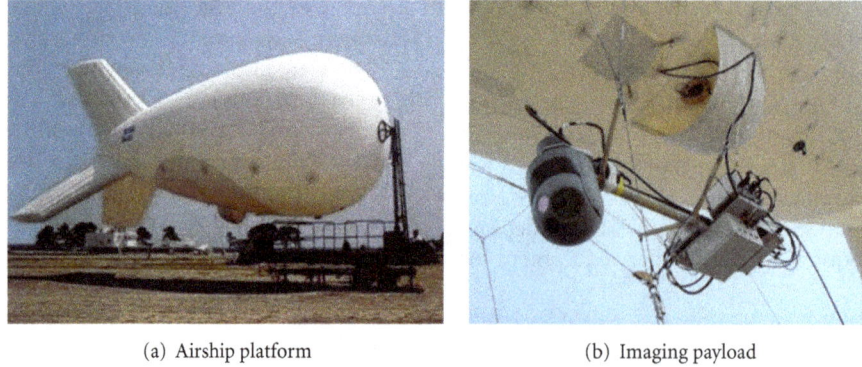

(a) Airship platform   (b) Imaging payload

FIGURE 1: RAID system of U.S troop.

equipped in 2003, and more than 60 sets of this system have been in service currently [26].

In the execution of the imaging reconnaissance tasks, the high-altitude airship maneuvers over the target area along with the cruising path and then begins the continuous observing in the hover-and-stare way. Usually, the imaging task has timeliness [27, 28] that is, each task has its time slot which is to reflect the requirements on the execution timing interval. The task execution must be finished within its time window. Otherwise, the task will lose its executive value or become invalid. Generally, imaging reconnaissance tasks are numerous and widely distributed in the battlefield. Hence, the high-altitude airship observes targets in different positions through multiple times cruising. However, due to the limitations of the cruising speed, the airship has to consume a lot of time in the cursing road. This may lead to parts of tasks that cannot be executed within their deadline. In order to maximize the overall observation benefit, it is essential to choose the observable tasks and determine the execution time interval for each airship based on the heterogeneity.

To facilitate description of this problem, we summarize the main notations which are used throughout this paper as follows:

$T_p = [t_{start}, t_{end}]$ is the task period of the airship; t_{start} denotes the starting time of the observation activity and t_{end} denotes the ending time of the observation activity;

Task = {task$_1$, task$_2$, . . . , task$_n$} is the set of the imaging tasks, where task$_i$ denotes the ith task and n denotes the number of the tasks;

$A = \{a_1, a_2, . . . , a_m\}$ is the set of the observation resources, where a_k denotes the kth airship and m denotes the number of the airships;

$o_k = (ox_k, oy_k)$ is the projective coordinates of a_k on the ground, where ox_k and oy_k denote the horizontal coordinate and vertical coordinate, respectively;

speed$_k$ is the average cursing speed of a_k;

RP_k is The resolution of the payload attached with a_k;

sit$_i = (tx_i, ty_i)$ is the position coordinates of task$_i$, where tx_i and ty_i denote the horizontal coordinate and vertical coordinate, respectively;

$[ta_i, td_i]$ is the time window of task$_i$, where ta_i denotes the allowable execution timing instant and td_i denotes the deadline;

tb_i is the beginning execution timing instant of task$_i$;

te_i is the ending execution timing instant of task$_i$;

t_i is the duration time (or called the task length) of task$_i$;

P_i is the value index of task$_i$;

u_i is the imaging resolution which needs to be satisfied in the execution of task$_i$;

$D = [d_{i,j}]_{n \times n}$ is The distance matrix, where $d_{i,j}$ denotes the distance from task$_i$ to task$_j$.

Due to the low-speed maneuverability and hover ability, the high-altitude airship is not restricted by the turning angle, climbing angle, dive angle, or other factors when performing the imaging observation activity. In addition, the common high-altitude airship is augmented by the efficient hybrid energy system, so as to enable it to be operated in long endurance [29]. However, this paper mainly focuses on the daily-scheduling problem with shorter task period. Therefore, the influence of the energy factors on the reconnaissance activities is put aside in the scheduling model.

Definition 1. If task$_i$ and task$_j$ are assigned to the same airship, and task$_j$ is arranged to be executed just next to task$_i$, then task$_i$ is called the preceding task of task$_j$, and task$_j$ is the following task of task$_i$.

Given a decision matrix $X = [x_{i,k}]_{n \times m}$, if task$_i$ is arranged to be executed on a_k, then $x_{i,k}$ equals "1"; otherwise, $x_{i,k}$ is "0". If $x_{i,k} x_{j,k} = 1$, and task$_i$ is the preceding task of task$_j$, then $y_{i,j}^k$ is "1"; otherwise, let $y_{i,j}^k$ equals "0".

In this paper, the main constraints are considered as follows.

Constraint 1. The airships execute observation activity within the task period, and each task only can be executed within its time window.

Constraint 2. If a task is executed, then the execution time of this task is no less than the required duration.

Constraint 3. Each task only can be assigned to one observation resource and just need to be done once.

Constraint 4. Only one preceding task and one following task of each task are allowable at most.

Constraint 5. The preemptive service in the task execution is prohibited. Once the execution starts, the process cannot be terminated until completion.

Constraint 6. Before the airship observes a new task, enough time should be given to change the observation position.

Constraint 7. The airship observes a task must meet its lowest imaging resolution.

Constraint 8. The airship only cruises among different tasks;

Constraint 9. With regard to any two tasks which are executed by the same observation resource, the execution should be in sequence.

Assume TB_k is the task benefit obtained by a_k and RD_k is its cruising distance. The primary optimization objective of the collaborative scheduling problem for the high-altitude airships is to maximize the task benefit:

$$\max z_1(X) = \sum_{k=1}^{m} TB_k. \tag{1}$$

The calculation method of TB_k is

$$TB_k = \sum_{task_i \in Task} x_{i,k} p_i. \tag{2}$$

On the basis of ensuring this objective, it is essential to shorten the cruising distance as far as possible.

$$\min z_2(X) = \sum_{k=1}^{m} RD_k. \tag{3}$$

If $x_{i,k} = 1$ and $\sum_{task_j \in Q} y_{j,i}^k = 0$, it means that $task_i$ is the first task to be executed on a_k; then, save the serial number of $task_i$ in a variable F_k. Furthermore, with any two points $A = (x_i, y_i)$ and $B = (x_j, y_j)$, we record the distance between them as $d(A, B) = \|A - B\|_2$. The calculation method of RD_k is expressed as follows:

$$RD_k = \sum_{task_i, task_j \in Task} \sum y_{i,j}^k d\left(sit_i, sit_j\right) + d\left(o_k, sit_{F_k}\right). \tag{4}$$

Assume R_0 is the resolution space of X. Let R_1 be the optimal solution set while $z_1(X)$ is optimized separately. Establish the constraint-programming model of the collaborative scheduling problem for high-altitude airships as follows:

$$z_1(X^*) = \max_{X \in R_0} z_1(X),$$

$$z_2(X^*) = \min_{X \in R_1} z_2(X),$$

$$\text{s.t.} \begin{cases} [tb_i, te_i] \subset [ta_i, td_i] \subset [t_{start}, t_{end}] \\ te_i - tb_i \geq t_i, & \text{if} \sum_{k=1}^{m} x_{i,k} = 1 \\ \sum_{k=1}^{m} x_{i,k} \leq 1 \\ \sum_{i=1}^{n} y_{i,j}^k \leq 1, \quad \sum_{j=1}^{n} y_{i,j}^k \leq 1 \\ [tb_i, te_i] \cap \left[tb_j, te_j\right] = \varnothing, & \text{if} \ x_{i,k} \cdot x_{j,k} = 1 \\ te_i + \dfrac{d_{i,j}}{speed_k} - K\left(1 - x_{i,j}^k\right) \leq tb_j \\ u_i \geq RP_j x_{i,j} \\ \sum_{k=1}^{m} y_{i,i}^k = 0 \\ y_{i,j}^k \cdot y_{j,i}^k = 0 \\ x_{i,k}, y_{i,j}^k \in \{0,1\} \\ \forall i, j \in \{1, 2, \ldots, m\}, \ k \in \{1, 2, \ldots, n\}, \end{cases} \tag{5}$$

where the former nine inequalities correspond to the aforementioned nine constraints, respectively, and the tenth inequality restricts the span of the decision variables.

3. Cooperative Scheduling Method

3.1. Designing of Scheduling Frame. There exist numerous constraints in the collaborative scheduling problem of the high-altitude airships, so it is hard to solve this problem directly.

As depicted in Figure 2, the hierarchical architecture is presented to convert the original problem into a main problem (namely, the task set partitioning) and a subproblem (namely, the cruising path selection).

(1) The Main Problem. The task set is divided into m portions according to the number of the observation resources. Each portion is called a cluster and denotes the assigning task set of an airship. The mapping relation between tasks and airships is constructed for decreasing the solution space of the original problem and facilitating the follow-up selection of cruising path.

Definition 2. Let $S_k = \{task_{k1}, task_{k2}, \ldots, task_{kq_k}\}$ be the cluster corresponding to a_k and regard the cluster set $S = \{S_1, S_2, \ldots, S_m\}$ as a feasible solution of the main problem, if the following conditions are satisfied:

$$\bigcup_{S_i \in S} S_i = S, \quad \forall S_i, S_j \in S, \ i \neq j, \ \text{if} \ S_i \cap S_j = \varnothing. \tag{6}$$

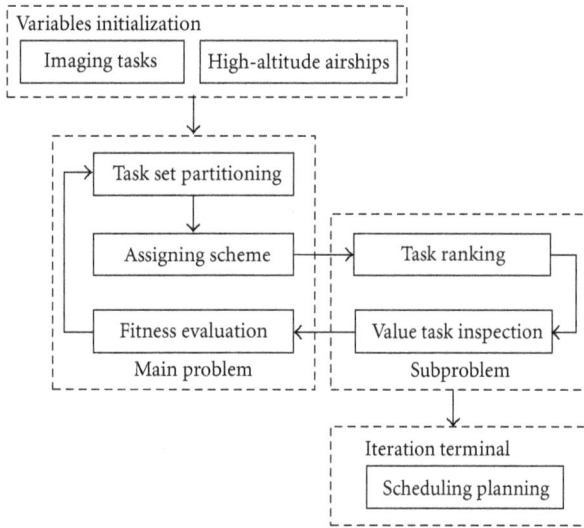

FIGURE 2: Architecture of cooperative scheduling algorithm.

Obviously, the task set partitioning problem belongs to the combinatorial optimization problem, and the scale of the solution space exponentially increases with the number of tasks and observation resources. At present, the global optimal solution algorithm with polynomial time complexity is nonexistent. Thus, this paper proposes a rapid algorithm named K-NGCA to solute this problem.

(2) The SubProblem. The priority for executing the sequence of each task is determined based on the solution of the main problem. The value tasks are selected according to the matching results between the observation capacity of airships and the requirement of tasks. In other words, the key nodes that each airship needs to fly through in sequence during the cruising are selected.

Definition 3. For any candidate task $task_i$ on a_k, if it can not be executed in its deadline, then $task_i$ is regarded as an invalid task on a_k; otherwise, $task_i$ is a value task on a_k.

The sub-problem mainly includes two aspects: one is to determine the priority execution sequence of each task (called the task ranking problem), which can be achieved by PA algorithm; the other is to select value task which should be observed (called the value task inspection problem), and it can be realized by KNSA algorithm.

3.2. K-NGCA Algorithm

Definition 4. Let $\sigma_k = (cx_k, cy_k)$ be the geometric position center point of the tasks in the cluster $S_k = \{task_{k1}, task_{k2}, \ldots, task_{kq_k}\}$. σ_k is called the core of S_k, and it can be calculated as follows:

$$cx_k = \frac{1}{\dim \|s_k\|} \sum_{task_{ki} \in S_k} tx_{ki},$$

$$cy_k = \frac{1}{\dim \|s_k\|} \sum_{task_{ki} \in S_k} ty_{ki},$$

(7)

where the operator dim $\| \cdot \|$ is used to calculate the element number of a set.

Definition 5. The nearness degree $r_{i,k}$ denotes the relative distance between $task_i$ and S_k, and it can be calculated as follows:

$$r_{ik} = \frac{\sqrt{(tx_i - cx_k)^2 + (ty_i - cy_k)^2}}{\sum_{j=1}^m \sqrt{(tx_i - cx_j)^2 + (ty_i - cy_j)^2}}.$$

(8)

In the initial construction of the cluster set, it is essential to avoid excessive elements which are assigned to a single cluster. This is beneficial to achieve the load balancing of the observation resources. Therefore, it is feasible to set a threshold variable Num to restrict the element number in a single cluster, which can be defined as

$$\text{Num} = \max_{a_i \in A} \left\{ \text{ceil} \left(\frac{\text{speed}_i}{\sum_{a_i \in A} \text{speed}_i} m \right) \right\},$$

(9)

where ceil $(\cdot)$ is a rounding function to be used to convert a decimal to an integer.

According to the position layout of the high-altitude airship, K-NGCA uses Greedy algorithm [30] to achieve the initiation of the cluster set. The main steps are shown as follows.

Step 1. All tasks in set task are sorted by their value indexes in descending order. The sequence is saved in set $STask$.

Step 2. The cluster set $S = \{S_1, S_2, \ldots, S_m\}$ is constructed, where $S_i \leftarrow NULL, \sigma_i \leftarrow o_i, i = 1, 2, \ldots, m$.

Step 3. For each task $task_i$ in set $STask$, let the set $F_i = \{a_{i1}, a_{i2}, \ldots, a_{ie}\}$ contain the airships which meet the imaging resolution constraint of $task_i$. Then, the subset $S' = \{S_{i1}, S_{i2}, \ldots, S_{ie}\}$ of set S is constructed.

Step 4. The nearness degrees between $task_i$ and each element in S' are calculated by (8), which are presented as $R_i = \{r_{i1}, r_{i2}, \ldots, r_{ie}\}$.

Step 5. Let $r_{ik} = \min_{r_{ij} \in R_i} \{r_{ij}\}$, and remove $task_i$ from $STask$ into S_{ik}.

Step 6. If $STask \neq \varnothing$, then proceed to Step 3; otherwise, the algorithm ends.

K-NGCA adjusts the cluster scheme by iteration method. The pseudocode of K-NGCA is outlined as Algorithm 1.

In the pseudocode of K-NGCA, the cluster set is initiated by Greedy (see line 1). If the number of iterations is less than the threshold value MaxNum, then the cluster is repartitioned again. Otherwise, K-NGCA ends (see line 3). In the iteration process, each cluster is set as null firstly. The element number of each cluster is stored in vector Q (see line 6). According to the executing priority sequence $STask$, each task $task_i$ is assigned in sequence (see lines 8–31). The observation resources which meet the lowest resolution of $task_i$ are selected and the corresponding clusters are saved.

```
01: Initialize S = {S_1, S_2, ..., S_m } by Greedy algorithm;
02: Calculate σ = {σ_1, σ_2, ..., σ_m} by (4) and Num
by (6);
03: while rep < MaxNum do
04:    rep ← rep + 1;
05:    Let each cluster S_i ← NULL in set S;
06:    Given a vector Q = {q_1, q_2, ..., q_n} with each
element q_i ← 0; /*Record the element quantity of each cluster*/
07:    Sort each task_i in Task by p_i, and
construction a set STask to record the new task sequence;
08:    for each task_i ∈ STask in sequence do
09:        Calculate R_i = {r_{i1}, r_{i2}, ..., r_{in}} by (5);
10:        find ← FALSE; step ← 0; backup ← NULL;
11:        while step ≤ n do
12:            Let v = {j | r_{ij} ≤ r_{ik}, ∀r_{ij}, r_{ik} ∈ R_i}; /*Find the nearest cluster S_v */
13:            step ← step + 1;
14:            if RP_v ≤ u_i then
15:                backup ← v; find ← TRUE;
16:                if q_v ≤ Num then
17:                    Add task_i to S_v;
18:                    q_v ← q_v + 1;
19:                    break;
20:                end if
21:            end if
22:            r_{iv} ← M; /*M is a tremendous number */
23:        end while
24:        if step > n then
25:            if find == TRUE then
26:                v ← backup; q_v ← q_v + 1;
27:                Add task_i to S_v;
28:            else
29:                Reject task_i due to all HAA are invalid;
30:            end if
31:        end if
32:    end for
33:    Update σ = {σ_1, σ_2, ..., σ_n} by (4);
34: end while
```

ALGORITHM 1: Pseudocodes of K-NGCA algorithm.

Search the cluster which has element number no more than Num and has the minimum nearness degree to task_i; then, task_i is add into this cluster (see lines 11–23). If such cluster is nonexistent, then find out the observation resources which meet the lowest resolution of task_i, but there is no restriction to the element number of the corresponding cluster. Select the cluster which has the minimum nearness degree to task_i and add task_i into this cluster (see lines 25–27). If such resource cannot be found, then task_i is rejected (see line 29). Update the cores of all clusters after iteration (see line 33).

3.3. PA Algorithm. The task ranking is a combination optimization issue, which is commonly solved by the intelligence algorithm (IA). The traditional IA involves simple operation and fast convergence, but it also has certain faults, such as inaccuracy search and early maturing. For instance, the classic PSO algorithm tends to terminate evolution in the iteration, because the particles tend to quickly gather to the historically optimized position of swarm. Similarly, while a superior chromosome is presented in the traditional GA algorithm, its genetic information may be rapidly spread to the swarm and make the algorithm being fast trapped into the local optimal solution. In this paper, a novel IA with the ability to avoid this prematurity is proposed.

PA is an evolutionary computation technique based on the propagation process of biological flock. The main principle of PA is summarized from animal swarm behavior, which indicates that individuals struggle to survive by competition and cooperation. In this algorithm, the survival activities of the flock are abstracted into three kinds of simple events, that is, search for food, breeding new individuals, and death. Similar to other swarm intelligent algorithms, each member in a flock is an independent individual, and its basic features include the search position and age. PA uses the iteration to implement the survival process and employs the minimum age unit of the population as iteration step. Each iteration is called a propagation and the aforementioned simple events are included. In order to find adequate food, all the individuals determine their new search positions in the next propagation based on their personal experience as well as information gained through inheritance or interaction. This process is accompanied with the feeble individual death and the new individual birth. Death means that several individuals are eliminated, and only the survivors can participate in the next propagation. The causes of individual death include

(1) inadequate food is found out due to the bad search position, so that the individuals die of hunger;

(2) the long survival time induces individuals to get aged, and elder individuals to die with the natural life gradually;

(3) the accidental death which is caused by other factors, for example, prey, disease, and accident.

It is essential to choose the individuals who are suitable to breed new members to participate in propagation. This operation is important to maintain the population scale and deliver the search position information of the excellent individuals. The main steps of PA are summarized as follows.

Step 1. The swarm is initialized with random search position and survival age of all individuals.

Step 2. Each search position is evaluated by fitness function, and the global optimal search positions of the individual and the population are stored, respectively.

Step 3. The hunger and the aging degree of each individual are analyzed, and then we can judge the dead ones in the swarm.

Step 4. If the propagation criterion is met, then go to Step 5; otherwise, go to Step 6.

Step 5. The suitable individuals are selected as parents to breed and initialize the child individuals.

Step 6. Move to the new search position and update the individual age.

Step 7. If the termination criterion is satisfied, then the iteration ends; otherwise, go to Step 2.

The aforementioned steps are the basic concepts of PA. There still needs to clear the encoding rule, search method, and propagation mechanism of this algorithm.

3.3.1. Encoding Rule. Decimal encoding rule is applied to represent search position of the individual with a Multidimensional vector, called the position vector, where the elements in the vector correspond to the task number. In the decoding operation, sort all tasks according to the value of the corresponding elements in a descending order, then the position vector can be converted into the task priority sequence.

For example, the task set $S_1 = \{task_1, task_2, \ldots, task_6\}$ is assigned to airship a_1. In encoding operation, the search position of each individual is denoted by six-dimensional vector. Assume $[0.27, 0.54, 0.95, 0.63, 0.15, 0.91]$ is a position vector, which denotes the execution priority sequence of $task_5$, $task_1$, $task_2$, $task_4$, $task_6$, and $task_3$. If this position vector is changed to $[0.96, 0.48, 0.81, 0.14, 0.37, 0.92]$, then the execution priority sequence is altered to $task_4$, $task_5$, $task_2$, $task_3$, $task_6$, and $task_1$.

Additionally, the twotuples $AI_i = \langle F_i, G_i \rangle$ are employed to denote fitness of individual in swarm, where F_i is the task benefit and G_i is the cruising distance of ith individual, respectively. The comparison method of fitness is given as follows:

$$
\begin{aligned}
AI_1 > AI_2, & \quad \text{if } F_1 > F_2, \\
AI_1 > AI_2, & \quad \text{if } F_1 = F_2, \quad G_1 < G_2, \\
AI_1 = AI_2, & \quad \text{if } F_1 = F_2, \quad G_1 = G_2, \\
AI_1 < AI_2, & \quad \text{otherwise.}
\end{aligned}
\tag{10}
$$

3.3.2. Search Method. In the process of searching a new position, the individuals are more inclined to move towards the global optimal search position where the population has appeared. However, there exists a certain distance between the current search position of the individuals and the global optimal search position of the swarm, so the individuals have to move many times to arrive at the destination. But the individuals may enter into the worse search position result in the deviation of movement direction. In order to reduce this risk, it is essential for the individual to consider the relationship among its own global optimal search position and current search position and the global optimal search position of the swarm in each movement.

Assume the population scale of algorithm is N. In the tth propagation, the global optimal position vector of the swarm is $GZ^t = (gz_1^t, gz_2^t, \ldots, gz_{q_i}^t)$; the current and the optimal position vectors of individual $agent_i$ are $Z_i^t = (z_{i,1}^t, z_{i,2}^t, \ldots, z_{i,q_i}^t)$ and $BZ_i^t = (bz_{i,1}^t, bz_{i,2}^t, \ldots, bz_{i,q_i}^t)$. If $agent_i$ can survive in the $t + 1$th propagation, Z_i^{t+1} can be calculated according to the following equations:

$$
\begin{aligned}
v_{i,j}^t &= z_{i,j}^t + R_1 \left(bz_{i,j}^t - z_{i,j}^t \right) + R_2 \left(gz_j^t - z_{i,j}^t \right), \\
z_{i,j}^{t+1} &= v_{i,j}^t + \Delta w_{i,j}^t, \quad i \in [1, N], \ j \in [1, q_i],
\end{aligned}
\tag{11}
$$

where $v_{i,j}^t$ is the jth element on the undisturbed position vector V_i^t in tth propagation, $R_1, R_2 \in (0, 1)$ is the random variable, and $\Delta w_{i,j}^t \in (0, 1)$ is the random disturbance variable of $v_{i,j}^t$.

3.3.3. Propagation Mechanism. In the iteration process, there are individuals which die of hunger, old, and accident reasons. The scale of population is set as a constant N in order to maintain the stable search ability of algorithm. If the individual number M is less than N, then select $N - M$ members whose survival time span is $[Age_1, Age_2]$ and food is adequate as the parents are to breed new individuals, so as to supplement swarm. If the number of suitable individuals is less than $N - M$, then take all individuals as parents, and the absent ones can be generated based on the random method (it can be regarded to acquire the absent parents externally). This operation that the new individuals are generated by the parents can refer to the choosing operation and the crossover operation in genetic algorithm (GA) [21].

In tth iteration, assume the aging degree of $agent_i$ is $OD_i(t)$ and the hunger degree is $HP_i(t)$. The probability of accidental death due to other reasons is expressed as the constant $a \in (0, 1)$, and the survival possibility of $agent_i$ after the tth propagation is given as follows:

$$LP_i(t) = (1 - a)[1 - OD_i(t)][1 - HP_i(t)]. \quad (12)$$

Calculate the survival possibility of all individuals in the swarm and determine the death by the roulette after normalization. The quantitative methods of aging degree and hunger degree are introduced as follows.

(1) Aging Degree. Usually, the search process of the traditional swarm intelligent algorithm easily drags into the local optimization. The main reason is that when the individuals with the dominant fitness appear in the swarm, their characteristic information will spread to other individuals rapidly and promote the swarm to converge towards the dominant individuals. In order to avoid the shortages of orthodox swarm intelligent algorithm, PA records the aging degree of each individual in every propagation and controls the expected survival time of each individual, so as to adjust the convergence speed of the algorithm.

Assume $age_i(t)$ is the age of $agent_i$ in the tth propagation. If $agent_i$ survives after the $t + 1$th propagation, let $age_i(t+1) = age_i(t) + 1$; otherwise, let $age_i(t + 1) = 1$. The aging degree of $agent_i$ in tth propagation is defined as

$$OD_i(t) = 1 - \exp\left(\frac{-age_i(t)}{EAge}\right), \quad (13)$$

where $EAge$ is the expected survival time of each individual.

In particular, if $age_i(t) = 0$ and $OD_i(t) = 0$, it shows that $agent_i$ is in a completely young state and that the death possibility of aging is nonexistent. If $age_i(t) \to \infty$ and $OD_i(t) = 1$, it shows that $agent_i$ will be in a continuous aging state with the increasing propagation time and that the death possibility infinitely approaches to 1.

(2) Hunger Degree. It is essential to reevaluate the hunger degree of each individual after propagation in order to analyze the performance of searching position. Generally speaking, the global optimal solution of the problem which we research cannot be obtained in advance, so the relative optimization is viewed as the criterion to measure the search position of the individuals. All individuals are sorted according to their fitness in a descending order after tth propagation, and the sequence is stored as $Fit_t = \{Fit_{t1}, Fit_{t2}, \ldots, Fit_{tN}\}$. Meanwhile, the first h individuals in Fit_t are assumed in hunger. The hunger degree of the individuals is given as follows:

$$HP_{ki}(t) = \begin{cases} \varepsilon, & \text{if } i \leq h, \\ 0, & \text{if } i > h, \end{cases} \quad (14)$$

where $\varepsilon \in (0, 1)$ is the death probability cause of hunger.

3.3.4. Parameters Analysis. The main parameters of PA include the population size, iteration number, hunger individual number, starvation probability, expected survival age, breeding age, and accidental probability. We will analyze these parameters in the following, and this work focuses on the parameter selection of PA.

(1) Population Size. The searching ability of PA is inadequate when the population size is small, so the iteration number should be increased to acquire a better solution. On the contrary, a large population size is conducive to a strong searching ability, but it could also lead to a low convergence speed. However, PA has the ability to jump the local optimal solution regardless of population size.

(2) Iteration Number. The iteration number is usually a constant, which is employed to adjust the searching time of algorithm. PA may not be able to find the ideal optimized solution with a small iteration number. If this parameter is enlarged, it is possible for PA to achieve a low searching efficiency as the searching time is increased.

(3) Hunger Individual Number and Starvation Probability. The product of hunger individual number and starvation probability is regarded as the expected number of dead one in a single iteration. The elimination rate of the inferior individuals is fast when this expected number is large this will help to improve the searching quality of swarm, but it will also incur a low convergence speed. On the contrary, a smaller value generates a faster convergence speed of PA because the swarm tends to keep the existing individuals, but prematurity is inevitable.

(4) Expected Survival Age. The expected survival age is proposed to control the convergence speed of swarm. The random searching of PA is presented in the solution space when this value is small. But a large value will also be invalid to acquire an ideal solution due to the rapid convergence of PA.

(5) Breeding Age. The position information of superior individuals is allowed to be spread to warm, which is similar to the genetic operation of GA. The difference is that PA controls the spreading velocity of the position information about the superior individuals by setting the breeding age. This parameter is assigned in the latter part of survival age because individuals are common in their best searching position at this time interval. Meanwhile, the little survival time of superior individuals prevents its position information from being spread excessively.

(6) Accidental Probability. The accidental probability is used to simulate the emergency death of swarm's evolution, which can decide the mutation probability of individual. Adaptive mutation can help the algorithm to jump the local optimal solution, but the extensive mutation will destroy the stable search of swarm in the solution space.

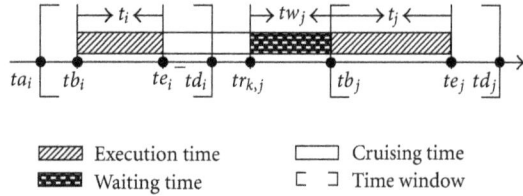

FIGURE 3: The waiting time between the successive key nodes.

3.4. KNSA Algorithm. According to the priority execution sequence of tasks acquired by PA, the key nodes are determined that each airship needs to fly through at a given constant speed and altitude. The cruising path optimization method between the successive key nodes can be learned from [24, 31, 32].

For any task, it can only be executed within its time window. If we can observe $task_i$ before allowable execution timing instant ta_i, then the waiting time should be introduced as is presented in Figure 3.

Figure 3 shows that $task_i$ and $task_j$ are two neighboring tasks on a_k; $tr_{k,j}$ is the preparation time of a_k before $task_j$ is executed; $tw_{k,j}$ is the waiting time of a_k before observing $task_j$. $tr_{k,j}$ and $tw_{k,j}$ can be calculated as follows:

$$tr_{k,j} = tb_i + t_i + d\left(\mathrm{sit}_i, \mathrm{sit}_j\right)\mathrm{speed}_k^{-1},$$
$$tw_{k,j} = \max\left\{tb_j - tr_{k,j}, 0\right\}. \tag{15}$$

With regard to the waiting time for an airship, it can be solved in many ways. For instance, the airship waits firstly and then reaches the location of the next task at the stipulated cruising speed or reaches location of the next task firstly and observes this task until the allowable moment.

The executing timing interval of $task_i$ can be easily determined as follows:

$$tb_j = tr_{k,j} + tw_{k,j},$$
$$te_j = tb_j + t_j. \tag{16}$$

Due to the limited maneuverability of the airship, some tasks may not be executed before their deadlines. Therefore, the execution value of a task decreases with the time advancement, and this trend is irreversible.

Assume $S = \{S_1, S_2, \ldots, S_m\}$ is a partitioning solution of the task set, where S_k is the task set assigned to a_k, and the elements in S_k have been sorted by the task priority in sequence. The method of detecting the value tasks and designing its execution timing interval is shown in Algorithm 2.

In Algorithm 2, the value tasks detection and the execution timing interval assignment are synchronized. The task sets assigned to various observation resources are scheduled by the algorithm, respectively. The variables including the preparation time, the task benefit, and the cruising path of each observation resources are initialized (see lines 2-3). According to the task priority execution sequence, the execution value of each task in set S_i is detected in order (see

lines 3-4). If $task_i$ can be executed in its time window, then $task_i$ is a value task. The algorithm calculates the execution time of $task_i$ and removes it to the key node set G_k (see lines 5–8); otherwise, $task_j$ is an invalid task which will not be arranged in the execution queue (see line 10). Set the decision variables by ranking result in G_k (see lines 13–18).

4. Experiment Designing

In this section, simulation experiment is conducted to illustrate the effectiveness of the proposed method. The proposed algorithms are implemented by Matlab 2007 on a laptop with Pentium IV 3.06 GHz CPU, 2 GB memory, and Windows XP operating system. As far as we know, there are no accepted benchmarks yet in cooperative scheduling problem of high-altitude airships, so the random models are used to construct the application scenario and simulate the battlefield area with $200 \times 300\,\mathrm{km}^2$.

4.1. Simulation Setup. Three high-altitude airships are tested in the experiment, and the task period is $0 \sim 24\,\mathrm{h}$. The main parameters for high-altitude airships simulation are listed in Table 1.

The task number varies from 50 to 300 as six instances, and they are deployed randomly within the battlefield. The task value indexes are changed from 1 to 10, and the imaging resolution is $0.3 \sim 0.6\,\mathrm{m}$. The equation of generating the time window for tasks is presented as follows:

$$td_i = ta_i + (1 + T\mathrm{Base}) \times t_i, \tag{17}$$

where $T\mathrm{Base}$ is used to adjust the tightness of the time window.

The setting of task parameters offers the flexibility to simulate the various workloads for high-altitude airships. Table 2 gives the configuration of task parameters employed in our experiment. We check the performance impact of parameters Task number, $T\mathrm{Base}$, and Task length by using the "once tuning one parameter (OTOP)" experiment method, which can be found in many researches [33–35]. The evaluation objects include task benefit and cruising distance.

In order to evaluate the effectiveness of K-NGCA algorithm, the Greedy algorithm which has been mentioned in the former section is applied in the task set partitioning. Furthermore, we verify the effectiveness of PA by comparing it with GA and PSO. We incorporate those algorithms to yield four new algorithms named K-NGCAPA, GreedyPA, K-NGCAGA, and K-NGCAPSO, respectively. The small-scale experiments have been finished to obtain the adaptive parameters of the PA, which are listed in Table 3.

4.2. Performance Impact of Task Number. In this experiment, we investigate the impact of task number on the performance of these algorithms. Figure 4 plots the scheduling results.

Figure 4(a) shows that K-NGCAPA obtains a higher task benefit than other algorithms (GreedyPA, K-NGCAGA, or K-NGCAPSO). In various task scales, the task benefit obtained by K-NGCAPA can be higher than that of GreedyPA, and K-NGCAGA, K-NGCAPSO reaching 0.384%, 2.92%, and

TABLE 1: Parameters for high-altitude airships simulation.

HAA	Initial position	Speed (km/h)	Imaging resolution (m)
a_1	$(20, 30)$	60	0.61
a_2	$(150, 170)$	75	0.36
a_2	$(280, 30)$	70	0.24

TABLE 2: Parameters for task simulation.

Parameters	Value (fixed)−(varied)
Task number	$(200)-(50, 100, 150, 200, 250, 300)$
Value index	$([1, 10])$
Imaging resolution (m)	$([0.3, 0.6])$
TBase	$(10)-(5, 10, 15, 20, 25, 30)$
Task length (min)	$([15, 30])-([0, 15], [15, 30], [30, 45])$

TABLE 3: Parameters for PA algorithm.

Parameters	Value (fixed)
Population size	30
Iteration number	200
Hunger individual number	5
Breeding age	$[5, 30]$
Expected survival age	35
Accidental probability	0.05
Starvation probability	0.1

2.71%, respectively, which shows a very high scheduling performance.

From Figure 4(a), we can observe that the task benefit increases gradually with the task number, but this tends to be flat while the variable is up to more than 200. This ascribes that the scale of value tasks increases gradually thus brings about the increasing number of the executable tasks and conduces to gain more task benefit. However, due to the limited observation capacity of the airships, when the task number ascends to a certain extent, the task benefit tends to increase slowly. It can be seen from Figure 4(b) that the cruising distance of the airships increases firstly and then decreases while the task number becomes larger. This cause of increasing task number will bring about the higher task density in the battlefield. In case the task density is low, a longer cruising distance needs to be flied if the airship is assigned to execute more tasks. While the task density reaches to a certain degree, the airships can execute more tasks just in a shorter scope. Therefore, the cruising distance declines.

4.3. Performance Impact of Time Window. The objective of this experiment is to investigate the performance impact of time window on the task guarantee ability. We divide TBase into six levels from 1 to 6. Figure 5 depicts the different scheduling results in various TBase levels.

The experimental results in Figure 5(a) show that the K-NGCAPA achieve higher guarantee ratios than GreedyPA, K-NGCAGA, and K-NGCAPSO. It can be seen that the task benefit increases with extending TBase. This ascribes that the increase of TBase brings about a loose time window. In

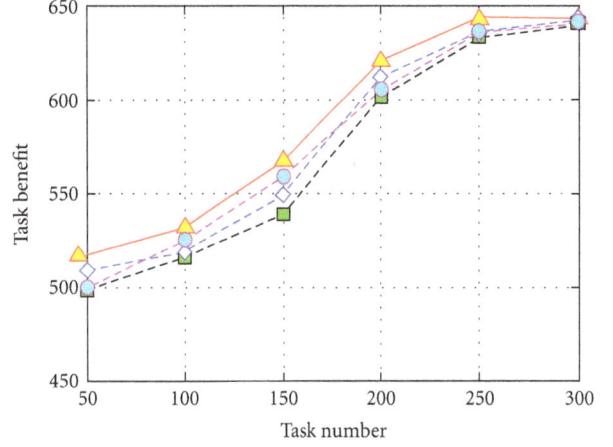

(a) Task benefit impact of task number

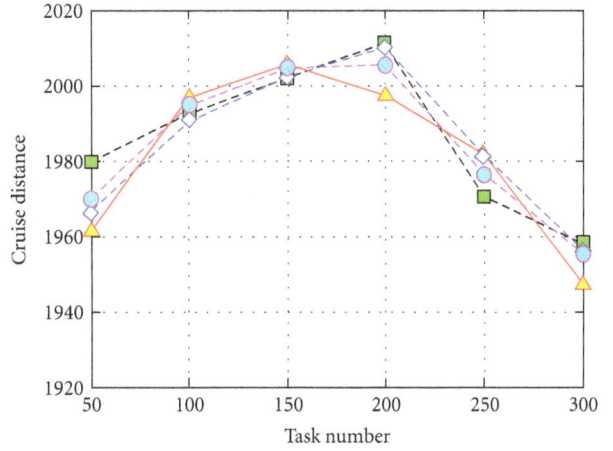

GreedyPA — K-NGCAGA
K-NGCAPA — K-NGCAPSO

(b) Cruising distance impact of task number

FIGURE 4: Scheduling results in various task number.

particular, the task deadline is extended, so more tasks can be executed timely. Hence, the task benefit increases gradually. From Figure 5(b), it can be seen that the cruising distance increases firstly and then decreases with the enlarging TBase. In the initial stage, the more tasks that airships execute are at the cost of longer cruising distance. Therefore, the cruising distance increases firstly. However, when the time window becomes loose, the tasks which are within the shorter distance range of the airships but once are unable to be executed timely become executable. Thus, the execution priority will be given to such tasks, and the scheduling scheme is adjusted by the new priority sequence which brings about the decline of the cruising distance.

4.4. Performance Impact of Task Length. To examine the performance impact of task length, three test configurations of task length can be found in Table 2. We assume the task length is satisfied with the uniform distribution. Figure 6 plots the scheduling results under short, middle, and long tasks.

(a) Task benefit impact of TBase

(b) Cruising distance impact of TBase

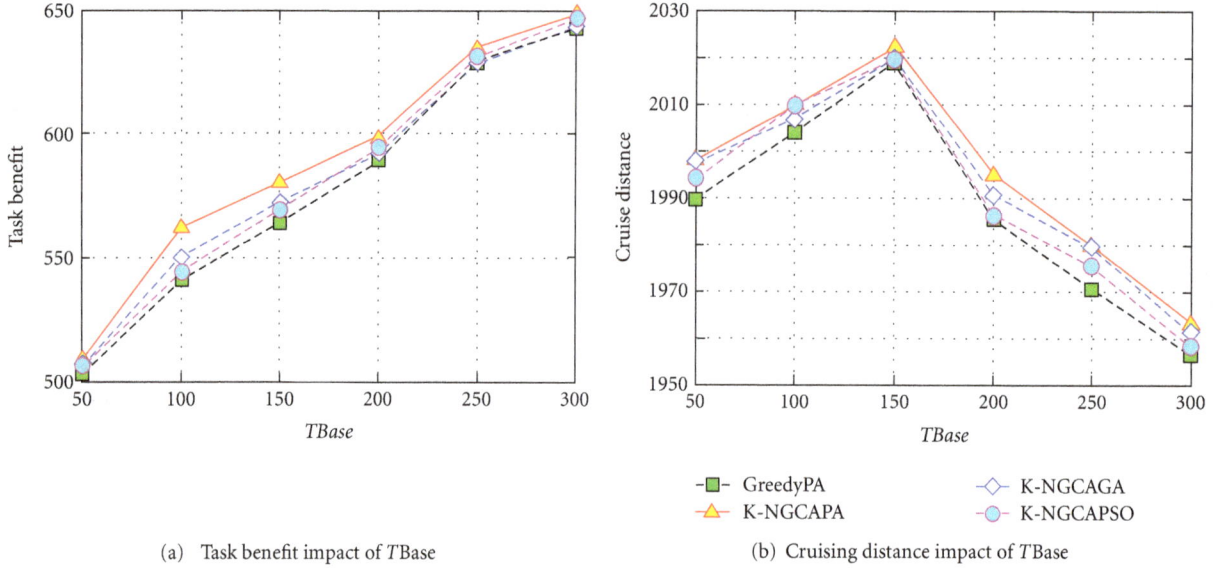

FIGURE 5: Scheduling results in various TBase.

```
01: for k ← 1 to m do
02:     idle_k ← 0; G_k ← NULL; /*Initialization*/
03:     A ← (ox_k, oy_k); TB_k ← 0; RD_k ← 0;
03:     for each task task_i ∈ S_k in sequence do
04:         B ← (x_i, y_i); /*Search the next key node */
05:         if idle_k + d(A,B)v_k^{-1} + t_i ≤ td_i then/*task_i is a key node */
06:             tb_i ← max(idle_i + d(A,B)v_k^{-1}, ta_i);
07:             te_i ← tb_i + t_i; idle_k ← te_i; A ← (x_i, y_i);
08:             TB_k ← TB_k + p_i; RD_k ← RD_k + d(A,B);
09:             Add task_i to the key node set G_k;
10:         else
11:             Reject task_i; /*task_i is a invalid task, and it can not been executed */
12:         end if
13:     end for
14:     for each task task_{gi} ∈ G_k in sequence do
15:         Delete task_{gi} from G_k, and set x_{gi,k} ← 1;
16:             if G_k ≠ NULL then
17:             Select the first task task_{gi} in G_k, and set y_{gi,gj}^k ← 1;
18:             end if
19:     end for
20: end for
```

ALGORITHM 2: Pseudocode of KNSA algorithm.

From Figure 6(a), it can be seen that the task benefit decreases gradually when the task length is extended. This is because that the required executing time of the single task become long, so that the tasks with later sequence cannot be executed timely. At the same time, the cruising distance is shortened with the decreasing of executable tasks. This trend can be observed from Figure 6(b).

Again, we can observe from Figures 4 and 6 that the effects of the task number and time window on the optimization objects are positive, that is, both the increase of task number or extension of time window will enlarge the task benefits. Meanwhile, the effect of task length on the

optimization objects is negative. Then, we can find the relationship between the aforementioned factors. In the given application scenarios, the task benefits are maintained by enlarging the time window or reducing the task length if the task number is reduced. On the contrary, while the time window is reduced or the task length is increased, we could also maintain the task benefits by increasing the task number.

5. Conclusions and Future Work

The cooperative scheduling of imaging observation tasks for high-altitude airships are a kind of complex combinatorial

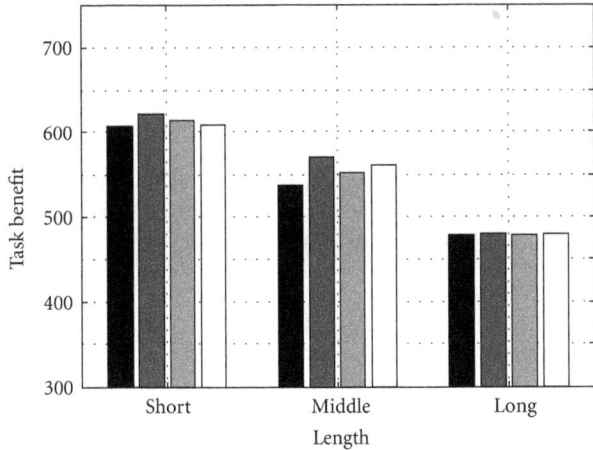

(a) Task benefit impact of task length

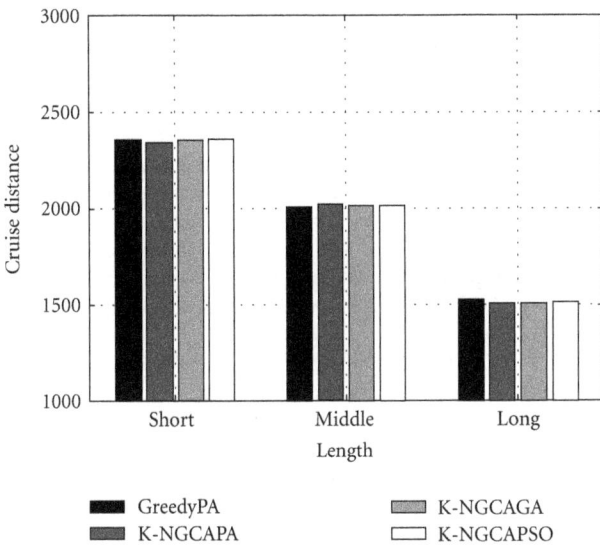

(b) Cruising distance impact of task length

FIGURE 6: Scheduling results in different task length.

the airships and tasks. The simulation results proved the effectiveness of this algorithm.

(4) A new swarm intelligent algorithm is presented, which is called PA algorithm. This algorithm can control the population convergence speed by adjusting the expected survival time of the individuals.

(5) The generation method of the application scenario for high-altitude airships is provided, and the influences of the parameters (such as the task number, the task length, and the time window) on the reconnaissance capability of high-altitude airships are analyzed by experiment.

Also for our future work, we plan to analyze the influences of the parameters (such as the expected survival time, the breeding age, and the accidental probability, etc.) on the convergence speed of PA algorithm. This work plays an important role in avoiding the defects that the traditional swarm intelligent algorithms are fast to fall into the local optimal solution effectively.

References

[1] Y. Li, M. Nahon, and I. Sharf, "Airship dynamics modeling: a literature review," *Progress in Aerospace Sciences*, vol. 47, no. 3, pp. 217–239, 2011.

[2] D. K. Schmidt, J. Stevens, and J. Roney, "Near-space station-keeping performance of a large high-altitude notional airship," *Journal of Aircraft*, vol. 44, no. 2, pp. 611–615, 2007.

[3] J. Wu and W. Zheng, "Adaptive fuzzy sliding mode control for robotic airship with model uncer-tainty and external disturbance," *Journal of Systems Engineering and Electronics*, vol. 23, no. 2, pp. 250–255, 2012.

[4] H. Wang, B. Song, and L. Zuo, "Effect of high-altitude airship's attitude on performance of its energy system," *Journal of Aircraft*, vol. 44, no. 6, pp. 2077–2080, 2007.

[5] A. K. Widiawan and R. Tafazolli, "High Altitude Platform Station (HAPS): a review of new infrastructure development for future wireless communications," *Wireless Personal Communications*, vol. 42, no. 3, pp. 387–404, 2007.

[6] P. Y. Bely, R. Ashford, and C. D. Cox, "High-altitude aerostats as astronomical platforms," in *Space Telescopes and Instruments*, vol. 2478 of *Proceedings of SPIE*, pp. 101–116, April 1995.

[7] B. Kevin, "How cargo lifter's airship will work?" http://www.howstuffworks.com/cargolifter.htm.

[8] J. M. Park, B. J. Ku, Y. S. Kim, and D. S. Ahn, "Technology development for wireless communications system using stratospheric platform in Korea," in *Proceedings of the 13th IEEE International Symposium on Personal, Indoor and Mobile Radio Communications, (PIMRC '02)*, pp. 1577–1581, September 2002.

[9] T. C. Hong, B. J. Ku, J. M. Park, D. S. Ahn, and Y. S. Jang, "Capacity of the WCDMA system using high altitude platform stations," *International Journal of Wireless Information Networks*, vol. 13, no. 1, pp. 5–17, 2006.

[10] W. Yao, Y. Li, W. J. Wang, and W. Zheng, "Thermodynamic model and numerical simulation of a stratospheric airship take-off process," *Journal of Astronautics*, vol. 28, no. 3, pp. 603–607, 2007.

optimization problem. This paper makes the following contributions in the study of this problem.

(1) The main constraints of the cooperative scheduling problem for high-latitude airships are presented. The timeliness of imaging observation tasks is proposed, and the influence of this feature on the reconnaissance activities is summarized. On the basis of the above analysis, we construct the constraint programming model.

(2) A hierarchical solution frame is developed to solve this cooperative scheduling problem. The original scheduling problem is converted into a main problem (task set partitioning) and a sub-problem (cruising path selection). It is available to simplify the solution process.

(3) The K-NGCA algorithm is proposed to partition the task set based on the position relationship between

[11] X. Fei and Y. Zhengyin, "Drag reduction for an airship with proper arrangement of propellers," *Chinese Journal of Aeronautics*, vol. 22, no. 6, pp. 575–582, 2009.

[12] G. Wang, M. Luo, and Z. Wu, "Size of high altitude long endurance airship affected by various technology guidelines," *Acta Aeronautica et Astronautica Sinica*, vol. 29, no. 1, pp. 66–69, 2008.

[13] C. Xiao, X. Wang, Z. Pu et al., "Recent studies in satellite observations of three-dimensional magnetic reconnection," *Science in China, Series E*, vol. 50, no. 3, pp. 380–384, 2007.

[14] W. J. Wolfe and S. E. Sorensen, "Three scheduling algorithms applied to the earth observing systems domain," *Management Science*, vol. 46, no. 1, pp. 148–168, 2000.

[15] F. Marinelli, S. Nocella, F. Rossi, and S. Smriglio, "A Lagrangian heuristic for satellite range scheduling with resource constraints," *Computers and Operations Research*, vol. 38, no. 11, pp. 1572–1583, 2011.

[16] P. Yan, M. Y. Ding, C. P. Zhou, and C. W. Zheng, "On-line real-time multiple-mission route planning for air vehicle," *Acta Aeronautica et Astronautica Sinica*, vol. 25, no. 5, pp. 485–489, 2004.

[17] C. Zheng, M. Ding, and C. Zhou, "Real-time route planning for unmanned air vehicle with an evolutionary algorithm," *International Journal of Pattern Recognition and Artificial Intelligence*, vol. 17, no. 1, pp. 63–81, 2003.

[18] Y. Xiao, G. Luo, J. Ma, and H. Li, "Research on modeling technology for dynamic modular design," in *Proceedings of the IEEE International Conference on Intelligent Computing and Intelligent Systems, (ICIS '09)*, pp. 681–685, November 2009.

[19] Y. Yin and S. Huang, "Optimization deployment of multi-sensor platforms in near-space based on adaptive genetic algorithm," in *Proceedings of the International Conference on Information Engineering and Computer Science, (ICIECS '09)*, December 2009.

[20] H. Y. Song, "A method of mobile base station placement for high altitude platform based network with geographical clustering of mobile ground nodes," in *Proceedings of the International Multiconference on Computer Science and Information Technology, (IMCSIT '08)*, pp. 869–876, October 2008.

[21] S. M. Han, S. W. Beak, K. R. Cho, D. W. Lee, and H. D. Kim, "Satellite mission scheduling using genetic algorithm," in *Proceedings of the International Conference on Instrumentation, Control and Information Technology (SICE '08)*, pp. 1226–1230, August 2008.

[22] S. Q. Chen, H. F. Wang, and B. F. Song, "Modeling and dynamic simulation study of big inertia pro-pulsion system of high altitude airship," in *Proceedings of the 2nd Artificial Intelligence, Management Science and Electronic Commerce*, pp. 4065–4068, 2011.

[23] Y. Ma and K. Sun, "Research on multi-power management system of high-altitude airship," in *Proceedings of the Asia-Pacific Power and Energy Engineering Conference, (APPEEC '10)*, pp. 1–4, March 2010.

[24] J. Rao, Z. Gong, J. Luo, Z. Jiang, S. Xie, and W. Liu, "Robotic airship mission path-following control based on ANN and human operator's skill," *Transactions of the Institute of Measurement and Control*, vol. 29, no. 1, pp. 5–15, 2007.

[25] N. Bessert and O. Frederich, "Nonlinear airship aeroelasticity," *Journal of Fluids and Structures*, vol. 21, no. 8, pp. 731–742, 2005.

[26] E. Aharon, "US Army orders hybrid airship for Afghan deployment," 2012, http://www.tgdaily.com/security-features/50238-us-army-orders-hybrid-airship-for-afghan-deployment.

[27] J. M. Wang, J. F. Li, and Y. J. Tan, "Study on heuristic algorithm for dynamic scheduling problem of earth observing satellites," in *Proceedings of the 8th ACIS International Conference on Software Engineering, Artificial Intelligence, Networking, and Parallel/Distributed Computing (SNPD '07)*, pp. 9–14, August 2007.

[28] S. Baolin, W. Wenxiang, and Q. Qianqing, "Satellites scheduling algorithm based on dynamic constraint satisfaction problem," in *Proceedings of the International Conference on Computer Science and Software Engineering, (CSSE '08)*, pp. 167–170, December 2008.

[29] T. G. Guzik, S. Besse, A. Calongne et al., "Development of the high altitude student platform," *Advances in Space Research*, vol. 42, no. 10, pp. 1704–1714, 2008.

[30] R. Chen, C. Li, J. Chen, and Z. Yu, "Optimization of near-space aerocraft track for regional coverage based on greedy algorithm," *Journal of Beijing University of Aeronautics and Astronautics*, vol. 35, no. 5, pp. 547–550, 2009.

[31] R. J. Szczerba, P. Galkowski, I. S. Glickstein, and N. Ternullo, "Robust algorithm for real-time route planning," *IEEE Transactions on Aerospace and Electronic Systems*, vol. 36, no. 3, pp. 869–878, 2000.

[32] N. Léchevin and C. A. Rabbath, "Decentralized detection of a class of non-abrupt faults with application to formations of unmanned airships," *IEEE Transactions on Control Systems Technology*, vol. 17, no. 2, pp. 484–493, 2009.

[33] C. He, X. M. Zhu, H. Guo et al., "Rolling-horizon scheduling for energy constrained distributed real-time embedded systems," *Journal of Systems and Software*, vol. 85, no. 4, pp. 780–794, 2012.

[34] Z. F. Yu and W. S. Shi, "Queue waiting time aware dynamic workflow scheduling in multicluster environments," *Journal of Computer Science and Technology*, vol. 25, no. 4, pp. 864–873, 2010.

[35] Z. Zong, A. Manzanares, X. Ruan, and X. Qin, "EAD and PEBD: two energy-aware duplication scheduling algorithms for parallel tasks on homogeneous clusters," *IEEE Transactions on Computers*, vol. 60, no. 3, pp. 360–374, 2011.

10

A Hybrid Metaheuristic DE/CS Algorithm for UCAV Three-Dimension Path Planning

Gaige Wang,[1,2] **Lihong Guo,**[1] **Hong Duan,**[3] **Heqi Wang,**[1] **Luo Liu,**[1,2] **and Mingzhen Shao**[1,2]

[1] *Changchun Institute of Optics, Fine Mechanics and Physics, Chinese Academy of Sciences, Changchun 130033, China*
[2] *Graduate School of Chinese Academy of Sciences, Beijing 100039, China*
[3] *School of Computer Science and Information Technology, Northeast Normal University, Changchun 130117, China*

Correspondence should be addressed to Lihong Guo, guolh@ciomp.ac.cn

Academic Editors: E. Acar and N. Avdelidis

Three-dimension path planning for uninhabited combat air vehicle (UCAV) is a complicated high-dimension optimization problem, which primarily centralizes on optimizing the flight route considering the different kinds of constrains under complicated battle field environments. A new hybrid metaheuristic differential evolution (DE) and cuckoo search (CS) algorithm is proposed to solve the UCAV three-dimension path planning problem. DE is applied to optimize the process of selecting cuckoos of the improved CS model during the process of cuckoo updating in nest. The cuckoos can act as an agent in searching the optimal UCAV path. And then, the UCAV can find the safe path by connecting the chosen nodes of the coordinates while avoiding the threat areas and costing minimum fuel. This new approach can accelerate the global convergence speed while preserving the strong robustness of the basic CS. The realization procedure for this hybrid metaheuristic approach DE/CS is also presented. In order to make the optimized UCAV path more feasible, the B-Spline curve is adopted for smoothing the path. To prove the performance of this proposed hybrid metaheuristic method, it is compared with basic CS algorithm. The experiment shows that the proposed approach is more effective and feasible in UCAV three-dimension path planning than the basic CS model.

1. Introduction

Unmanned combat aerial vehicles (UAVs) are remotely piloted or self-piloted aircrafts that can carry many different types of accessories such as cameras, sensors, and communications equipment. They have a very wide range of applications that include both civil and military areas. Some important features that make them very popular are their low cost, smaller size, and their extended maneuver capability because of absence of a human pilot [1]. In particular, UCAV is one of the inevitable trends of the modern aerial weapon equipment, which develop in the direction of unmanned attendance and intelligence. Research on UCAV directly affects battle effectiveness of the air force and is a fundamental and significant research related to safeness of a nation. Trajectory generation and path planning is one of the key technologies in cooperative UCAV combatting. The flight path planning in a large mission area is a typical large-scale optimization problem, a series of algorithms have

been proposed to solve this complicated multiconstrained optimization problem, such as differential evolution [2], genetic algorithm [3], ant colony algorithm [4] and its variant [5, 6], chaotic artificial bee colony [7], and intelligent water drops optimization [8]. However, those methods can hardly solve the contradiction between the global optimization and excessive information.

In 1995, Storn and Price firstly proposed a novel evolutionary algorithm (EA): differential evolution (DE) [9, 10], which is a new heuristic approach for minimizing possibly nonlinear and nondifferentiable continuous space functions. It converges faster and with more certainty than many other acclaimed global population-based optimization methods [11]. This new method requires few control parameters, which makes DE more robust, easy to implement, and lends itself very well to parallel computation.

Cuckoo search (CS) is an optimization algorithm developed by Yang and Deb in 2009 [12, 13], which was inspired by the obligate brood parasitism of some cuckoo species by laying their eggs in the nests of other host birds (of other species) [14]. Each egg in a nest represents a solution, and a cuckoo egg represents a new solution. The aim is to use the new and potentially better solutions (cuckoos) to take the place of a not-so-good solution in the nests. In the simplest form, each nest has one egg. An important advantage of CS algorithm is its simplicity. In principle, comparing with other population-based metaheuristic algorithms such as particle swarm optimization and harmony search, there is essentially only a single parameter p_a in CS (apart from the population size). Therefore, it is very easy to implement [15].

However, in the field of UCAV path planning, no application of CS algorithm exists yet. In this work, the Differential Evolution (DE) algorithm is combined with CS algorithm, which uses the DE mutation and crossover operator instead of Lévy flights to form the new cuckoo egg updating strategy, in order to reduce the number of exact evaluations of candidate solutions. The candidate paths are modeled in the physical space and evaluated with respect to the task space. A smooth path is essential for a real UCAV, because nonsmooth path cannot satisfy the turning constraint. In the UCAV community, most researchers apply the Dubins algorithm to generate a smooth path [16]. In this paper, to improve the quality of the paths, we used a computationally efficient path-smoothing method called B-Spline curve smoothing strategy [17]. B-Spline curve is used for path line modeling, and complicated paths can be produced with a small number of control variables. To verify the feasibility and effectiveness of our proposed approach, the series experiments conducted under complicated combating environment demonstrate that our hybrid metaheuristic approach with B-Spline curve path smoothing can generate a feasible optimal three-dimension path of UCAV more quickly than the basic CS algorithm.

The remainder of this paper is structured as follows. Section 2 describes the mathematical model in UCAV three-dimension path planning problem. In Section 3, preliminary knowledge of DE and CS algorithm is introduced. Then, an improved CS algorithm for UCAV three-dimension path planning is presented in Section 4 and the detailed implementation procedure is also described. Subsequently, a B-Spline curve method for UCAV path smoothing is described in Section 5. The simulation experiments are conducted in Section 6. Finally, Section 7 concludes the paper and discusses the future path of our work.

2. Mathematical Model in UCAV Three-Dimension Path Planning

As a key component of mission planning system [18], path planning for UCAV is the design of optimal flight route to meet certain performance requirements according to the special mission objective and is modeled by the constraints of the terrain, data, threat information, fuel, and time. The goal for three-dimension path planning is to calculate the optimal or near-optimal flight route for UCAV within the appropriate time, which enables the UCAV to break through the enemy threat environments and self-survive with the

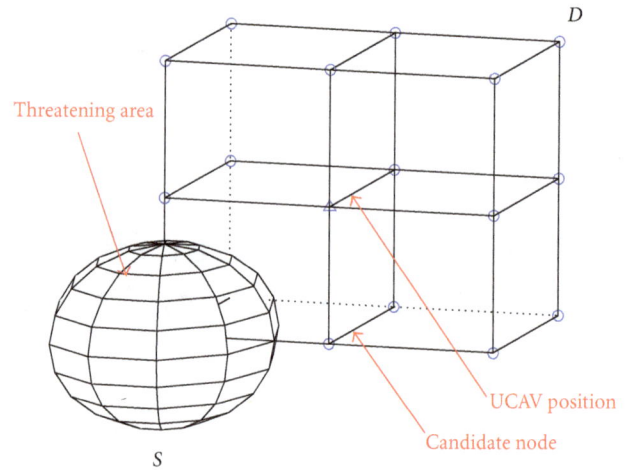

FIGURE 1: Typical UCAV battle field model in three-dimension.

perfect completion of mission. In our work, we use the mathematical model for UCAV 3-dimension path planning described as follows [5].

In order to simplify the UCAV three-dimension path planning problem, the UCAV task region can be divided into three-dimensional mesh, thus forming a three-dimensional network diagram connecting the starting point and end point. In this way, the problem of UCAV optimal three-dimension path planning is the general path optimization problem essentially. The typical UCAV battle field model in three-dimension can be shown in Figure 1.

In Figure 1, suppose the flight task for UCAV is from node S to node D. There are some threatening areas in the task region. We divide the space into m subcubes equally, so there are n nodes in the area, which can be labeled with $L_1, L_2, \ldots, L_n$. Let $L_i(x_i, y_i, z_i)$ be the ith node. It is obvious that there are 26 candidate nodes which could be chosen at most by the UCAV in each step. The nodes in the vertical direction of current point are unaccepted, so the number of the candidate nodes decreases to 24. Then, all the selected nodes could be connected one by one as the step going on until getting the target. In this way, the path from the starting node to the end node can be described as follows:

$$\text{Path} = \{S, L_1(x_1, y_1, z_1), L_2(x_2, y_2, z_2), \ldots, \\ L_{m-1}(x_{m-1}, y_{m-1}, z_{m-1}), D\}. \tag{1}$$

A performance indicator of three-dimension path planning for UCAV mainly contains the completion of the mandatory safety performance indicator, fuel performance indicator, and height performance indicator, that is, indicators with the least threat, the least fuel, and optimal height.

Minimum of performance indicator for threat

$$\min J_t = \int_0^L w_t dl, \quad L \text{ is the length of the path.} \tag{2}$$

Minimum of performance indicator for fuel

$$\min J_f = \int_0^L w_f dl, \quad L \text{ is the length of the path.} \tag{3}$$

Minimum of performance indicator for height

$$\min J_h = \int_0^L w_h dl, \quad L \text{ is the length of the path.} \quad (4)$$

Then the total performance indicators for UCAV route

$$\min J = k_1 J_t + k_2 J_f + (1 - k_1 - k_2) J_h \quad (0 \le k_1, k_2 \le 1), \quad (5)$$

where w_t, w_f, and w_h are the threat cost, fuel cost, and height cost for each point on the path that depend on path length, respectively. The choice of k_1 and k_2 all between 0 and 1 gives the designer certain flexibility to dispose relations among the threat exposition degree, the fuel consumption, and the height information. When k_1 is more approaching 1, more attention is paid to the radar's exposed threat, and it requires avoiding the threat as far as possible at the sacrifice of the trajectory length and flight height. Similarly, when k_2 is more approaching 1, a shorter path is needed to be planned regardless of the cost of other two factors.

When the UCAV is flying along the subpath L_{ij}, the total threat cost generated by N_t threats is calculated as follows:

$$w_{t,L_{ij}} = \int_0^{L_{ij}} \sum_{k=1}^{N_t} \frac{t_k}{\left[(x-x_k)^2 + (y-y_k)^2\right]^2} dl. \quad (6)$$

A computationally more efficient and acceptably accurate approximation to the exact solution is to calculate the threat cost at several locations along an edge and take the length of the edge into account. In this work, the threat cost was calculated at five points along each edge, as shown in Figure 2. To simplify the calculations, each path segment is discretized into five subsegments and the threat cost is calculated on the end of each subsegment. If the distance from the threat point to the end of each subsegment is within threat radius, we can calculate the responding threat cost according to

$$w_{t,L_{ij}} = \frac{L_{ij}^5}{5} \sum_{k=1}^{N_t} t_k \left(\frac{1}{d_{0.1,k}^4} + \frac{1}{d_{0.3,k}^4} + \frac{1}{d_{0.5,k}^4} + \frac{1}{d_{0.7,k}^4} + \frac{1}{d_{0.9,k}^4} \right), \quad (7)$$

where L_{ij} is the length of the subsegment connecting node i and node j; $d_{0.1,k}$ is the distance from the 1/10 point on the subsegment L_{ij} to the kth threat; t_k is threat level of the kth threat. Moreover, we can simply consider the fuel cost w_f to L, and height cost $w_{h,i}$ equals to H which is the flight height of the UCAV when the speed is a constant. The total cost for traveling along the path comes from a weighted sum of the threat and fuel costs shown as in (5).

3. Preliminary Knowledge

3.1. Differential Evolution. The differential evolution (DE) algorithm, proposed by Storn and Price [9, 10], is a simple evolutionary algorithm (EA), which generates new candidate solutions by combining the parent individual and a few other individuals of the same population. A candidate substitutes

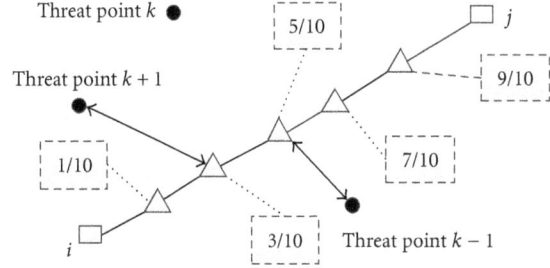

FIGURE 2: Modeling of the UCAV threat cost [5].

the parent only if it has better fitness. This is a rather greedy selection scheme, which often overtakes traditional EAs. Advantages of DE are easy implementation, simple structure, speed, and robustness. Due to these advantages, it has many real-world applications, such as power dispatch, parameters estimation, economic emission load dispatch, and neural network training.

The mainframe of the original DE algorithm is described in Algorithm 1, where D is the number of decision variables. NP is the size of the parent population P. F is the mutation scaling factor. CR is a constant for crossover operator. $X_i(j)$ is the jth variable of the solution X_i. U_i is the offspring. $\lceil NP * rand \rceil$ is a uniformly distributed random integer number between 1 and NP. And rand is a uniformly distributed random real number in interval (0, 1). Different types of strategies of DE have been proposed depending on the target vector selected and the number of difference vectors used. We use the DE/rand/1/bin scheme shown in Algorithm 1. From Algorithm 1, we can see that there are only three control variables in this algorithm, which are NP, F, and CR.

3.2. Cuckoo Search (CS). Cuckoo has a smart reproduction strategy that involves the female laying her fertilized eggs in the nest of another species so that the replaced parents unwittingly raise her brood. Sometimes the cuckoo's eggs in the nest are discovered and the surrogate parents throw them out or leave the nest and start their own brood elsewhere [14].

Cuckoo search (CS) is a new metaheuristic algorithm for solving optimization problems, which is based on the obligate brood parasitic behavior of some cuckoo species in combination with the Lévy flight behavior of some birds and fruit flies. In the case of CS, the walking steps of a cuckoo are determined by the Lévy flights.

A Lévy flight is a random walk in which the steps are defined in terms of the step-lengths, which have a certain probability distribution, with the directions of the steps being isotropic and random. Lévy flights is a class of random walk in which the jumps are distributed according to a power law, that is,

$$y = x^{-\beta}, \quad (8)$$

where $1 < \beta < 3$ and therefore has an infinite variance.

Barthelemy et al. [19] had reported the relationship between light, and Lévy flights has subsequently been applied

```
Begin
    Step 1: Initialization. Set the generation counter t = 1; randomly generate a population of NP
            individuals P; set the weighting factor F and a crossover constant CR
    Step 2: Evaluate the fitness for each individual in P
    Step 3: while the halting criteria is not satisfied do
                for i = 1 to NP do
                    Select uniform randomly r₁ ≠ r₂ ≠ r₃ ≠ i
                    randj = ⌈NP * rand⌉
                    for j = 1 to D do
                        if rand ≤ CR or j == randj then
                            Vᵢ(j) = X_{r₁}(j) + F × (X_{r₂}(j) − X_{r₃}(j))
                        else
                            Vᵢ(j) = Xᵢ(j)
                        end if
                    end for
                end for
                for i = 1 to NP do
                    Evaluate the offspring Uᵢ
                    if Uᵢ is better than Pᵢ then
                        Pᵢ = Uᵢ
                    end if
                end for
                Memorize the best solution achieved so far
                t = t + 1
    Step 4: end while
End.
```

ALGORITHM 1: Algorithm of DE with DE/rand/1/bin scheme.

to improve and optimize searching. In the case of CS, the walking steps of a cuckoo are determined by the Lévy flights.

For simplicity in describing cuckoo search in [12], Yang and Deb used the following three idealized rules.

(1) Each cuckoo lays only one egg at a time, and places its egg in a selected nest at random.

(2) The best nests with high quality of eggs will carry over to the next generation.

(3) The number of available host nests is fixed, and the egg laid by a cuckoo is discovered by the host bird with a probability $p_a \in [0, 1]$. In this case, the host bird can either throw the egg away or leave the nest, and build a fully new nest. For simplicity, this last assumption can be approximated by the fraction p_a of the n nests which are displaced by new nests (with new random solutions) [15].

Based on these three rules, the basic steps of the CS can be summarized as shown in Algorithm 2.

In CS, each egg in a nest represents a solution, and a cuckoo egg represents a new solution. The aim is to use the new and potentially better solutions (cuckoos) to replace a not-so-good solution in the nest. In the simplest form, each nest has one egg. The algorithm can be extended to more complicated cases in which each nest has multiple eggs representing a set of solutions.

When generating new solutions $x^{(t+1)}$ for, say, a cuckoo i, a Lévy flight is performed

$$x_i^{(t+1)} = x_i^{(t)} + \alpha \oplus \text{Lêvy}(\lambda), \tag{9}$$

where $\alpha > 0$ is the step size which should be related to the scales of the problem of interests. In most cases, we can use $\alpha = 1$. The above equation is essentially the stochastic equation for random walk. In general, a random walk is a Markov chain whose next status/location only depends on the current location (the first term in the above equation) and the transition probability (the second term). The product $\oplus$ means entrywise multiplications. This entrywise product is similar to those used in PSO, but here the random walk via Lévy flight is more efficient in exploring the search space as its step length is much longer in the long run.

The Lévy flights essentially provides a random walk, while the random step length is drawn from a Lévy distribution

$$\text{Lêvy}(\lambda) \sim u = t^{-\lambda} \quad (1 < \lambda \le 3), \tag{10}$$

which has an infinite variance with an infinite mean. Here the steps essentially form a random walk process with a power-law step-length distribution with a heavy tail. Some of the new solutions should be generated by Lévy walk around the best solution obtained so far; this will speed up the local search. However, a substantial fraction of the new solutions should be generated by far field randomization and whose locations should be far enough from the current best

```
Begin
    Step 1: Initialization. Set the generation counter G = 1; initialize the population P of n host
            nests randomly and each egg in a nest corresponding to a potential solution to the
            given problem; set the discovery rate p_a.
    Step 2: While the termination criteria is not satisfied or G < MaxGeneration do
            Sort the population/nest from best to worst.
            Get a cuckoo randomly (say, i) and replace its solution by performing Lévy flights.
            Evaluate its quality/fitness F_i.
            Choose a nest among n (say, j) randomly.
            if (F_i < F_j)
                Replace j by the new solution.
            end if
            A fraction (p_a) of the worse nests is abandoned and new ones are built.
            Keep the best solutions/nests.
            Sort the population/nest from best to worst and find the current best.
            Pass the current best to the next generation.
            G = G + 1.
    Step 3: end while
    Step 4: Post-processing the results and visualization.
End.
```

ALGORITHM 2: The algorithm of cuckoo search (CS) via Lévy flights.

solution; this will make sure the system will not be trapped into a local optimum.

4. Differential Evolution/Cuckoo Search: DE/CS

Generally speaking, the standard DE algorithm is adept at exploring the search space and locating the region of global optimal value, but it is not relatively good at exploiting solution. On the other hand, standard CS algorithm is usually quick at the exploitation of the solution though its exploration ability is relatively poor. Therefore, in this paper, a hybrid metaheuristic algorithm by integrating differential evolution into cuckoo search, so-called DE/CS, is used to solve the three-dimension path planning for UCAV. The difference between DE/CS and CS is that the mutation and crossover of DE is used to replace the original CS selecting a cuckoo. In this way, this method can explore the new search space by the mutation of the DE algorithm and exploit the population information with CS and therefore can conquer the lack of the exploitation of the DE algorithm. In the following, we will show the algorithm DE/CS, which is a variety of DE and CS.

4.1. Mainframe of DE/CS. The critical operator of DE/CS is the hybrid differential evolution selecting cuckoo operator, which embeds the differential evolution into the CS. The core idea of the proposed differential evolution selecting cuckoo operator is based on two considerations. First, the mutation operator of DE can add diversity of the population to improve the search efficiency. Second, the mutation operator of DE can improve the exploration of the new search space. Pseudocode of hybrid differential evolution selecting cuckoo operator can be described as in Algorithm 3. In Algorithm 3, D is the number of decision variables. NP is the size of the parent population P. F is the mutation scaling factor. CR is a constant for crossover operator. $X_i(j)$ is the jth variable of the candidate solution X_i. X_u is the offspring. $\lceil NP * rand \rceil$ is a uniformly distributed random integer number between 1 and NP. And rand is a uniformly distributed random real number in interval (0, 1). We use the DE/rand/1/bin scheme shown in Algorithm 3.

By incorporating above-mentioned hybrid differential evolution selecting cuckoo operator into original CS algorithm, the DE/CS has been developed as a new algorithm. DE/CS algorithm is given as in Algorithm 4, where a fraction of worse nests are discovered with a probability p_a. K is a status matrix with NP $\times$ D whose value is logical value 0 or 1, meaning the egg in the nest discovered or not, and $K(i,:)$ represents the ith row elements in the status matrix K. The Hadamard product of two matrices $\mu \odot v$ is defined as the entrywise product, that is, $[\mu \odot v] = \mu_{ij} v_{ij}$. In the real world, if a cuckoo's egg is very similar to host's eggs, then this cuckoo's egg is less likely to be discovered, thus the fitness should be related to the difference in solutions. Therefore, it is a good idea to do a random walk in a biased way with some random step sizes. Vector *Step* is the step size that determines how far a random walker can go for a fixed number of iterations. P_1 and P_2 are the copy of the population P; Y_i and Z_i are the individuals in the population P_1 and P_2, respectively. From Algorithm 4, we can see that there are only four control parameters in this algorithm, which are NP, F, CR, and p_a.

4.2. Algorithm DE/CS for UCAV Three-Dimension Path Planning. In essence, UCAV three-dimension path planning is to reach minimum value for the objective function shown as in (5). For a minimization problem, the quality or fitness of a solution can simply be inversely proportional to the value of the cost function (5). For simplicity, we can use

```
Begin
    for i = 1 to NP do
        Select uniform randomly r₁ ≠ r₂ ≠ r₃ ≠ i
        X_v = X_{r_1} + F × (X_{r_2} − X_{r_3})
        r₄ = ⌈NP * rand⌉
        for j = 1 to D do
            if rand ≤ CR or j == r₄ then
                X_u(j) = X_v(j)
            else
                X_u(j) = X_i(j)
            end if
        end for
        if F(X_u) < F(X_{r_4}) then
            X_u = X_{r_4}
            F(X_{r_4}) = F(X_u)
        end if
    end for
End.
```

ALGORITHM 3: Algorithm of selecting cuckoo for DE/CS.

the following simple representations that each egg in a nest represents a solution, and a cuckoo egg represents a new solution; the aim is to use the new and potentially better solutions (cuckoos) to replace a not-so-good solution in the nests. For this present work, we will use the simplest approach where each nest has only a single egg. In this case, there is no distinction between egg, nest, or cuckoo, as each nest corresponds to one egg which also represents one cuckoo. Therefore, in the following, we do not distinguish the egg, nest, and cuckoo all of which represent a candidate solution.

Let NP cuckoos be in the starting node; the cuckoos will choose the next nodes in the grid network diagram according to the selecting cuckoo rule shown as in Algorithm 3 instead of the Lévy flights used in CS shown in (8). A cuckoo lays an egg in a nest which may be found by the hosting bird; if then, the egg would be discarded, and then it is replaced by another cuckoo's egg. Thus cuckoo birds are always looking for a better place in order to decrease the chance of their eggs to be discovered. The process can be approximated by the fraction p_a of the NP nests which are displaced by new nests (with new random solutions). Consequently, it will enhance the original quality of the candidate solution. Thus, the more cuckoos a UCAV path is passed by, the bigger possibility that a path can be selected by the other cuckoos. This process can guarantee nearly all cuckoos walk along the shortest UCAV path in the end.

Based on the above analysis, the pseudocode of improved CS-DE/CS for UCAV three-dimension path planning is described as follows (Algorithm 5).

5. Path-Smoothing Strategies

The generated UCAV optimal three-dimension path using the proposed hybrid metaheuristic method DE/CS is usually hard for exact flying. There are some turning points on the

optimized path [20, 21]. In this section, we adopt a class of dynamically feasible trajectory smooth strategy called B-Spline curves smoothing strategy [17]. B-Splines are adopted to define the UCAV desired path, providing at least first-order derivative continuity. B-Spline curves are well fitted in the evolutionary procedure; they need a few variables (the coordinates of their control points) in order to define complicated curved paths. Each control point has a very local effect on the curve's shape and small perturbations in its position produce changes in the curve only in the neighborhood of the repositioned control point.

B-Spline curves are parametric curves, with their construction based on blending functions [22]. Their parametric construction provides the ability to produce nonmonotonic curves. If the number of control points of the corresponding curve is $n + 1$, with coordinates $w_0(x_0, y_0, z_0), \ldots, w_n(x_n, y_n, z_n)$, the coordinates of the B-Spline curve may be written as

$$x(u) = \sum_{i=1}^{n} x_i \cdot N_{i,p}(u),$$

$$y(u) = \sum_{i=1}^{n} y_i \cdot N_{i,p}(u), \qquad (11)$$

$$z(u) = \sum_{i=1}^{n} z_i \cdot N_{i,p}(u),$$

where u is the free parameter of the curve, $N_{i,p}(u)$ are the blending functions of the curve, and p is its degree, which is associated with curve's smoothness ($p + 1$ being its order). Higher values of p correspond to smoother curves.

The blending functions are defined recursively in terms of a *knot* vector $U = \{u_0, \ldots, u_m\}$, which is a nondecreasing sequence of real numbers, with the most common form being the *uniform nonperiodic* one, defined as

$$u_i = \begin{cases} 0 & \text{if } i < p + 1, \\ i - p & \text{if } p + 1 \leq i \leq n, \\ n - p + 1 & \text{if } n < i. \end{cases} \qquad (12)$$

The blending functions $N_{i,p}$ are computed, using the knot values defined above, as

$$N_{i,0} = \begin{cases} 1 & u_i \leq u \leq u_{i+1}, \\ 0 & \text{otherwise,} \end{cases}$$

$$N_{i,p}(u) = \frac{u - u_i}{u_{i+p} - u_i} N_{i,p-1}(u) + \frac{u_{i+p+1} - u}{u_{i+p+1} - u_{i+1}} N_{i+1,p-1}(u). \qquad (13)$$

If the denominator of either of the fractions is zero, that fraction is defined to have zero value. Parameter u varies between 0 and $(n - p + 1)$ with a constant step, providing the discrete points of the B-Spline curve. The sum of the values of the blending functions for any value of u is always 1.

The use of B-Spline curves for the determination of a flight path provides the advantage of describing complicated nonmonotonic 3-dimensional curves with controlled

```
Begin
    Step 1: Initialization. Set the generation counter G = 1; initialize the population P of NP
            host nests randomly and each egg in a nest corresponding to a potential solution to
            the given problem; set the mutation scaling factor F and crossover constant CR.
    Step 2: While the termination criteria is not satisfied or G < MaxGeneration do
            Sort the population/nests from best to worst.
            Store the best nests to KeepNest.
            Get a cuckoo (say, i) and replace its solution by performing Algorithm 3.
            K = rand(NP, D) > p_a.
            P_1 = P; P_2 = P.
            for i = 1 to NP do
                Step = rand * (Y_i − Z_i);
                X_new = X_i + Step ⊙ K(i, :);
            end for
            for i = 1 to NP do
                if F(X_new) < F(X_i) then
                    X_new = X_i; F(X_new) = F(X_i)
                end if
            end for
            Keep the best solutions/nests.
            Sort the population/nest from best to worst and find the current best.
            Replace the worst nests with the best nests KeepNest stored.
            Pass the current best to the next generation.
            G = G + 1.
    Step 3: end while
    Step 4: Post-processing the results and visualization.
End.
```

ALGORITHM 4: The main procedure of DE/CS.

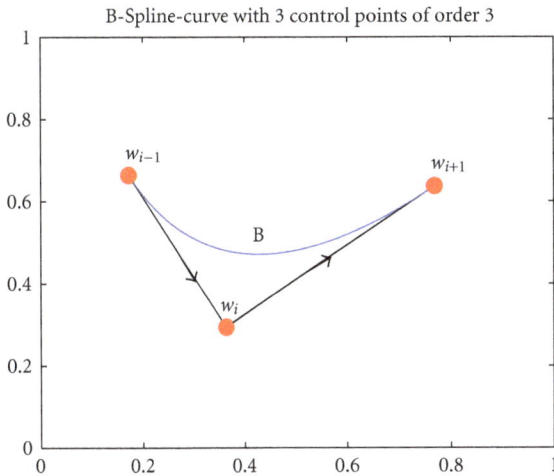

FIGURE 3: A quadratic ($p = 2$) 2-dimensional B-Spline curve, produced using a uniform nonperiodic knot vector, and its control polygon.

smoothness with a small number of design parameters, that is, the coordinates of the control points. Another valuable characteristic of the adopted B-Spline curves is that the curve is tangential to the control polygon at the starting and end points. This characteristic can be used in order to define the starting or end direction of the curve, by inserting an extra fixed point after the starting one, or before the end control point. Figure 3 shows a quadratic 2-dimensional B-Spline curve ($p = 2$) with its control points and the corresponding control polygon.

After this process, the original path $\overline{w_{i-1}w_i} \rightarrow \overline{w_iw_{i+1}}$ could be replaced by the path $\widehat{w_{i-1}B} \rightarrow \widehat{Bw_{i+1}}$. In this way, the optimized path can be smoothed for feasible flying. This trajectory smoothing algorithm has a small computational load and can be run in real time.

6. Simulation Experiments

In this section, we look at the performance of the proposed hybrid metaheuristic DE and CS to UCAV three-dimension path planning through a series of experiments conducted under complex combat field environment.

To allow a fair comparison of running times, all the experiments were implemented on a PC with a Pentium IV processor running at 2.0 GHz, 512 MB of RAM, and a hard drive of 160 Gbytes. Our implementation was compiled using MATLAB R2012a (7.14) running under Windows XP3. No commercial CS tools or other population-based optimization tools were used in the following experiments.

To our knowledge, parameter setting has a great effect on the performance of optimization method. According to simulation experiments [12], Yang and Deb found that population size NP = 15 to 40 and discovery rate $p_a = 0.25$ are sufficient for most optimization problems. Their results

Begin
    Step 1: Initializing. Set the generation counter $G = 1$; randomly generate UCAV path
        (population/nest) of n individuals and each egg in a nest corresponding to a
        potential optimal path to the given problem; set discovery rate p_a; set the mutation
        scaling factor F and crossover constant CR.
    Step 2: Evaluate the population of P according to (5).
    Step 3: While the termination criteria is not satisfied **or** $G <$ MaxGeneration **do**
            Sort the UCAV path (population/nest) from best to worst.
            Store the Keep best nests to KeepNest.
            Get a cuckoo (say, i) and replace its solution by performing Algorithm 3.
            Evaluate its cost (fitness) J_i according to (5).
            Choose a path among n (say, j) randomly.
            if $(J_i > J_j)$
                Replace j by the new solution.
            end if
            $K = \text{rand}(NP, D) > p_a$.
            $P_1 = P; P_2 = P$.
            for $i = 1$ to NP **do**
                Step $= \text{rand} * (Y_i - Z_i)$;
                $X_{\text{new}} = X_i + \text{Step} \odot K(i, :)$;
            end for
            for $i = 1$ to NP **do**
                if $F(X_{\text{new}}) < F(X_i)$ **then**
                    $X_{\text{new}} = X_i; F(X_{\text{new}}) = F(X_i)$
                end if
            end for
            Keep the best solutions/paths.
            Sort the population/nest/paths from best to worst and find the current best.
            Replace the Keep worst nests with the Keep best nests KeepNest stored.
            Pass the current best to the next generation.
            $G = G + 1$.
    Step 4: end while
    Step 5: Inversely transform the coordinates in final optimal path into the original coordinate,
        and output
End.

ALGORITHM 5: Algorithm of DE/CS for UCAV three-dimension path planning.

and analysis also illustrate that the convergence rate, to some degree, is insensitive to the parameters selected. This means that we do not need to fine-tune parameters for any given problems. Therefore, in all experiments, we will use the same set of CS algorithm parameter, which are step size $\alpha = 1$, discovery rate $p_a = 0.25$, population size NP = 30, and maximum generation Maxgen = 200.

Figure 4 shows the UCAV path planning results comparison between basic CS and the proposed hybrid metaheuristic CS and DE algorithm in three-dimension and two-dimension space with NP = 30, $p_a = 0.25$, and the curve path comparison by the smooth algorithm, and also the evolution curves comparison. The symbol "$\bigcirc$" denotes the starting point, the cone denotes the threaten area, while the symbol "$\square$" denotes the end point. And the green line is the path generated by the basic CS, while the red one is generated by the improved CS.

The values of each optimal solution searched by the different algorithm could be given by the value of the "shortest length," which can be shown in Table 1. Table 1 shows the results found by basic CS and improved CS algorithm over 100 Monte Carlo runs. From the experimental results presented in Figure 4 and Table 1, it is apparent that the proposed hybrid metaheuristic CS and DE method can find feasible and optimal three-dimension path for the UCAV very quickly and can effectively solve the three-dimension path planning of UCAV in complicated combating environments. This method provides a new way for three-dimension path planning of UCAV in real application.

7. Conclusion and Future Work

This paper presented a hybrid metaheuristic CS and DE algorithm for UCAV three-dimension path planning in complicated combat field environments. A novel type of CS model has been described for single UCAV path planning, and DE is applied to optimize selecting cuckoo operator during the process of egg updating in nest. Then, the UCAV can find the safe path by connecting the chosen nodes while avoiding the threat areas and costing minimum fuel. This

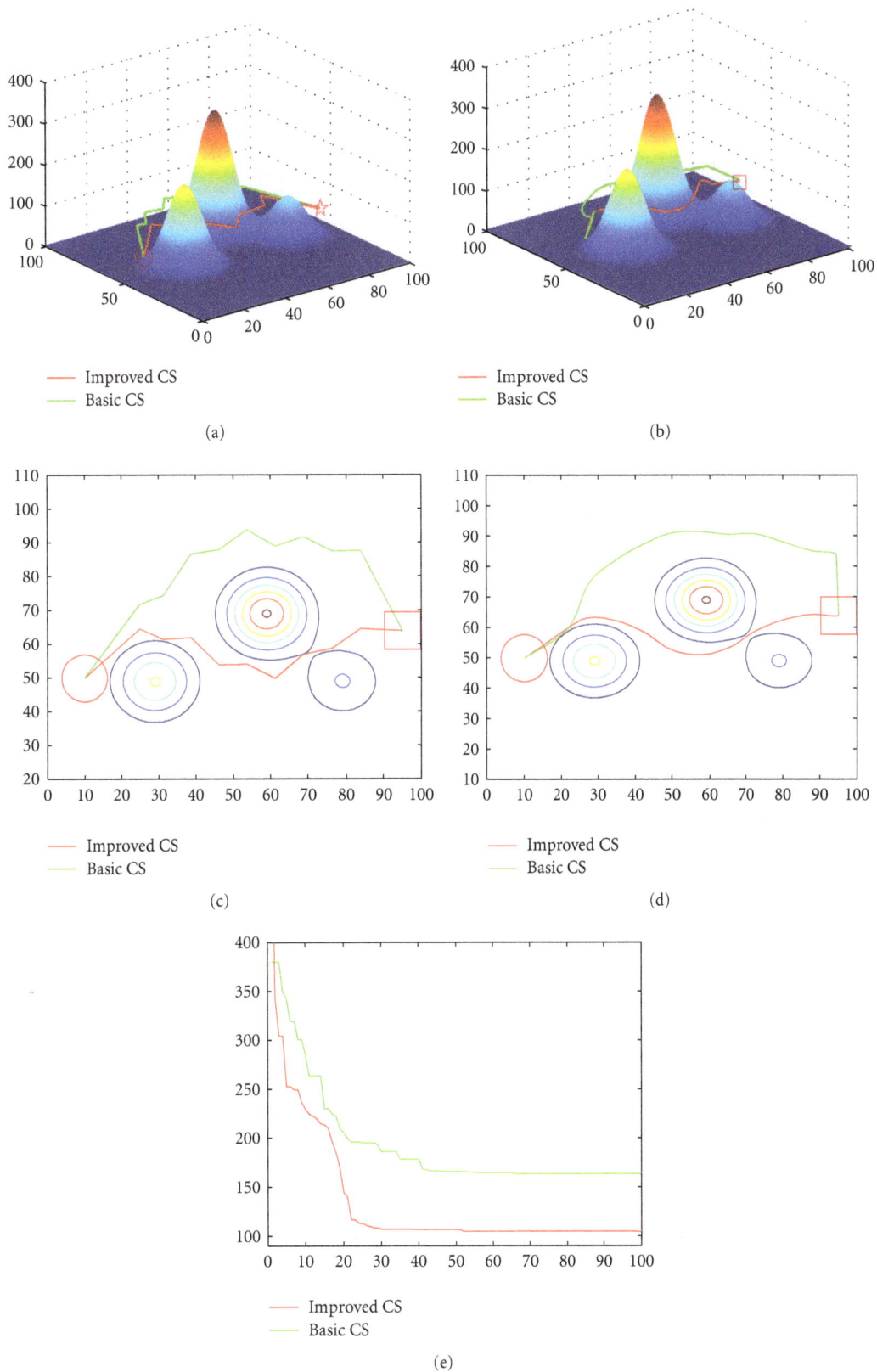

FIGURE 4: Parameter values were NP = 30, p_a = 0.25. (a) Path-planning original results comparison between basic CS and improved CS in a three-dimension space. (b) Route comparison after using the smoothing strategy in a three-dimension space. (c) Path-planning original results comparison between basic CS and improved CS in a two-dimension space. (d) Route comparison after using the smoothing strategy in a two-dimension space. (e) Evolution curves comparison between the basic CS and the improved CS.

TABLE 1: Shortest length comparison between the basic CS and the improved CS in UCAV three-dimension path planning problem. The numbers shown are the results found after 100 Monte Carlo simulations of each algorithm.

	Basic CS	Improved CS
Mean	193.1960	165.5124
Std	34.9554	25.0811
Best	127.1038	105.1268
Worst	282.3626	207.9992
Time (sec)	75.6491	26.1248

new approach can accelerate the global convergence speed while maintaining the strong robustness of the basic CS. The detailed implementation procedure for this metaheuristic approach is also described. In order to make the optimized UCAV path more feasible, the B-Spline curve is adopted for smoothing the path, and this trajectory smoothing algorithm has a small computational load and can be run in real time. Compared with the basic CS algorithm, the simulation experiments show that this method is a feasible and effective way in UCAV path planning. It is also flexible, in complicated dynamic battle field environments, and pop-up threats are easily incorporated.

In the algorithm of UCAV three-dimension path planning, there are many issues worthy of further study, and efficient route planning method should be developed depending on the analysis of specific combat field environments. Currently, the hot issue contains self-adaptive route planning for a single UCAV and collaborative route planning for a fleet of UCAVs. As the important ways of improving aircraft survivability, adaptive route planning should analyze real-time data under the uncertain and dynamic threat condition, it can even remodify preplanned flight path to improve the success rate of completing mission. The difficulty of the collaborative route planning for a fleet of UCAVs exists in coordination between the various UCAVs, including the fleet formation, target distribution, arrival time constraint, and avoidance conflict, each of which is a complicated question worthy of further study. Our future work will focus on the two hot issues and develop new methods to solve problem in UCAV 3-dimension path planning.

Acknowledgments

This work was supported by State Key Laboratory of Laser Interaction with Material Research Fund under Grant no. SKLLIM0902-01 and Key Research Technology of Electric-discharge Non-Chain Pulsed DF Laser under Grant no. LXJJ-11-Q80.

References

[1] S. Kurnaz, O. Cetin, and O. Kaynak, "Adaptive neuro-fuzzy inference system based autonomous flight control of unmanned air vehicles," *Expert Systems with Applications*, vol. 37, no. 2, pp. 1229–1234, 2010.

[2] H. B. Duan, X. Y. Zhang, and C. F. Xu, *Bio-Inspired Computing*, Science Press, Beijing, China, 2011.

[3] Y. V. Pehlivanoglu, "A new vibrational genetic algorithm enhanced with a Voronoi diagram for path planning of autonomous UAV," *Aerospace Science and Technology*, vol. 16, pp. 47–55, 2012.

[4] W. Ye, D. W. Ma, and H. D. Fan, "Algorithm for low altitude penetration aircraft path planning with improved ant colony algorithm," *Chinese Journal of Aeronautics*, vol. 18, no. 4, pp. 304–309, 2005.

[5] H. Duan, Y. Yu, X. Zhang, and S. Shao, "Three-dimension path planning for UCAV using hybrid meta-heuristic ACO-DE algorithm," *Simulation Modelling Practice and Theory*, vol. 18, no. 8, pp. 1104–1115, 2010.

[6] H. B. Duan, X. Y. Zhang, J. Wu, and G. J. Ma, "Max-min adaptive ant colony optimization approach to multi-UAVs coordinated trajectory replanning in dynamic and uncertain environments," *Journal of Bionic Engineering*, vol. 6, no. 2, pp. 161–173, 2009.

[7] C. Xu, H. Duan, and F. Liu, "Chaotic artificial bee colony approach to Uninhabited Combat Air Vehicle (UCAV) path planning," *Aerospace Science and Technology*, vol. 14, no. 8, pp. 535–541, 2010.

[8] H. Duan, S. Liu, and J. Wu, "Novel intelligent water drops optimization approach to single UCAV smooth trajectory planning," *Aerospace Science and Technology*, vol. 13, no. 8, pp. 442–449, 2009.

[9] R. Storn and K. Price, "Differential evolution—a simple and efficient adaptive scheme for global optimization over continuous space," Technical Report, International Computer Science Institute, Berkeley, Calif, USA, 1995.

[10] R. Storn and K. Price, "Differential evolution—a simple and efficient heuristic for global optimization over continuous spaces," *Journal of Global Optimization*, vol. 11, no. 4, pp. 341–359, 1997.

[11] V. V. D. Melo and G. L. C. Carosio, "Evaluating differential evolution with penalty function to solve constrained engineering problems," *Expert Systems With Applications*, vol. 39, pp. 7860–7863, 2012.

[12] X. S. Yang and S. Deb, "Cuckoo search via Lévy flights," in *Proceedings of the World Congress on Nature and Biologically Inspired Computing (NABIC '09)*, pp. 210–214, December 2009.

[13] X. S. Yang and S. Deb, "Engineering optimization by cuckoo search," *International Journal of Mathematical Modelling and Numerical Optimization*, vol. 1, no. 4, pp. 330–343, 2010.

[14] R. B. Payne, M. D. Sorenson, and K. Klitz, *The Cuckoos*, Oxford University Press, Oxford, UK, 2005.

[15] Wikipedia, Cuckoo search, 2009, http://en.wikipedia.org/wiki/Cuckoo_search.

[16] G. Yang and V. Kapila, "Optimal path planning for unmanned air vehicles with kinematic and tactical constraints," in *Proceedings of the 41st IEEE Conference on Decision and Control*, pp. 1301–1306, Las Vegas, Nev, USA, December 2002.

[17] I. Nikolos, E. Zografos, and A. Brintaki, "UAV path planning using evolutionary algorithms," in *Innovations in Intelligent Machines*, S. C. Javaan, C. J. Lakhmi, M. Akiko, and S. I. Mika, Eds., pp. 77–111, Springer, Berlin, Germany, 2007.

[18] W. Ye and H. D. Fan, "Research on mission planning system key techniques of UCAV," *Journal of Naval Aeronautical Engineering Institute*, vol. 22, no. 2, pp. 201–207, 2007.

[19] P. Barthelemy, J. Bertolotti, and D. S. Wiersma, "A Lévy flight for light," *Nature*, vol. 453, no. 7194, pp. 495–498, 2008.

[20] T. Kito, J. Ota, R. Katsuki et al., "Smooth path planning by using visibility graph-like method," in *Proceedings of the IEEE International Conference on Robotics and Automation*, vol. 3, pp. 3770–3775, Taipei, Taiwan, September 2003.

[21] J. Miura, "Support vector path planning," in *Proceedings of the International Conference on Intelligent Robots and Systems (IROS '06)*, pp. 2894–2899, Beijing, China, October 2006.

[22] G. Farin, *Curves and Surfaces for Computer Aided Geometric Design, a Practical Guide*, Academic Press, San Diego, Calif, USA, 2th edition, 1990.

11

New Sensors and Techniques for the Structural Health Monitoring of Propulsion Systems

Mark Woike,[1] **Ali Abdul-Aziz,**[1] **Nikunj Oza,**[2] **and Bryan Matthews**[2]

[1] *National Aeronautics and Space Administration, Glenn Research Center, Cleveland, OH 44135, USA*
[2] *National Aeronautics and Space Administration, Ames Research Center, Moffett Field, CA 94035, USA*

Correspondence should be addressed to Mark Woike; mark.r.woike@nasa.gov

Academic Editors: D. Greatrix, C. Hajiyev, A. Kalfas, Y.-H. Li, and L. Massotti

The ability to monitor the structural health of the rotating components, especially in the hot sections of turbine engines, is of major interest to aero community in improving engine safety and reliability. The use of instrumentation for these applications remains very challenging. It requires sensors and techniques that are highly accurate, are able to operate in a high temperature environment, and can detect minute changes and hidden flaws before catastrophic events occur. The National Aeronautics and Space Administration (NASA), through the Aviation Safety Program (AVSP), has taken a lead role in the development of new sensor technologies and techniques for the in situ structural health monitoring of gas turbine engines. This paper presents a summary of key results and findings obtained from three different structural health monitoring approaches that have been investigated. This includes evaluating the performance of a novel microwave blade tip clearance sensor; a vibration based crack detection technique using an externally mounted capacitive blade tip clearance sensor; and lastly the results of using data driven anomaly detection algorithms for detecting cracks in a rotating disk.

1. Introduction

The development of in situ measurement technologies and fault-detection techniques for the structural health monitoring of gas turbine engines is of major interest to NASA's Aviation Safety Program and the aeronautical community. The rotating components of modern gas turbine engines operate in severe environmental conditions and are exposed to high thermal and mechanical loads. The cumulative effects of these loads often lead to high stresses, structural deformities, cracks, and eventual component failure. Current structural health monitoring practices involve periodic inspections and schedule-based maintenance of engine components to ensure their integrity over the lifetime of the engine. However, these methods have their limitations, and failures are experienced leading to unscheduled maintenance and unplanned engine shutdowns. To prevent these failures and enhance aviation safety, the NASA Glenn Research Center has investigated new sensor technologies and techniques for the in situ structural health monitoring and detection of defects in gas turbine engines.

Much of the research effort to date has dealt with the development of low technology readiness level (TRL) structural health concepts with the intent of validating these concepts and transitioning them to higher TRLs for usage on actual aero engine hardware. In this study, microwave blade tip clearance and blade tip timing sensor technology is being investigated as a means of making high temperature noncontact structural health measurements in the hot sections of gas turbine engines. It is specifically being targeted for use in the high pressure turbine (HPT) and high pressure compressor (HPC) sections to directly monitor the structural health of the rotating components. The capability to make in situ health measurements is a need that has been identified by the aero engine community as blade damage in the HPT and HPC sections account for 12 percent of the inflight engine shutdown events and 32 percent of the damage events that caused engine removal for unscheduled maintenance [1].

Currently there are no off-the-shelf blade tip clearance sensors that are used in commercial turbine engines for in-situ structural health monitoring, as there are challenges with the sensors being able to operate in and survive the harsh high temperature environment. Microwave sensor technology is appealing in that it is accurate, it has the ability to operate at extremely high temperatures, and is unaffected by contaminants that are present in turbine engines. In addition to its use for structural health monitoring, this type of sensor also has parallel usage in improving engine performance through its use in active turbine tip clearance control schemes. As a means of better understanding the issues associated with the microwave sensors, a series of evaluation experiments were conducted to evaluate their performance on aero turbine engine like hardware.

Along with the development and evaluation of new types of blade tip clearance sensors, efforts have been placed into developing techniques that utilize these sensors for the structural health monitoring of engine components. The vibration based crack detection experiment that will be discussed involved introducing a notch to simulate a crack on a subscale turbine engine disk and monitoring its vibration response as the disk was rotated at speeds up to 12 000 rpm. The vibration response was characterized by using externally mounted capacitive blade tip clearance sensors to measure the combined disk-rotor system's whirl amplitude and phase during operation. Testing was performed on a clean undamaged baseline disk and a disk with a 50.8 mm long notch machined into the disk to simulate a crack. The responses were compared and evaluated against the theoretical models to investigate the applicability and success of detecting the notch.

And finally, in parallel with the vibration based crack detection investigation an experiment using data driven anomaly detection algorithms was undertaken as a means of evaluating whether these data mining techniques could be used to detect a fault or an anomaly in a rotating engine disk. In this experiment blade tip clearance data sets were acquired on an undamaged subscale engine rotor disk as it was operated at speeds up to 10 000 rpm. This baseline data was used to train three different data mining algorithms. Data was then acquired from a damaged disk with a known crack and the data mining algorithms were used to detect the presence of the crack.

This paper presents a summary of key results and findings obtained from three different structural health monitoring approaches that have been investigated. This includes evaluating the performance of a novel microwave blade tip clearance sensor; a crack detection technique using externally mounted blade tip clearance sensors; and lastly discussing the results of a data driven anomaly detection technique for sensing cracks in a rotating rotor disk.

2. Microwave Blade Tip Clearance/Blade Tip Timing Sensors

The ability to monitor the structural health of the rotating components, especially in the hot sections of turbine engines, is of major interest to aero community in improving engine safety and reliability [2]. In addition, the active control and minimization of the gap between the rotating turbine blades and the stationary case of gas turbine engines is being sought as a means of increasing engine efficiency, reducing fuel consumption, reducing emissions, and increasing engine service life [3]. The use of instrumentation for these applications in gas turbine engines requires sensors that are highly accurate and can operate in a high temperature environment. To address this need, microwave sensor technology is being investigated as a means of making high temperature non-contact blade tip clearance and tip timing measurements for use in structural health monitoring and active clearance control applications in turbine engines. This technology is appealing due to its high accuracy and its potential to operate at extremely high temperatures that are present in turbine engines. It is intended to use blade tip clearance to monitor blade growth and wear and blade tip timing to monitor blade vibration and deflection.

2.1. Sensor Background and Theory. NASA has worked with Radatec (now Meggitt Inc.) through the Small Business Innovation Research (SBIR) program for the development of microwave sensor technology for high temperature noncontact blade tip clearance and blade tip timing measurements. The initial development of the technology was accomplished through a phase II SBIR contract awarded in 2002. Further development of the technology was accomplished in 2004 through 2005 as part of NASA's Ultra Efficient Engine Technology (UEET) Program. A prototype first generation 5.8 GHZ system was delivered as part of a phase III SBIR commercialization contract in 2007 and a second generation 24 GHZ system along with upgraded electronics was delivered as part of subsequent follow-on contracts in 2009 and 2010, respectively.

The microwave blade tip clearance sensor operates essentially as a field disturbance device. The tip clearance probe contains both a transmitting and receiving antenna. The sensor emits a continuous microwave signal and measures the signal that is reflected off a rotating blade. The sensor measures the changes in the microwave field due to the blade passing through the field. The motion of the blade phase modulates the reflected signal and this reflected signal is compared to an internal reference. Changes in amplitude and phase directly correspond to the distance to the blade. The time interval of when the blade passes through the field is measured to provide blade tip timing. More detailed information on the sensor's theory of operation can be found in [4–6]. The microwave blade tip clearance probes are made of high temperature material and are designed to operate in temperatures up to 900°C. The first generation probes (Figure 1(a)) operate at 5.8 GHz and can measure clearance distances up to one-half the radiating wavelength which is 25 mm. The second generation probes (Figure 1(b)) operate at 24 GHz and can measure clearance distances up to 6 mm. The 5.8 GHz sensor is targeted for use on large rotating machinery such as land based power turbines or in the fan sections of aero gas turbine engines. The 24 GHz sensor is being targeted for use in smaller rotating machinery applications such as the turbine and compressor sections of aero engines. This technology has an ultimate goal of obtaining

(a)   (b)

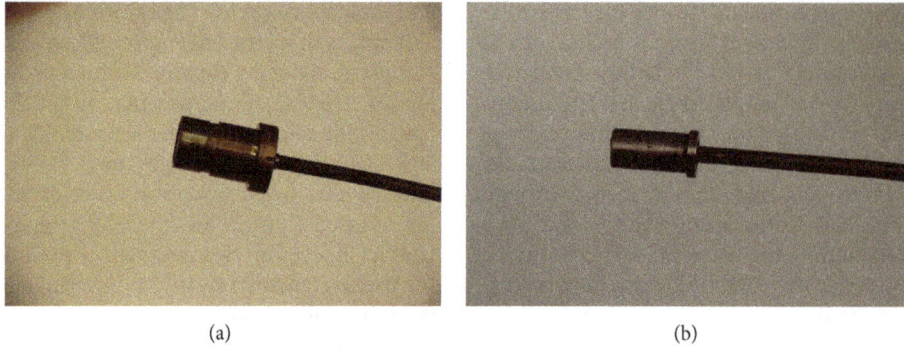

FIGURE 1: Microwave blade tip clearance probe (Meggitt). (a) 5.8 GHz probe. (b) 24 GHz probe.

FIGURE 2: Axial vane fan located at the 10 × 10 SWT.

FIGURE 3: NASA Turbofan.

clearance measurement accuracies approaching +/−25 μm. A frequency response of up to 5 MHz is typical, with up to 25 MHz being possible with this technology which lends itself for use in structural health monitoring applications for the measurement of blade deflection and vibration.

2.2. Experimental Results. NASA's primary goal is to demonstrate this microwave blade tip clearance sensor technology on an actual gas turbine engine in a relevant high temperature environment. However, the use of microwave sensors for this application is an emerging concept. Techniques on their use and calibration needed to be understood and developed. In addition, the microwave sensor's accuracy and ability to make blade tip clearance and deflection measurements had to be assessed prior to use on actual engine hardware. As a means of better understanding the issues associated with the microwave sensors, a series of experiments were conducted to evaluate the sensor's performance on aero engine type applications. A summary of these experiments and their results are as follows.

The first generation 5.8 GHz microwave sensors were used to a make blade tip clearance measurements on a large axial vane fan and a subscale NASA Turbofan. The purpose of the test on the large Axial Vane Fan (Figure 2) was to develop the infrastructure required for the calibration of the sensors and the techniques for their use in the field to make clearance measurements on large rotating machinery. The motivation behind their use on the NASA Turbofan (Figure 3)

was to evaluate the first generation sensor's ability to acquire blade tip clearance data on an aero engine size test article and blades. Blade tip clearance data sets were acquired for several test runs of the NASA Turbofan. Data was acquired at a variable sampling rate that was synchronized to the fan's speed. Each measurement consisted of two revolutions of data with 10 000 samples taken per revolution. Figure 4 shows the individual blade clearances measured for several fan speeds. It is clearly noted from the polar plot that the tip clearances decreased as the fan speed is increased.

This result is expected and is due to the growth and expansion of the turbofan's composite blades as the fan operates at higher speeds. An average decrease of 0.22 mm was observed as the rig speed was increased to 8 875 rpm. The change in clearance detected in this experiment was within the range predicted for these blades. In addition, the change in tip clearances measured by the microwave sensors was nearly identical to previously recorded values obtained with capacitive clearance sensors. In previous test entries, changes in tip clearances of up to 0.22 mm were also noted when the turbofan was operated over the same speed range. These two initial experiments served as test beds for the development of techniques and infrastructure required for the calibration of the sensors and successfully demonstrated the microwave clearance sensor's ability to make measurements on aero engine size hardware.

The second generation 24 GHz sensors were used to make measurements on a 32 blade subscale turbine engine like

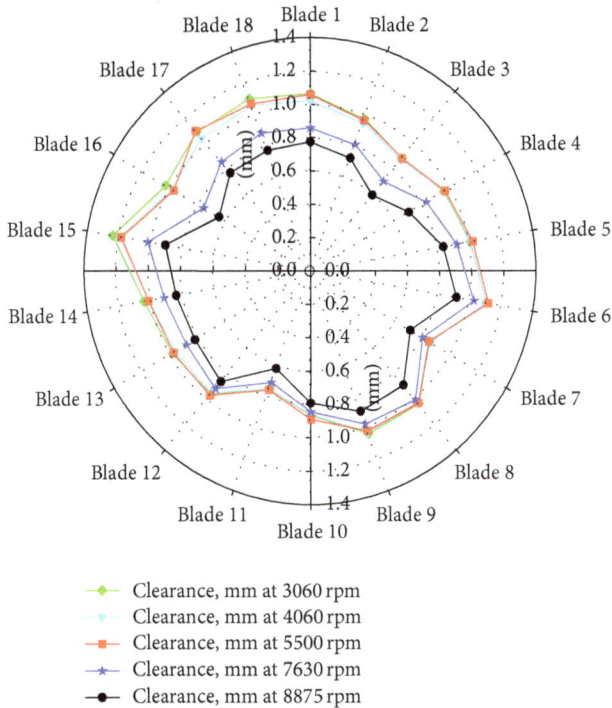

Clearance, mm at 3060 rpm
Clearance, mm at 4060 rpm
Clearance, mm at 5500 rpm
Clearance, mm at 7630 rpm
Clearance, mm at 8875 rpm

FIGURE 4: Blade tip clearance polar plot for probe #1, 90° location—NASA Turbofan experiment.

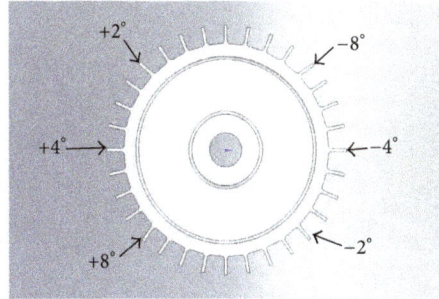

FIGURE 5: Simulated turbine disk with prebent blades for blade tip deflection testing.

disk and on an actual compressor stage from a small aero engine. These experiments were conducted in a laboratory environment using a calibration spin rig, a High Precision Spin Rig, and a High Temperature Spin Rig for the purposes of evaluating the second generation sensor's capability of making both clearance and timing measurements on small aero engine hardware. The 32 blade subscale engine disk (Figure 5) was manufactured to have 6 equally spaced blades that were bent at predefined angles or deflection distances. The purpose of this experiment was to evaluate the sensors ability to generate blade tip clearance and blade tip timing measurements simultaneously. During this experiment it was successfully demonstrated that the sensors were able to reliably make absolute measurements down to a clearance level of 100 μm. This was an improvement over the previously demonstrated minimum measurement of 500 μm made with the first generation sensors. This ability to measure very low clearance ranges is critical for use in closed loop clearance control applications. In addition, the sensor was able to simultaneously detect the minimum deflection of 0.70 mm (2 degrees) that was machined onto the disk. It is planned to further explore this capability by fabricating a disk that has smaller deflections in order to see the minimum deflection that can be reliably measured.

The investigation using a compressor stage from a small aero engine was conducted for the purposes of calibrating the sensors for future use on a small compressor and evaluating their ability to measure blade tip clearances on actual engine hardware over the very low clearance ranges typically associated with aero engines. For this experiment

clearances were successfully measured over a range from 0.10 to 1.50 mm with a maximum observed error of +/− 0.021 mm. These results were encouraging in that the sensor was able to make measurements on actual aero engine hardware over a relatively low clearance range within the desired accuracy of +/− 0.025 mm. However, it is acknowledged that these results were achieved in the ideal set-up of a laboratory and that the results may differ in an actual engine due to noise and other environmental effects.

In summary, the evaluation testing that was accomplished on the microwave blade tip clearance technology has shown that sensors are a viable option for propulsion health monitoring and clearance control applications. Techniques for the calibration, integration, and use of the sensors to make measurements on aero engine hardware were successfully developed and both the first and second generation sensors have been successfully demonstrated on rotating machinery and aero engine hardware. A demonstration test of these sensors along with other advanced technologies on an engine ground test is being planned to occur in the 2013-2014 timeframe.

3. Vibration Based Crack Detection Technique

This crack-detection methodology involved introducing a notch on a simulated turbine engine disk and monitoring its vibration response as the disk was rotated at speeds up to 12 000 rpm. The vibration response was characterized by monitoring the disk-rotor system's whirl amplitude and phase during operation. The whirl amplitude and phase were derived from the blade tip clearance profile that was measured using externally mounted blade tip clearance sensors. This type of sensor was chosen as it represents the type of sensor that is most likely to be installed on a commercial engine due to its minimal installation impact, low overhead, and parallel benefits that it can bring to improve engine performance through its use in active turbine tip clearance control. The testing was performed on a clean undamaged baseline disk and on a damaged disk with a 50.8 mm long notch machined into the disk to simulate a crack. The experiment was conducted at the NASA Glenn Research Center's High Precision Spin Rig, a rig which is ideally suited for the development and validation of low technology readiness level (TRL) concepts before implementation on more

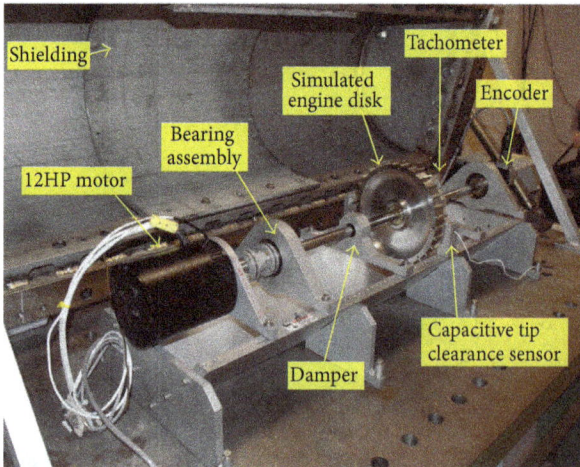

FIGURE 6: High Precision Spin Rig.

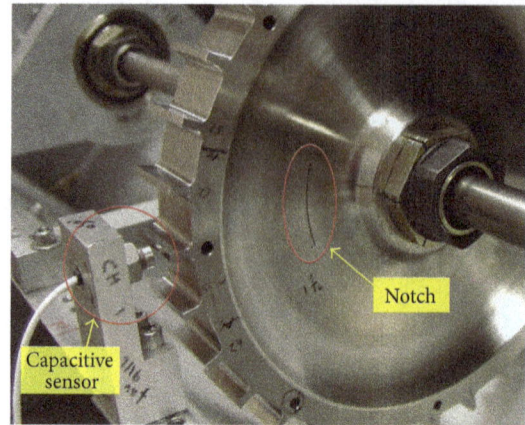

FIGURE 7: Capacitive blade tip clearance sensor and subscale turbine engine disk with notch.

expensive and complicated rotating machinery. A description of the High Precision Spin Rig, the baseline theory behind this technique, the experimental setup, and the results are discussed.

3.1. High Precision Spin Rig Description. Figure 6 shows the Rotordynamics Laboratory's High Precision Spin Rig, which can accommodate simulated engine rotor disks of up to 235 mm in diameter. It has a stainless steel shaft with a length of 781 mm and diameter of 20 mm. The shaft is supported by precision contact ball bearings on each end and has adjustable dampers that were positioned along the length of the shaft for these experiments. An encoder mounted on the end of the shaft was used by the control system to provide closed-loop control of rig speed. A secondary optical tachometer was used to record the speed into the data system and to synchronize the data to the rig's rotation. A 12-hp custom-built, brushless direct-current (dc) motor was used to rotate the spin rig and the subscale engine disks at speeds up to 12 000 rpm.

The rig was set up to acquire two channels of radial blade tip clearance data from the simulated engine disk using capacitive displacement sensors (Figure 7). These sensors were developed as part of a NASA Small Business Innovation Research (SBIR) contract and were different from traditional capacitive sensors in that their operation was based on a dc offset technique instead of the typical modulation technique. A National Instruments System was used to acquire data from the capacitive displacement probes at a fixed sampling rate of 1 MHz. This system and its application software were delivered as part of the SBIR contract for the capacitive blade tip clearance sensors. The system used custom data acquisition and processing applications that were tailored for acquiring and processing data from the blade tip clearance sensors.

3.2. Experimental Theory and Setup. The theory behind the crack-detection methodology that was investigated was based on previous theoretical and experimental work performed by Abdul-Aziz et al. [7, 8], Gyekenyesi et al. [9, 10], and

Haase and Drumm [11]. The goal of this experiment was to determine if the crack-detection methodology investigated in these earlier studies could be validated by using the spin rig to conduct tests on simulated engine rotor disks with a notch introduced to replicate a crack.

The detection methodology is based on monitoring the vibration response of rotating disks to determine if a crack has developed. The theory implies that a defect, such as a crack, creates minute deformations in the disk as it is rotated. The deformation, in turn, creates a speed-dependent shift in the disk's center of mass. It was theorized that this shift could be detected by analyzing the vibration whirl amplitude and phase as measured by the blade tip clearance profile. The system's behavior was modeled after a 2-degree-of-freedom Jeffcott rotor. The model predicts that the vibration amplitude peaks when a clean, undamaged disk goes through the first critical speed but heads to a lower steady-state value as the speed is increased above the critical speed. Correspondingly, the phase shifts 180° when going through the first critical speed and then stabilizes to a steady-state value as the speed is increased past critical. At this point, the rotor is rotating about the combined system's center of mass and has stabilized. However, as the speed of a cracked disk is increased, centrifugal forces open the crack. This, in turn, deforms the disk and shifts its center of mass. At speeds above critical, the crack-induced shift in the disk's center of mass starts to grow and dominate the overall system's vibration response. The modeling predicts that, instead of heading toward a steady-state value, the vibration amplitude will change as a second-order function of rotational speed as it is increased beyond the first critical speed. This can be detected by analyzing the vibration response of the disk-rotor system, particularly the amplitude and phase of the first harmonic, as the system is operated over a range of speeds.

The testing approach used in this experiment was to spin a subscale engine disk over a speed range from 0 to 12 000 rpm and simultaneously record its vibration response. The vibration response was acquired as previously described using capacitive displacement probes to measure the blade tip clearance profile. Two subscale simulated engine turbine disks were tested. The first disk was undamaged and was used to acquire the baseline vibration response data. The disk had an outside diameter of 235 mm, a bore thickness of 25.4 mm, and an outside rim thickness of 31.75 mm. The thinnest portion of the disk's web was 2.54 mm. Thirty-two teeth to simulate blades were evenly spaced around the circumference of the disk. Each simulated blade had a cross section of 31.75 mm by 3.30 mm and a height of 8.38 mm. The disk was made of a nickel base alloy, Haynes X-750 (Haynes International, Inc.) and had a weight of 4.88 kg.

The damaged disk is shown in Figure 7. It was identical to the baseline disk with the exception that a 50.8 mm long notch had been introduced in its mid-span region to imitate a crack in the disk. The mid-span region was selected since prior finite-element analysis had shown that this area experienced high stress levels during operation. The data acquired from this disk were compared with the data from the undamaged baseline disk to determine if the simulated crack could be detected by analyzing the vibration response. Supportive analytical calculations were made using finite-element analysis to complement the experimental work and determine the expected radial growth of the disk. The finite-element analysis predicted a radial growth on the order of ~0.075 mm at the highest operating speed, 12 000 rpm. This was within the detection limit of the blade-tip-clearance instrumentation. However, it should be noted that the major effect that was to be monitored was how the notch changed the combined center of mass of the disk-rotor system and its vibration amplitude and phase, not the radial blade tip growth of the disk.

3.3. Experimental Results.

The experimentation consisted of operating the baseline and notched disks at several speed profiles and recording their vibration responses using the capacitive blade tip clearance sensors. The first harmonic component of the tip-clearance profile for each revolution was then analyzed to determine if the amplitude and phase exhibited the expected characteristics associated with a crack in the disk. Approximately 15 test runs were conducted for each disk. A typical test run or cycle consisted of ramping up the disk's speed from 0 rpm to a predefined maximum speed, remaining on condition for 0 to 60 s, then ramping back down to 0 rpm. Blade tip clearance data were acquired over the entire cycle. Data sets were acquired at ramp rates of 60 and 100 rpm/s for peak operating speeds of 5 000, 10 000, and 12 000 rpm. In addition, complex mission profiles were conducted where the speed was ramped to various power levels in an attempt to simulate the rigorous loading conditions that an engine would experience during a typical mission.

The run profile presented in this paper is shown in Figure 8. It is typical of the profiles that were used in the investigation. For this test case, the speed of the disk was increased from 0 to 10 000 rpm at a ramp rate of 100 rpm/s, cycled between 10 000 and 5 000 rpm for three cycles, then ramped down from 10 000 to 0 rpm. This was done to simulate a mission profile from start-up to full power and

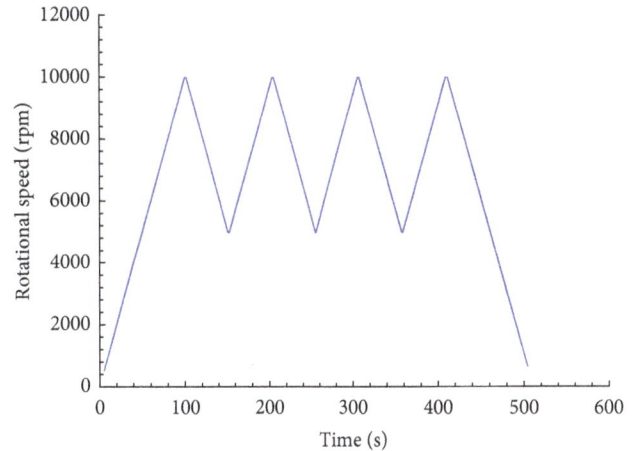

FIGURE 8: Simulated mission profile.

to observe the effects of cyclical loading on the disks. This ramping up and down also maximized the amount of data that could be acquired over a wide range of speeds, which was critical for the validation of this concept.

Figure 9 shows the results from this multiple-cycle test run for both the baseline and notched disks. The figure shows the analysis for the last cycle, starting at 10 000 rpm and ending at 0 rpm. As previously discussed, the presence of a crack is indicated if a second-order speed-dependent variation can be observed in the vibration amplitude as it is operated in the postcritical speed regime. In this test case, a speed-dependent rise was observed in the amplitude data for the notched disk. The area of interest showing the speed-dependent rise is highlighted within the circle of Figure 9(b). A curve-fit analysis, shown in Figure 10, was conducted on the vibration data in this region and it was found to closely follow a second-order polynomial fit, thus following the predicted behavior and potentially indicating the presence of the simulated crack.

Although the vibration data looked promising in this region, it was also observed that the second-order speed-dependent variation was not consistent over the entire range and that it appeared to reset itself at a speed of ~9500 rpm. Moreover, this relatively large amplitude variation was not consistently observed in other test runs, which cast some doubt on whether the effect was entirely due to the notch or to some other factor that is yet to be determined. However, this case yielded the best data so far and showed positive indications of being able to detect a defect such as a crack by analyzing the disk's vibration response as it is operated over a range of speeds. It is theorized that what made this case different, and more promising than other test runs, is that this test was conducted towards the end of the test series and many cycles had been placed on the disk prior to this test run. In addition, the disk experienced more loading because of the cyclical nature of the test profile. Overall, the results were promising and plans are in place to further investigate and refine this crack-detection technique as part of future studies.

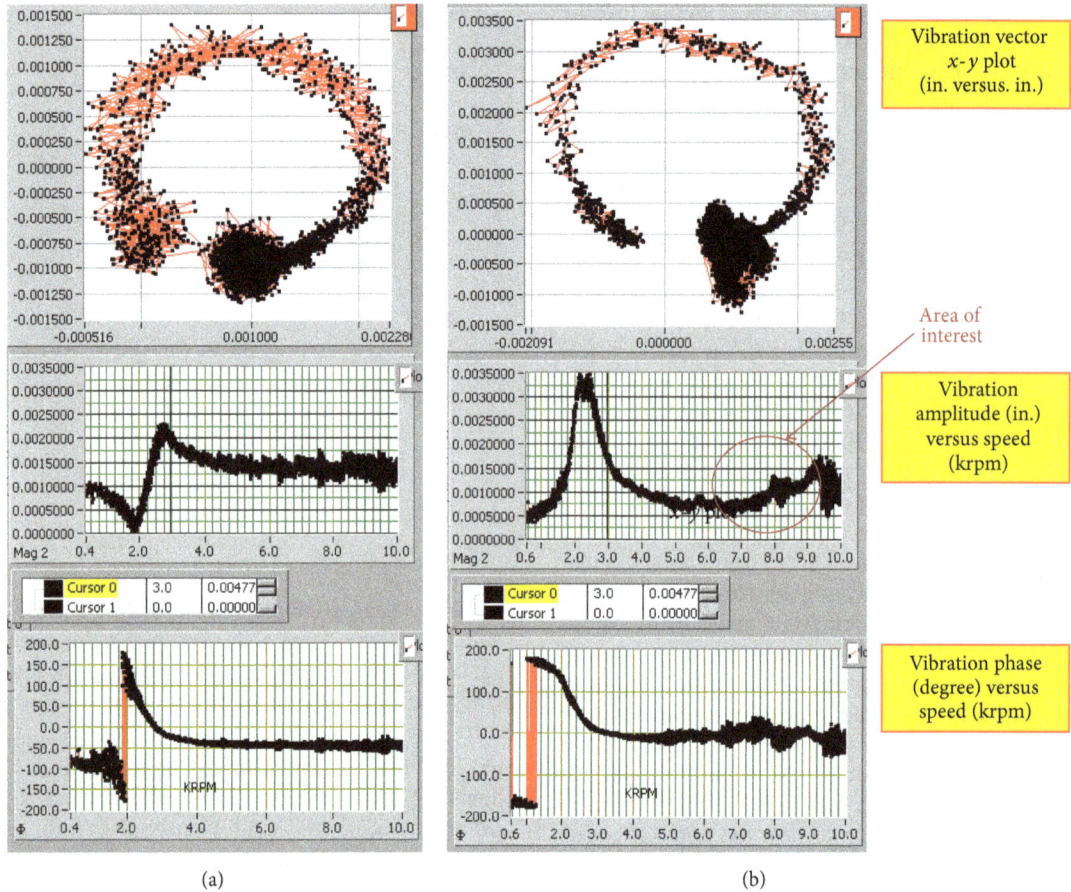

FIGURE 9: 10 000 rpm multiple-cycle run comparison. (a) Baseline disk. (b) Notched disk.

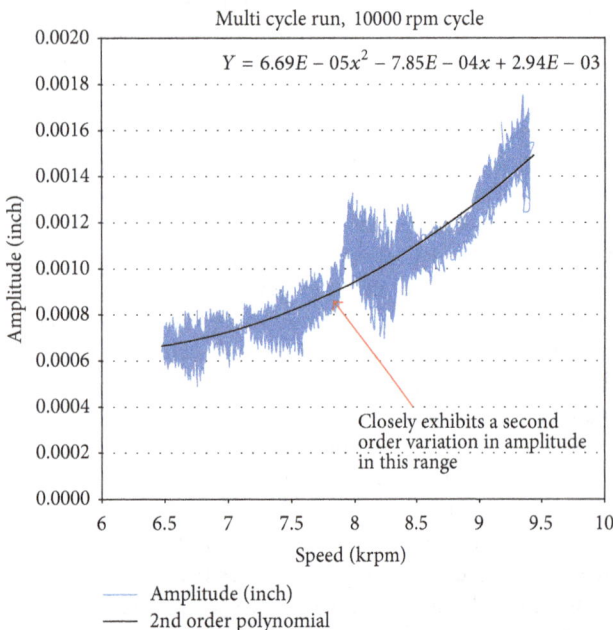

FIGURE 10: Detailed vibration amplitude plot for 10 000 rpm multiple-cycle run.

4. Data Driven Anomaly Detection Technique

In addition to the experimental work previously described, parallel health monitoring assessments have been conducted using data mining algorithms in order to detect flaws, such as cracks, in a rotating engine disk. In this technique blade tip clearance data were acquired on an undamaged subscale engine disk using the same 10 000 rpm multi-cycle run profile that was used for the previously described vibration based technique. These baseline data were used to train three different data mining algorithms: Orca, Inductive Monitoring System (IMS), and One-Class Support Vector Machine (OCSVM) [12–15]. These algorithms were then used to analyze the blade tip clearance data that was acquired from a damaged disk which had a notch machined into it to simulate a crack in order to evaluate their performance and determine if the algorithms could detect the presence of the notch due to changes in the disks operating characteristics as measured by the blade tip clearance sensors.

As stated previously, three different techniques were investigated: Orca, Inductive Monitoring System (IMS), and One-Class Support Vector Machine (OCSVM). These methodologies provide a technique that can monitor the health of

TABLE 1: Multiple 10 000 rpm cycle test results.

	Algorithm	Correct detection rate	False alarm rate	Accuracy	Area under ROC
5% false alarm rate	Orca	100%	5%	97.55%	1.00
	IMS	93%	5%	94.02%	0.93
	OCSVM	100%	5%	97.55%	0.99
90% correct detection	Orca	90%	0.58%	94.62%	1.00
	IMS	90%	0.36%	94.73%	0.93
	OCSVM	90%	1.26%	94.28%	0.99

a system with fidelity by training the model to identify normal system parameters from abnormal ones. This is implemented by defining groups of consistent system parameter data using nominal system data. These training data are then used to model the system. Upon learning how the system should behave under nominal operating conditions, these models are then used to identify abnormal behavior, such as a crack or flaw in the disk which causes a change in the known operation of the system.

4.1. Detection Algorithms

4.1.1. Orca. Orca is an outlier detection algorithm which uses a Euclidean distance nearest neighbor based approach to determine outliers. For computational efficiency, it employs a modified pruning technique which allows it to perform in near linear time. For each point in the test data set, where a point is a row in the data set consisting of measurements taken at a single point in time, Orca calculates the nearest neighbor points from the reference data set. The output from Orca is a distance score which represents the average distance to its k-nearest neighbors; the more anomalous the point is the higher the score, since the nearest neighbors are farther away. More information about this algorithm can be found in [13].

4.1.2. Inductive Monitoring System (IMS). IMS is a cluster based modeling method. The algorithm is given a set of nominal data points and builds a model by agglomerative clustering of the data points. The resulting model is used to generate anomaly scores for new data. For each test data point, IMS finds its distance to the nearest cluster's boundary. The score that is reported is the sum of the squares of the distances from the test data point to the dimensional bounds of the closest cluster. If a data point falls entirely within the cluster bounds, the point is expected to represent normal behavior and it is assigned a score of zero. More information about IMS can be found in [12].

4.1.3. One-Class Support Vector Machine (OCSVM). OCSVM is a one-class nonlinear kernel based algorithm that maps the training data to a higher-dimensional feature space and then linearly separates nominal data from anomalies in that feature space. The idea is that such a model corresponds to a nonlinear model in the original data space but still maintains the benefit of a linear model in that it is guaranteed to return the model with the lowest error over the training set.

The algorithm identifies a subset of the training data, called the "support vectors," which is used to generate a hyperplane model. The anomaly score that is reported is the distance from the test data point to the hyperplane as measured in the feature space. More information on OCSVM can be found in [14, 15].

4.2. Analysis-Results. For the multiple-cycle test run previously shown in Figure 8 blade tip clearance data were recorded for both the undamaged and damaged, notched, disk. For the analysis, the 32 individual blade tip clearance measurements acquired for each revolution were utilized in training and evaluating the three anomaly detection algorithms. The undamaged disk's data was randomly divided in half. The first half was used for training of the algorithm and second half was used for validation. The means and standard deviations for all 32 channels were calculated for the training data and used to normalize both the training and testing data sets.

The plots of Figure 11 are the global anomaly scores for each of the algorithms for both the validation undamaged data set and the damaged data set over time. The anomaly scores are portrayed as positive values and any nominal points are represented as zero. This was done to allow for similar comparison across the algorithms. It is important to note that OCSVM allows some nominal sample points in the training set to be classified as anomalous. This percentage is governed by the Nu parameter, which is set by the user, but for which we set currently at the default value of 10%. Due to this characteristic of the algorithm 10% of all nominal data tested may result in anomalous classification. In analyzing the effectiveness of the three techniques in Table 1 the threshold for correct detection was set to 90% to make the comparisons fair between all three algorithms, since OCSVM, in this case, was optimized for a correct detection of 90%.

Figure 11 shows all three algorithms and they appear to have performed quite well in differentiating between the nominal undamaged disk and the damaged, notched disk test runs. Orca and OCSVM seem to show slightly better performance across a few of the metrics. In Table 1 the metrics used for comparison are correct detection, false alarm rate, accuracy, and area under the Receiver Operating Characteristic (ROC) curve. The ROC curve is a plot of false positives versus true positives. The ideal curve is one that has a 90° bend in it shooting straight up with the false positive = 0 and holding the true positive = 1 across resulting in an area of 1.00. When fixing the false alarm rate at 5%, Orca and

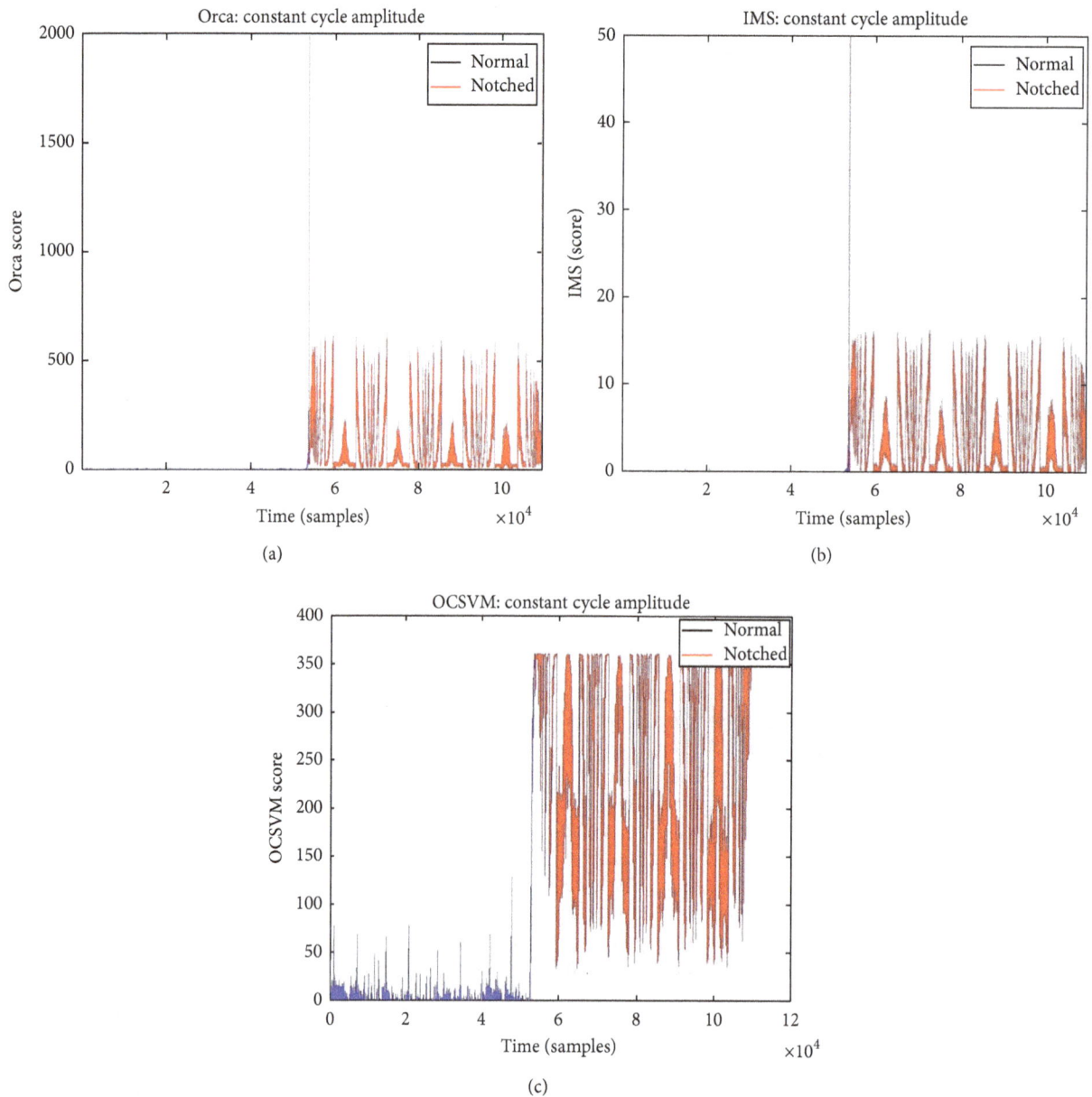

FIGURE 11: Results of the data driven anomaly detection techniques. (a) Orca. (b) IMS. (c) OCSVM.

OCSVM both have 100% correct detection rate of the notched anomalous data, where IMS's correct detection rate is at 93%. When fixing a threshold so that the correct detection rate is 90%, all three show very good false alarm rates. An additional metric, that is independent of choosing a threshold, is measuring the area under the ROC curve. When comparing this metric across the three algorithms, all three methods are reporting very good areas with Orca and OCSVM doing slightly better than IMS. The results obtained showed that the detection algorithms are capable of predicting anomalies in the rotor disk with very good accuracy. Each detection scheme performed differently under the same experimental conditions and each delivered a different level of precision in terms of detecting a fault in the rotor.

5. Conclusion and Future Plans

The work presented in this paper represents a summary overview of the on-going research activities that are being conducted at the NASA Glenn Research Center for the development of new sensors and fault detection techniques for the structural health monitoring of rotating turbine engine components. The main focus of these efforts is to first develop and validate low technology readiness level (TRL) structural health monitoring concepts and then mature them for further use on more complicated aero engine hardware.

The microwave blade tip clearance sensor technology was found to be a promising option for use in structural health monitoring applications for gas turbine engines. The sensors have been successfully demonstrated on aero engine like hardware. The first generation 5.8 GHz sensors were used on a large axial vane fan and on a NASA Turbofan for the purposes of evaluating the sensor's ability to acquire blade tip clearance data on an aero engine size test article and blades. The second generation 24 GHz sensors were used to make basic blade tip deflection and low range clearance measurements on a simulated engine and on a small aero engine's compressor disk in order to evaluate the sensor's capability of making both clearance and timing measurements on small aero engine hardware. A ground test with these sensors installed on a full scale engine is being planned for the 2013-2014 timeframe.

The vibration based crack detection experimental results showed that this technique has the potential and merit to be used for detecting cracks on a rotating disk. The data acquired from the 10 000 rpm multiple-cycle test run showed a speed-dependent rise in the vibration amplitude, indicating that a crack-induced shift in the disk's center of mass was present. Other test cases equally showed that the system was on the verge of being able to detect the simulated crack using this technique, but not all of the expected trends were consistently observed. Nevertheless, this technique looks very promising and appears to have the sensitivity to detect faults in the rotor disk system. However, additional work remains warranted to further validate and refine this crack detection concept.

Additional health monitoring was performed through combined experimental and data-driven anomaly detection techniques. Three different automatic data-driven detection algorithms, Orca, OCSVM, and IMS, were used. These techniques were utilized to analyze blade tip clearance data acquired from the rotor operating under undamaged and damaged conditions to check the viability of the detection methodology and to evaluate the performance of each approach. The results obtained showed that the detection algorithms are capable of predicting anomalies in the rotor disk with very good accuracy. Each detection scheme performed differently under the same experimental conditions and each delivered a different level of precision in terms of detecting a fault in the rotor. Overall rating showed that both the Orca and OCVSM performed better than the IMS technique, but in the end all of the algorithms showed promise and will be further refined for eventual use in detecting flaws in rotating components of an engine.

References

[1] U.S. Department of Transportation, Federal Aviation Administration, "Engine Damage-Related Propulsion System Malfunctions," DOT/FAA/AR-08/24, 2008.

[2] T. Kurtoglu, K. Leone, M. Revely, and C. Sandifer, "A study on current and emerging technologies and future research requirements for integrated vehicle health management," Internal NASA Report, 2008.

[3] S. B. Lattime and B. M. Steinetz, "Turbine engine clearance control systems: current practices and future directions," NASA TM 2002-211794, AIAA-2002-3790, 2002.

[4] T. A. Holst, T. R. Kurfess, S. A. Billington, J. L. Geisheimer, and J. L. Littles, "Development of an optical-electromagnetic model of a microwave blade tip sensor," AIAA 2005-4377, 2005.

[5] J. L. Geisheimer, S. A. Billington, and D. W. Burgess, "A microwave blade tip clearance sensor for active clearance control applications," AIAA 2004-3720, 2004.

[6] T. A. Holst, Analysis of spatial filtering in phase-based microwave measurements of turbine blade tips [M.S. thesis], Georgia Institute of Technology, Atlanta, Ga, USA, 2005.

[7] A. Abdul-Aziz, M. R. Woike, J. D. Lekki, and G. Y. Baaklini, "Health monitoring of a rotating disk using a combined analytical-experimental approach," NASA/TM 2009-215675, 2009.

[8] A. Abdul-Aziz, M. R. Woike, G. Abumeri, J. D. Lekki, and G. Y. Baakilini, "NDE using sensor based approach to propulsion health monitoring of a turbine engine disk," in Health Monitoring of Structural and Biological Systems, Proceedings of SPIE, pp. 9–12, San Diego, Calif, USA, March 2009.

[9] A. L. Gyekenyesi, J. T. Sawicki, and G. Y. Baaklini, "Vibration based crack detection in a rotating disk, part 1—an analytical study," NASA/TM 2003-212624, 2003.

[10] A. L. Gyekenyesi, J. T. Sawicki, R. E. Martin, W. C. Haase, and G. Y. Baaklini, "Vibration based crack detection in a rotating sisk, part 2—experimental results," NASA/TM 2005-212624/PART2, 2005.

[11] W. Haase and M. Drumm, "Detection, discrimination and real-time tracking of cracks in rotating disks," IEEE 0-7803-7321-X/01, 2002.

[12] D. L. Iverson, "Inductive system health monitoring," in Proceedings of the International Conference on Artificial Intelligence, Proceedings of the International Conference on Machine Learning, Models, Technologies & Applications (MLMTA '04), vol. 2, June 2004.

[13] S. D. Bay and M. Schwabacher, "Mining distance-based outliers in near linear time with randomization and a simple pruning rule," in Proceedings of the 9th ACM SIGKDD International Conference on Knowledge Discovery and Data Mining (KDD '03), pp. 29–38, Association for Computing Machinery, August 2003.

[14] D. M. J. Tax and R. P. W. Duin, "Support vector domain description," Pattern Recognition Letters, vol. 20, no. 11–13, pp. 1191–1199, 1999.

[15] B. Schölkopf, J. C. Platt, J. Shawe-Taylor, A. J. Smola, and R. C. Williamson, "Estimating the support of a high-dimensional distribution," Neural Computation, vol. 13, no. 7, pp. 1443–1471, 2001.

12

Nondestructive Damage Evaluation in Ceramic Matrix Composites for Aerospace Applications

Konstantinos G. Dassios,[1] Evangelos Z. Kordatos,[1] Dimitrios G. Aggelis,[2] and Theodore E. Matikas[1]

[1] *Department of Materials Science & Engineering, University of Ioannina, 45110 Ioannina, Greece*
[2] *Department of Mechanics of Materials and Constructions, Vrije Universiteit Brussel, Pleinlaan 2, 1050 Brussels, Belgium*

Correspondence should be addressed to Theodore E. Matikas; matikas@cc.uoi.gr

Academic Editors: D. Ouinas and K. N. Shivakumar

Infrared thermography (IRT) and acoustic emission (AE) are the two major nondestructive methodologies for evaluating damage in ceramic matrix composites (CMCs) for aerospace applications. The two techniques are applied herein to assess and monitor damage formation and evolution in a SiC-fiber reinforced CMC loaded under cyclic and fatigue loading. The paper explains how IRT and AE can be used for the assessment of the material's performance under fatigue. IRT and AE parameters are specifically used for the characterization of the complex damage mechanisms that occur during CMC fracture, and they enable the identification of the micromechanical processes that control material failure, mainly crack formation and propagation. Additionally, these nondestructive parameters help in early prediction of the residual life of the material and in establishing the fatigue limit of materials rapidly and accurately.

1. Introduction

Owing to their unique properties such as damage tolerance, fracture toughness, wear and corrosion resistance with respect to monolithic ceramics, and crack growth resistance, CMCs can withstand severe thermomechanical loading conditions [1] and are used today in many aerospace applications as braking systems, structural components, nozzles, and thermal barriers. SiC fibres are the number one candidate reinforcement for such composites as they offer high strength and modulus, thermal stability, and good mechanical performance under high temperatures [2].

The importance of monitoring the structural safety of aerospace structures is imperative. Prevention of catastrophic failure as well as safe and economical management of the structures can be achieved by early assessment of material conditions before the appearance of large-scale fracture. Regular observation of the structures for signs of damage or deterioration will enable the realization of proper repair actions which, in turn, will help extend the useful life span of the component. Among the highly sought-after nondestructive methods capable of monitoring the structural integrity of aerospace structures in an efficient and economical manner, infrared thermography and acoustic emission stand out for being fast, straightforward, and highly reliable. Today both the National Aeronautics and Space Administration (NASA) and Astrium, the European space company, use IRT and AE to detect defects in shuttle wings, rudders and tails, thruster chamber assemblies, and other composite components [3–8]. While IRT captures the thermal energy emissivity of the specimen which is directly related to the damage mechanisms that form and develop during material fracture, the idea behind AE monitoring is that any fracture incident inside a material releases energy which propagates in the form of elastic waves and can be captured at the surface of the material by appropriate sensors.

While IRT and AE have been successfully applied to detect flaws in CMCs, little information is available on their potential to capture and follow the formation of subsurface cracks. Moreover, it is extremely interesting to investigate the advantages of combined application IRT and AE and to evaluate complementary input that these two techniques can give about CMC damage.

In the present work, IRT and AE are combined to monitor the formation and development of damage during cyclic and fatigue loading of SiC-fiber reinforced barium osumilite (barium, magnesium, aluminium, and silicate (BMAS)) glass-ceramic matrix composites. IRT was used to identify the most critical, with respect to fracture, damage mechanisms as well as to monitor crack propagation under cyclic and dynamic loads and to predict the composite's residual life. State-of-the-art IR lock-in thermography was used in a unique manner to rapidly and precisely assess the fatigue limit of the CMC, using data from a single specimen test. AE parameters were very powerful in identifying and quantifying real time damage in the CMC. The significance of a large number of IRT and AE indices with respect to mechanical performance and damage evaluation is discussed and explained in the text.

2. Experimental

2.1. Materials and Mechanical Testing. SiC/BMAS laminates were provided as 3 mm thick plates. The BMAS glass matrix consisted of 50 wt% SiO_2, 28 wt% Al_2O_3, 7 wt% MgO, and 15 wt% BaO and was reinforced by SiC Tyranno fibers stacked and hot-pressed at 1200°C for 10 min in a symmetric $(0/90)_{4s}$ orientation. During hot-pressing, a chemical reaction between the fiber and the oxides of the matrix is known to result in the formation of a weak carbon-rich interphase [9] responsible for large-scale bridging and pull-out phenomena during composite fracture [10]. Rectangular specimens of dimensions $105 \times 12 \times 3$ mm^3 were prepared in a vertical CNC with fiber orientation in the external plies set to 0° with respect to the specimen's longitudinal axis. Double-edge notch (DEN) specimens of initial notch-to-width ratios of 0, 0.2, and 0.35 were prepared using a diamond wafering blade intended for cyclic tension testing as shown in Figure 1. Dogbone specimens were prepared for monotonic tensile testing as well as for fatigue loading (Figure 1).

All mechanical testing was performed at ambient temperature on an Instron 8800 servohydraulic frame equipped with a ±100 kN load cell. Specimens were gripped with a pressure of 4 MPa and were tested without end tabs at a nominal gauge length of 50 mm. Static tensile testing, both monotonic and cyclic, was performed under crosshead displacement control with a rate of 0.2 mm/min corresponding to an initial strain rate of 4.0×10^{-3} min^{-1} within the 25 mm gauge length of the external, knife-edge-mounted axial extensometer. In cyclic tension experiments with unloading/reloading loops, unloading commenced at 10^{-3} strain and repetitions occurred with a step of 1.5×10^{-3} strain. The composites were unloaded to full relaxation before reloading.

Fatigue step loading until fracture was conducted on dogbone specimens. The first loading step was set to 10% σ_{ULT} and endured for 6000 cycles. The commencing load level was chosen to be low in order to capture the whole mechanical response of the material. The subsequent four loadings, up to 60% σ_{ULT}, occurred with a step of 10% σ_{ULT}. At 60% σ_{ULT} the fatigue loading step was decreased to 5% σ_{ULT} and remained

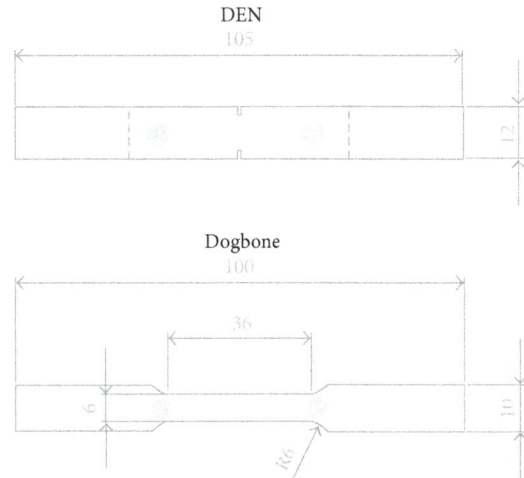

FIGURE 1: Double-edge-notch and dogbone specimen configurations with marked AE monitoring locations (grey circles).

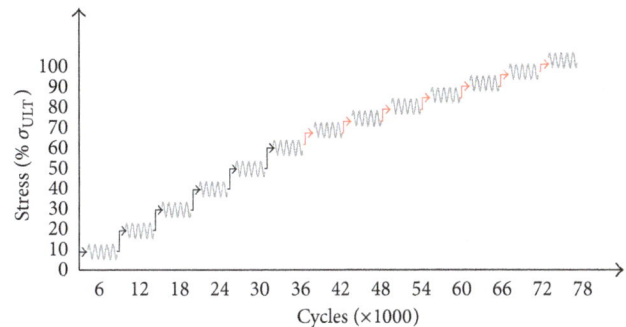

FIGURE 2: Schematic of fatigue loading protocol.

such until specimen fracture. The frequency of the sinusoidal fatigue load was 10 Hz and R was 0.1. The fatigue loading protocol is represented schematically in Figure 2.

2.2. Acoustic Emission Monitoring. Acoustic emission activity was monitored throughout mechanical testing on all specimens using two "Pico" microminiature AE sensors tape-mounted at a separation of 40 mm on the central part of the specimen (Figure 1). In static tests (monotonic and cyclic tension) the sensors were mounted on the same side of the specimen as the extensometer. This manner of mounting always provided an instrument-free face on the specimen that could be monitored by IRT in real time. Acoustic coupling between the sensors and the specimen was provided by application of silicon grease. The broadband frequency response of the sensors, 50–800 kHz, enabled signal acquisition from a wide range of damage mechanisms. On the other hand, their small size facilitated geometric location of event sources along the specimen. AE sampling was conducted on a PCI-2 board (Physical Acoustics Corporation, Princeton, NJ, USA) with a sampling rate of 5 MHz, an amplification of 40 dB, and a threshold of 45 dB that enabled exclusion of ambient noise from the recorded signal.

2.3. Thermography. Throughout testing, temperature variations due to the applied loading, on the AE/extensometer-free face of the specimen, were monitored by an infrared thermography camera (CEDIP, MIW). The camera featured a cooled indium antimonide (InSb) detector (3–5 μm), a focal plane array (FPA) with pixel format of 320 (H) × 240 (V), and a temperature sensitivity of 20 mK. Temperature was recorded with a sampling rate of 100 Hz. Aliasing was avoided by recording the baseline emissivity of the material prior to load application by capturing the IR fingerprint of the surface with the thermal camera.

Throughout cyclic loading of DEN specimens of the SiC/BMAS composite, thermographs were recorded from the high-stress concentration area between the notches. For lock-in thermography measurements during fatigue loading, specimens were spray-coated with a matte black varnish in order to achieve uniform high-level surface emissivity. Optimal field of view (FOV) conditions were achieved by positioning the camera at approximately 40 cm in front of the gripped specimen. The IR camera was connected to the lock-in amplifier which, in turn, was connected to the servohydraulic controller. This enabled the synchronization of the lock-in amplifier and the testing machine frequencies and capturing of lock-in images and data during fatigue loading. The IR camera was used to measure the amount of energy emitted as infrared radiation, which is a function of the temperature and emissivity of the specimen. According to a previous study, the measured energy corresponds to the intrinsically dissipated energy while the fatigue limit is located at the break of the intrinsic dissipation regime of the loaded specimen [11].

3. Results and Discussion

3.1. Mechanical Response under Static Tension. The stress-strain response of notched and dogbone SiC/BMAS specimens under cyclic and monotonic tension, respectively, is presented in Figure 3. Unnotched specimens exhibited a triple regime behavior consisting of a linear initial part followed by a regime of gradually decreasing tangent modulus and a final regime of apparent stiffening. The second regime (regime "II") is associated mainly with interfacial damage, most importantly interfacial debonding but also with progressive matrix cracking evidenced as decreasing material stiffness (average slope of unloading/reloading loops). In the third regime (regime "III") an increase in material stiffness and tangent modulus coupled with an almost linear stress-strain relationship are apparent. In this ultimate regime, the mechanisms of interfacial debonding and matrix cracking have reached a saturated state; hence, material damage is not governed by the interface or matrix anymore; but by a mechanism of superior strength, essentially load bearing by intact fibers [12]. Similar triple regime phenomena with prefailure macroscopic stiffening and linear stress-strain relationships have been encountered before [13]. As observed in the curves of Figure 3(b), Regime III is absent from the mechanical behaviors of notched specimens. This is probably due to premature fiber—hence also composite—stemming from stress concentration in the vicinity of the notch roots.

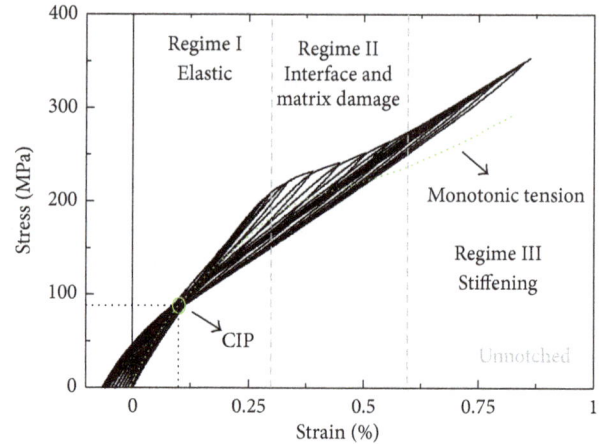

(a)

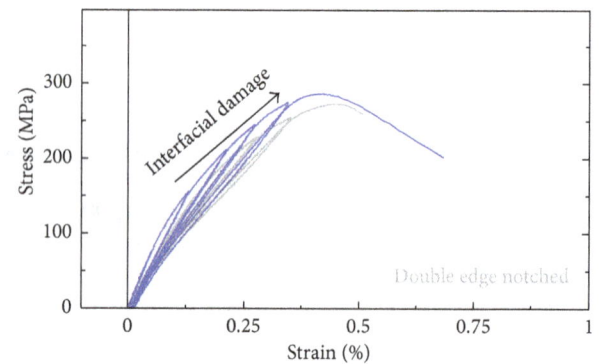

Notch ratio 0.35
Notch ratio 0.2

(b)

FIGURE 3: Stress-strain response of SiC/BMAS under monotonic loading (dotted line in (a)) and cyclic tension with unloading/reloading loops for (a) un-notched specimens and (b) notched specimens with various notch lengths.

In notched specimens instead, the material failed soon after the maximum load was attained, giving minimal "tail" effects.

Most importantly, the stress-strain curves of unnotched specimens are defined with unique precision, a common intersection point (CIP) of unloading-reloading curves in the first quadrant of the stress-strain curve in the tension domain. The coordinates of the CIP, 0.001 strain and 90 MPa stress, are directly related to the axial residual stress state of the composite [14, 15]. While a thorough analysis of the CIP feature for the particular composite has been the subject of a previous work [16], it is interesting to repeat here that a self-assembled CIP had never before been encountered experimentally.

Comparing the monotonic and cyclic tension curves for the SiC/BMAS composite (Figure 3(a)), it can be concluded that cyclic loading results in an increase by 20% in attainable material stress, calculated at fracture. If this increase is due to higher amounts of energy dissipated at damage mechanisms such as interfacial debonding, matrix cracking, and load bearing by intact fibers [10], it is then suggested that cyclic

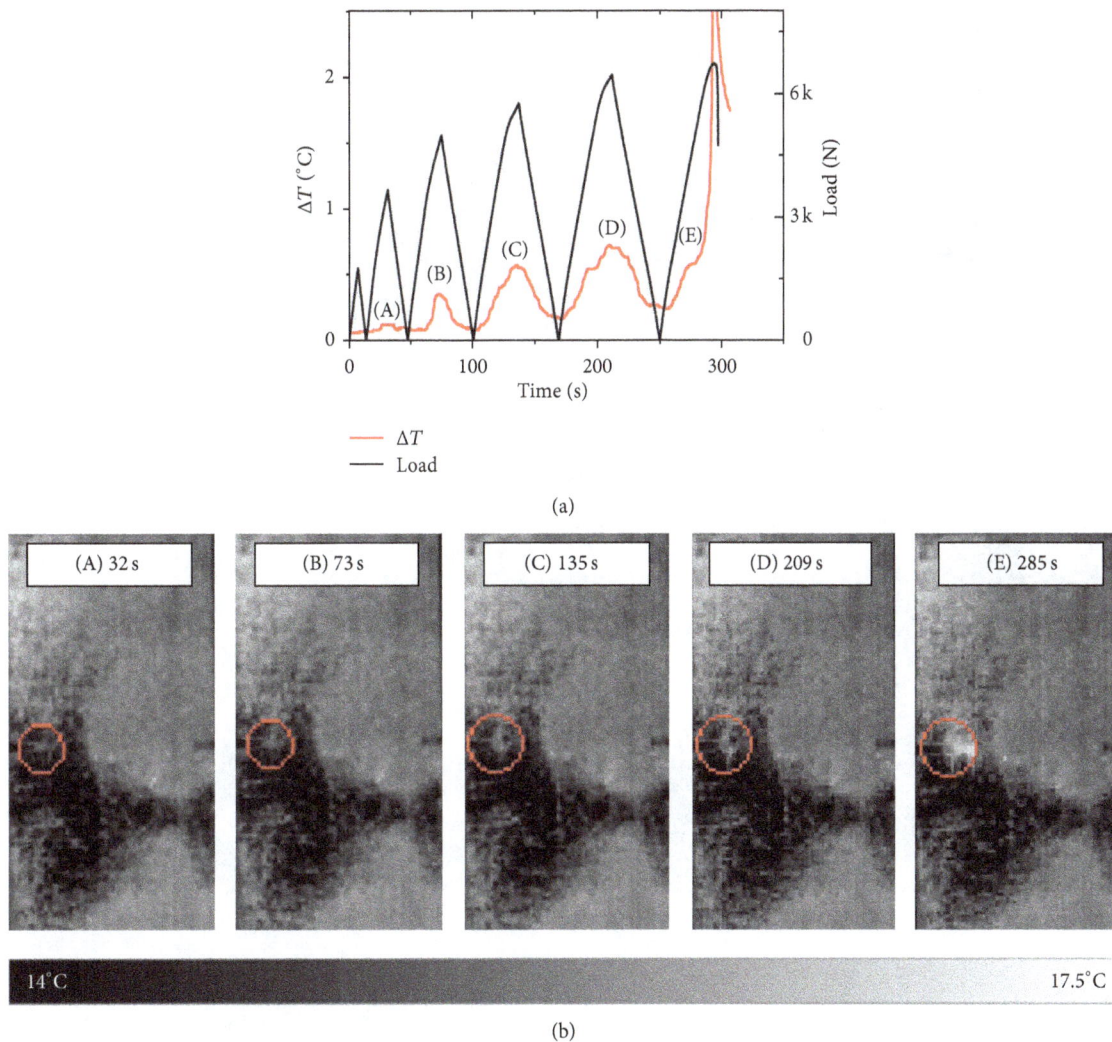

FIGURE 4: (a) ΔT and load versus time for a 0.35 notch-to-width ratio specimen loaded in cyclic tension. Sequential alphabet letters indicate instance of thermographs presented in (b).

loading by itself improves the energy dissipation capacity of the material. The existence of another energy dissipation mechanism, pull out, anticipated by the weak interfacial bond discussed in the experimental section, was verified after the end of the tests: failed specimens had not separated in two pieces after removal from the grips, with the frame still indicating small load values of the order of a few Newtons. This meant that fibers had failed within the matrix environment and had pulled out noncompletely before removal of the specimens from the grips [17].

Composite strength and modulus appeared to increase with decreasing notch length. Unnotched specimens enjoyed average strengths and moduli of 355 MPa and 151 GPa, respectively. The corresponding values for the 0.2 and 0.35 notched-to-width length were 280 MPa/119 GPa and 270 MPa/108 GPa, respectively.

3.2. Thermography

3.2.1. Static/Cyclic Loading. Temperature variation as measured by IRT, ΔT, and load is shown in Figure 4(a) as a function of time for a DEN specimen with a 0.35 notch-to-width ratio. Indices "A" to "E" denote the instances of the thermographs shown in Figure 4(b), collected at the notched ligaments of the composites. It is observed that peak ΔT location coincides with the location of maximum load for every cycle, whereas peak ΔT magnitude increases with progressing loading, hence also material damage. At the ultimate cycle, the ΔT trace appears to follow a completely different pattern than those in previous cycles, wherein temperature appears to drastically increase, indicating that the specimen is heading for catastrophic fracture.

In Figure 4(b), the locations of crack initiation, as identified by IRT, are indicated by a red circle mark. The apparent high-temperature area located outside and to the right of the circle mark is a baseline pattern that exists even before load application and remains constant until catastrophic failure. It is associated with the specimen's surface emissivity, not with material damage. It should not be ignored that IRT is concerned with temperature *variations* as a result of progressive damage, not with *absolute* values. Under this rationale,

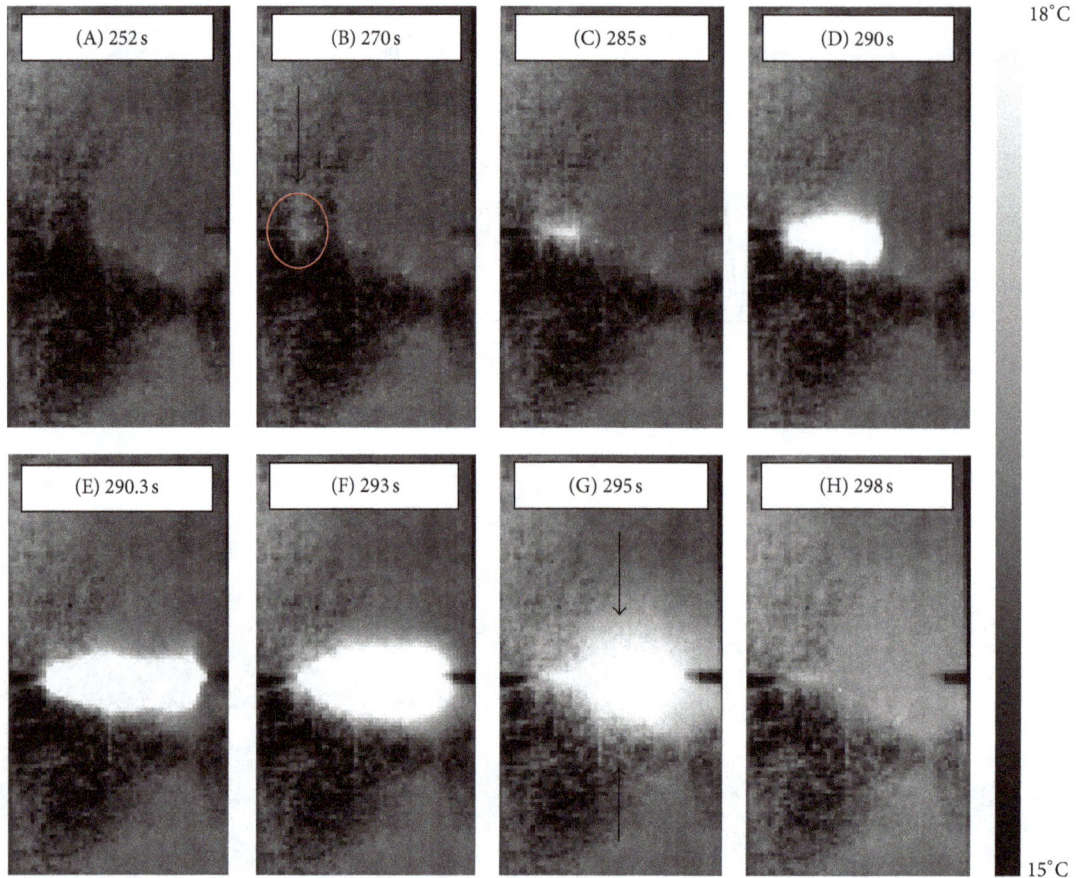

(a)

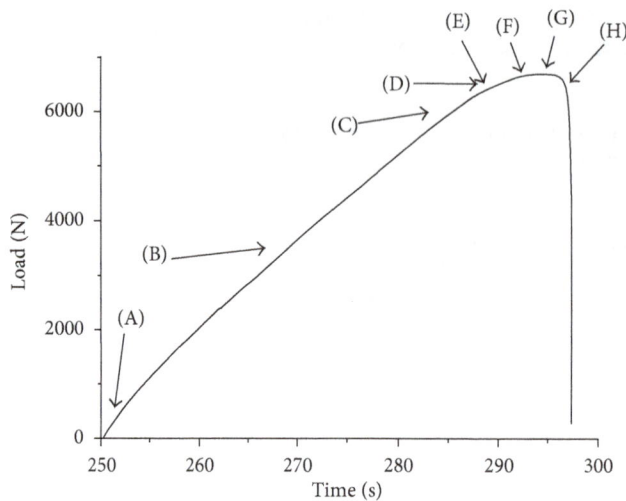

(b)

FIGURE 5: Final loading cycle of 0.35 notch-to-width ratio DEN specimen. (a) Thermographs showing crack propagation and (b) load versus time curve.

no noticeable change in temperature is seen up to 32 sec experimental time (thermograph (A) of Figure 4(b)). 73 sec within testing (thermograph (B) of Figure 4(b)), very small temperature variations can be observed within the marked

(circle) area. Temperature increases become more obvious in thermographs (C) and (D) of Figure 4(b), 135 and 209 into loading, respectively. It is indicated that the damage is extending in area and magnitude. In the last loading cycle, 285 sec

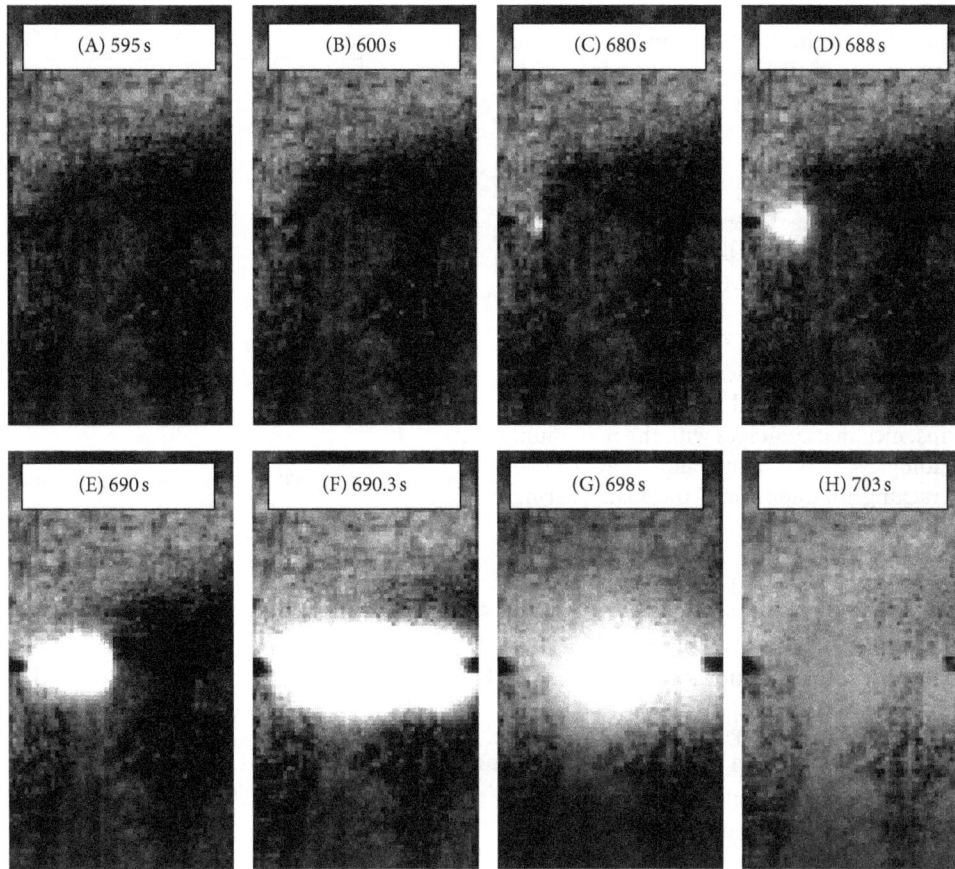

(a)

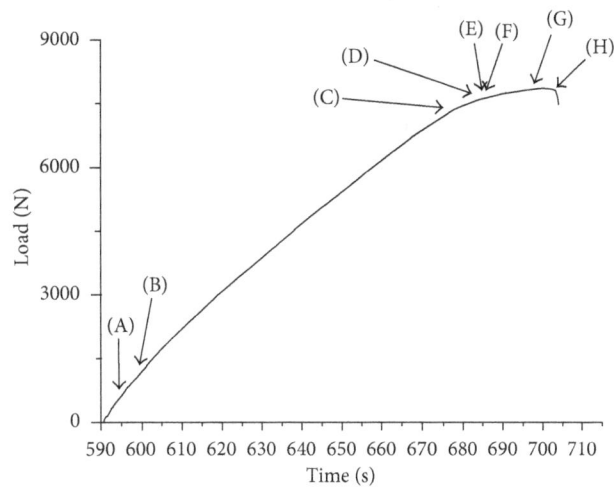

(b)

FIGURE 6: ((a), (b)) Thermographs of crack propagation and diagram of load versus time of final loading cycle (0.2 notch-to-width ratio specimen).

in the test, a dramatic increase in temperature throughout the whole notched ligament signifies that material failure is imminent.

The thermographic behaviour within the ultimate loading cycle, of the same 0.35 notch-to-width ratio specimen, is demonstrated in Figure 5 and analyzed in the following.

Indices "A" to "H" presented in Figure 5(b) define the instances of the thermographs in Figure 5(a) (note the notch roots). In thermograph (A) of Figure 5(a), no hot appears exist within the notched ligament. In the next instance, thermograph (B) of Figure 5(a), a red arrow indicated what appears to be crack initiation. At 285 sec, thermograph

(C) of Figure 5(a), a significant temperature difference is observed, which coincides with the change in the slope of the mechanical response curve. It is believed that from this instance on, subsurface crack starts propagating from the left notch root with direction to the right. It is important to establish this instance as precisely as possible, as this will facilitate early prediction of the final fracture. It is noted that the associated time (285 sec) corresponds to 73% of the total duration of this ultimate loading cycle. 5 sec later (thermograph (D) of Figure 5(a) at 290 sec) the subsurface crack appears to span half of the notched ligament while only another 300 msec later (thermograph (E) of Figure 5(a)) it propagates abruptly and unstably towards the right notch. The maximum temperature is attained (thermograph (F) of Figure 5(a)) at an instance that coincides with the maximum load of the final loading cycle. This temperature is associated with the matrix cracking saturation and the load bearing completely by the reinforcing fibers. Temperature starts decreasing at the left notch in the next thermograph, (G) of Figure 5(a), while it increases above and under the subsurface crack, as indicated by black arrows. This increase is due to the fiber failure under the critical level of applied load. Failed fibers pull out giving rise to the frictional thermal energy evidenced in thermograph (H) of Figure 5(a).

Similar trends were observed for DEN composites with smaller notches, as in specimens with 0.2 notch-to-width ratios; the thermographic behaviour within the ultimate loading cycle of such a specimen is demonstrated in Figure 6 and analyzed in the following. Again, indices "A" to "H" presented in Figure 6(b) define the instances of the thermographs in Figure 6(a). After 600 sec of testing time, Figure 6(a)(B), no "warm" damage areas are seen in the thermographs. Crack initiation appears at 680 sec, Figure 6(a)(C), which compares favorably with the instance of slope change in mechanical behavior of the material. It is hence possible to foresee early fracture still at 80% of the final cycle duration. In the subsequent thermographs Figure 6(a)(D) and (E), the subsurface crack propagates from the left notch towards the middle of the notched ligament. The crack then propagates abruptly and unstably towards the right notch. A temperature variation profile compatible with pull out is seen again in the last thermograph, Figure 6(a)(H), where the specimen has failed completely.

Peak ΔT is shown in Figure 7 for the two notch lengths used in the current study. It is observed that specimens with shorter notches exhibit ΔT of 15°C at fracture while the ones with longer notches exhibit ΔT values around 10°C. It is believed that in specimens with larger notched ligaments (smaller notches), damage evolves over a wider material region throughout testing; hence peak temperature at the critical load is not high. On the other hand, damage is accumulated and relieved not so drastically in a specimen with less material available within the notched region.

3.2.2. Fatigue.
Lock-in thermography was applied during fatigue loading of SiC/BMAS dogbone specimen. The intrinsically dissipated energy as monitored by the IR camera for 10 different stress levels ranging from 30% to 90% σ_{ULT} is plotted as a function of % σ_{ULT} in Figure 8.

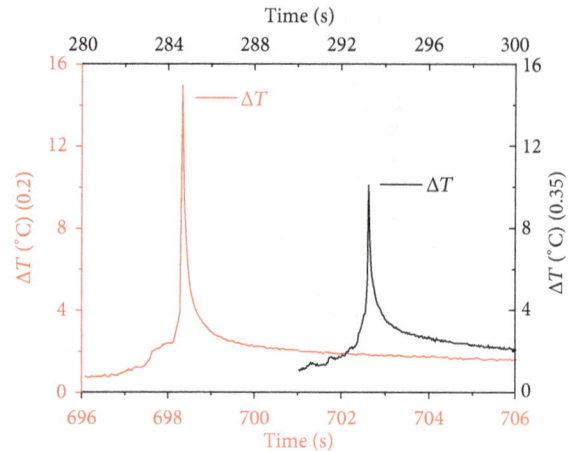

FIGURE 7: Peak ΔT for the two notch lengths used in the DEN specimens.

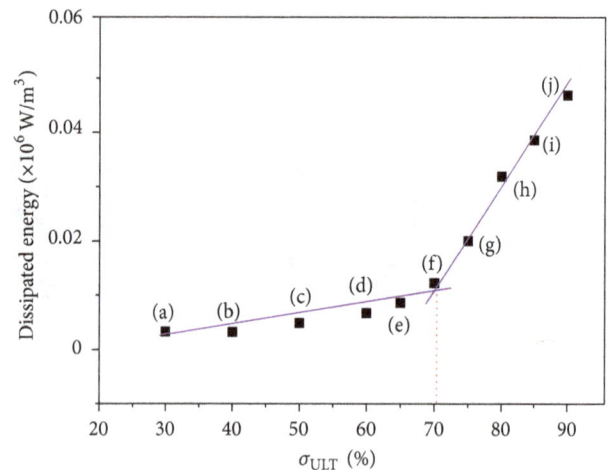

FIGURE 8: Dissipated energy versus % σ_{ULT} for fatigue-loaded SiC/BMAS.

The curve exhibits two distinct slopes, visualized by the two linear regressions seen in Figure 8. At low stress levels, 30% to 60% σ_{ULT}, the dissipated energy increases with a low rate, while from 70% σ_{ULT} and upwards it rises considerably more rapidly. The intersection point of the two lines defines the fatigue limit of the material. The value of fatigue limit calculated for the cross-ply SiC/BMAS through the thermographic approach of this study is 70% σ_{ULT} or 205 MPa.

An examination of the thermographic pattern of cross-ply SiC/BMAS at different stress levels is of particular interest in view of the established fatigue limit value. This information is presented in Figures 9(a)–9(j), wherein the 70% σ_{ULT} fatigue level which corresponds to thermograph Figure 9(f). Two distinct cases are made obvious by examination of this figure: (i) thermographs Figures 9(a)–9(d) depict low energies in cold (blue) color coding in the initial loading stages associated with minimal material damage and (ii) thermographs Figures 9(e)–9(j) capture progressive damage accumulation which is captured by increasingly warmer colors (high energy). In the first four thermographs, up to

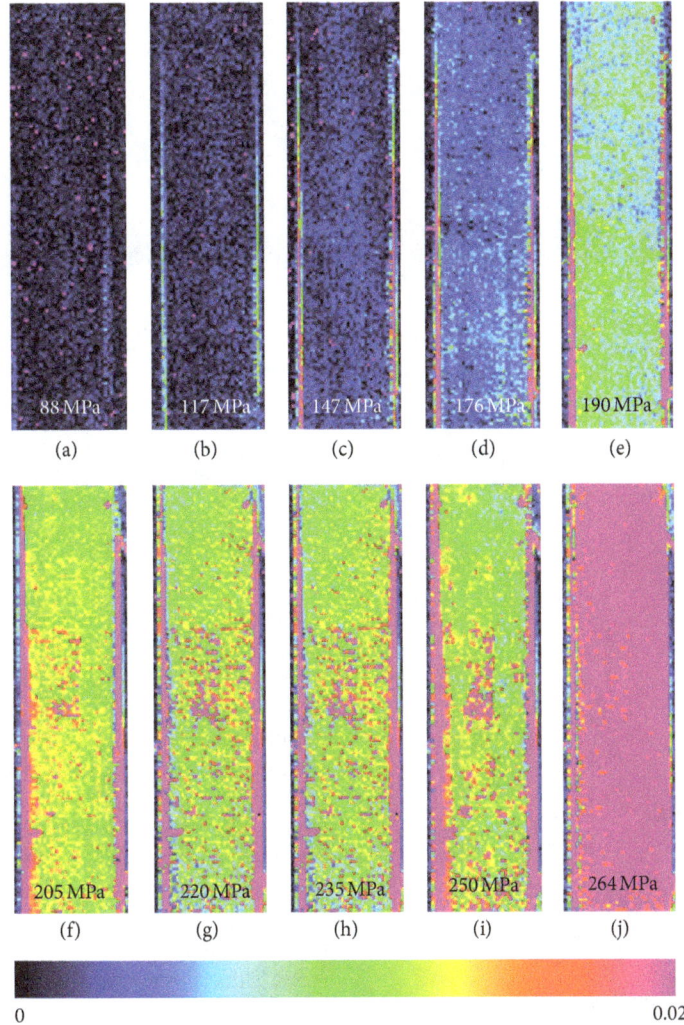

FIGURE 9: Thermographic pattern during fatigue loading of a SiC/BMAS composite.

176 MPa applied stress, there is practically no appreciable change in the dissipated energy. In the fourth thermograph, Figure 9(e), a slight change in color can be attributed to the saturation of elastic energy accumulation on the onset of appearance of fatigue. A totally dissimilar energy distribution pattern appears in thermograph Figure 9(f) due to the unfolding of internal energy dissipation phenomena such as interfacial damage, delamination, and fiber sliding across the debonded interface [18]. At thermographs Figures 9(g)–9(j) (75%–90% of σ_{ULT}), a raise in energy can be noticed indicated by the increment of the magenta spots until fracture. At the particular loading level, this energy can be attributed to fiber bridging, fiber failure, and pull out.

3.3. Acoustic Emission

3.3.1. Static/Cyclic Loading. Cumulative AE signal history collected during cyclic loading of a DEN specimen with 0.35 notch-to-width ratio is shown in Figure 10 alongside with strain. The rate of AE acquisition exhibits fluctuations according to the cyclic loading protocol. Specifically, AE rate increases as the load increases to the maximum within

each cycle. After these maxima points, the AE rate decreases without however being completely eliminated, at the cycle's minimum load. The total activity was of the order of 4000 signals.

Apart from the cumulative activity, which counts the separate acquisitions of the sensors, different AE descriptors help to distinguish the severity of the condition according to loading level. Two of them are the ASL and RMS. ASL is the average signal level defined as the average amplitude of samples of the rectified waveform while RMS is the square root of the average of the squares of all points of a waveform (root mean square) [19].

They are given by

$$\text{ASL (Average Signal Level): ASL} = \frac{1}{n}\left(x_1 + x_2 + \cdots + x_n\right)$$

$$\text{RMS (root mean square, } X_{\text{rms}}\text{): } X_{\text{rms}}$$

$$= \sqrt{\frac{1}{n}\left(x_1^2 + x_2^2 + \cdots + x_n^2\right)},$$

$$(1)$$

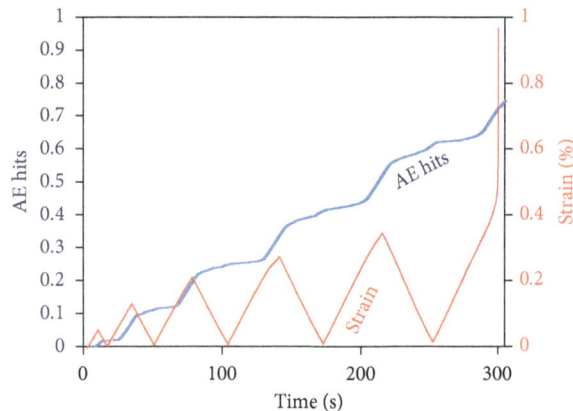

FIGURE 10: Strain and AE cumulative history for specimen B.

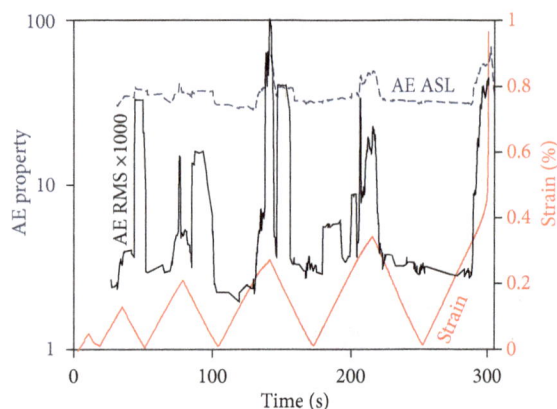

FIGURE 12: Strain history and AE peak frequency for a double-edge notched specimen with notch-to-width ratio of 0.35. The solid line is the sliding average of recent 20 points.

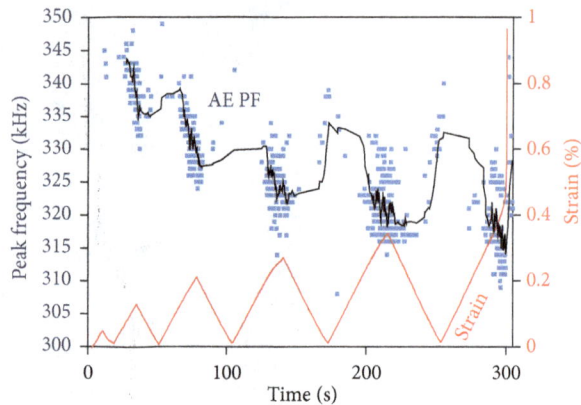

FIGURE 11: Strain history and AE amplitude parameters for a double-edge notched specimen with notch-to-width ratio of 0.35.

where n is the number of samples (waveform points) and x_i stands for the rectified amplitude of the waveform samples.

Therefore, both of them are indicative of the AE signal emitted by the fracture. Figure 11 shows the sliding average (window of 20 points) for both AE parameters. Focusing on the ASL first, local peaks are indicated at the moments of local maxima of strain, especially for the last three cycles. This means that at those periods of high strain, stronger fracture incidents take place which are recorded as waveforms with different intensity characteristics. The same holds for the RMS which again gives a measure of the elastic energy recorded by the sensor being proportional of the pressure wave that impinged on the sensor's surface. The values of RMS were multiplied by a thousand for visibility reasons in the graph. These parameters show that monitoring of AE and specific qualitative AE features allows the recording of the high stressing moments and the whole loading history until the material is brought to failure.

Apart from the amplitude or energy-related parameters, significant information can be derived by the frequency content of the emitted waves. In mechanical materials it has been shown that a drop of frequency indicates increase of

damage accumulation and is linked with the shift between fracture modes (e.g., initial tensile matrix cracking to ultimate shearing) [20, 21]. The index that is displayed here is peak frequency (PF) and is the frequency of the highest peak of the FFT of the AE waveforms. Using the specific AE sensors typical peak frequencies average around 450 kHz which is the maximum sensitivity of the sensors, while different bands from 100 kHz up to approximately 800 kHz can also be recorded. Figure 12 focuses on a specific family of emissions which initially exhibit frequencies between 330 and 350 kHz. It can be easily seen that this group of emissions are registered near the maxima of the strain cycles. However, one certain trend is that the main frequency of this family steadily drops in frequency after each cycle. This drops is of the order of 5 to 10 kHz in each cycle resulting in an average of 315 kHz at the last cycle just before failure. This drop of frequency is a result of the shifting between fracture mechanisms within the material since the monitoring conditions (sensors, separation distance) were constant throughout the experiment duration.

The observed behavior can be due to the increasing number of interfacial F that is expected to happen at the higher strain levels of the cycles. This allegation is supported by visual evidence of fiber bundle sliding and existence of off-axis layers seen in the postmortem side view of a specimen's notched ligament, microphotograph in Figure 13. It is observed that crack opening is approximately 500 μm. It is also observed that fibers are bridging the crack sides being pulled out of the matrix. While rupture of a large number of individual fibers is evident, fiber bundles have not failed completely since the specimen was removed in one piece after the test.

3.3.2. Fatigue Loading. During fatigue loading continuous monitoring by AE was applied, as mentioned earlier. Apart from the activity (number of emissions) all different AE parameters were recorded. As the fatigue life proceeds and damage is being accumulated, apart from the larger number of signals which are emitted, the nature (waveform shape) of the AE incidents starts to change. One of the indicative AE

FIGURE 13: Stereoscope image of the side of a fractured specimen.

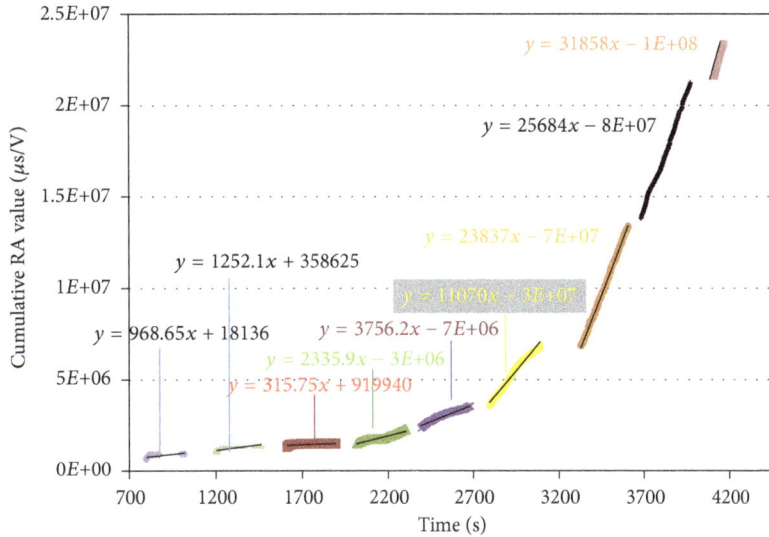

FIGURE 14: Cumulative RA value for the different fatigue loading stages.

descriptors is the RA value which is the inverse of the rising angle of the waveform and is given by

$$RA = \frac{RT}{A}, \qquad (2)$$

where RT is the rise time of the waveform (or delay between the onset and the highest peak in μs) and A is the amplitude of the highest peak in V. In general this feature increases as damage develops along with the shift of the failure mechanisms from matrix cracking to shearing phenomena like interfacial sliding and debonding [18, 19]. Therefore, RA values are progressively increasing until the final failure of the specimen. Figure 14 shows the cumulative RA value for the different stages of fatigue loading. As mentioned above, at the early stages, RA values accumulation is low. At the block of loading starting at approximately 2000 s (70% of UTS) the RA values seem to accumulate at a higher rate, which is increasing for each successive step.

The rate of RA accumulation is shown in Figure 15. This allows a more clear visualization of the trends between the AE and RA at different stages. It is concluded that the curve can be fitted by two straight lines. The change of slope between the two lines occurs below 75% load which is a certain indication that significant changes take place after that load level. At lower loads, the events emit notably lower RA values showing

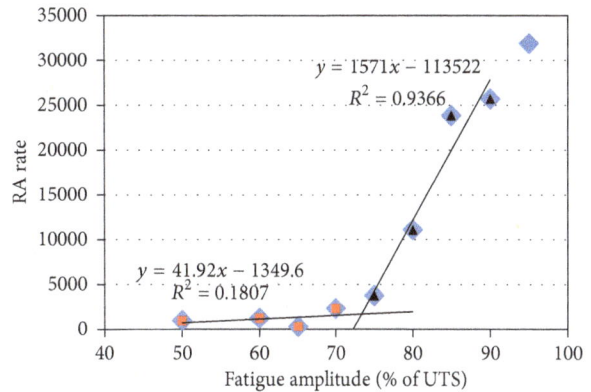

FIGURE 15: RA increasing rate for different fatigue loading amplitudes.

less intensity compared to levels higher than 75%. The intersection of the two lines is between 70% and 75% and coincides well with the change of slope in the corresponding thermal dissipation curve presented earlier. This shows that both techniques can register small but delicate changes of the failure processes, depending on the fatigue loading. It seems that although incidents that, emit acoustic emission and heat are induced by low loading, these exhibit distinct values from corresponding events at loads higher than the fatigue limit which eventually will lead to the failure of the specimen.

4. Conclusions

Thermography and acoustic emission were used to capture the initiation and evolution of damage in SiC-fiber reinforced glass-ceramic matrix composites under static and fatigue loading. Infrared thermography results helped identify the intact fiber population as the mechanism that control ultimate material failure and that under the presence of notches the composite fails shortly after the attainment of a saturated matrix cracking state. Infrared thermography (IRT) was also used to monitor, both in location and in time, the crack propagation path during mechanical testing in cyclic tension and fatigue. The technique also enabled early prediction of the residual life of the material, as early as at 73% of the duration of the final loading cycle. Successful application of the technique under such dynamic conditions where the surface changes with usage is close to real-life scenarios found in aerospace applications.

A novel infrared lock-in thermographic methodology was used for the determination of the fatigue limit of the ceramic matrix composites (CMCs). The limit was unconventionally rapidly assessed by the thermographic technique at 70% σ_{ULT} (i.e., 205 MPa). The outcome makes lock-in IRT a new, versatile, and accurate method that overcomes the limitations of Wöhler's curve approach, as it significantly reduces experimental time and requires testing of a single sample only for obtaining the fatigue limit of the material.

Furthermore, acoustic emission (AE) monitoring enables monitoring fracture behavior in real time. Apart from the increase of AE acquisition for higher load and damage accumulation, energy- and frequency-related parameters help discern the moments higher stress. Descriptors like the root mean square (RMS) and average signal level (ASL) increase their values at high stresses, while peak frequency shows the inverse trend, being continuously downgraded for the successive loading steps. Concerning fatigue, AE showed the capability of detecting the different intensity of the fracture incidents. Waveform shape parameters like the RA exhibit changes as the load increases above the fatigue level of the material. The results are benchmarked by the heat dissipation curves offered by thermography and allow the determination of the fatigue limit of the material by using only one specimen.

References

[1] A. G. Evans, "The mechanical performance of fiber-reinforced ceramic matrix composites," *Materials Science and Engineering A*, vol. 107, pp. 227–239, 1989.

[2] J. J. Brennan and K. M. Prewo, "Silicon carbide fibre reinforced glass-ceramic matrix composites exhibiting high strength and toughness," *Journal of Materials Science*, vol. 17, no. 8, pp. 2371–2383, 1982.

[3] J. G. Sun, M. J. Verrilli, R. Stephan, T. R. Barnett, and G. Ojard, "Nondestructive Evaluation of Ceramic Matrix Composite Combustor Components," NASA/TM-2003-212014, April 2003.

[4] E. I. Madaras, W. P. Winfree, W. H. Prosser, R. A. Wincheski, and K. E. Cramer, "Nondestructive evaluation for the space shuttle's wing leading edge," in *Proceedings of the 41st AIAA/ASME/SAE/ASEE Joint Propulsion Conference and Exhibit*, July 2005, AIAA 2005-3630.

[5] W. P. Winfree, E. I. Madaras, K. E. Cramer et al., "NASA langley inspection of rudder and composite tail of American Airlines Flight 587," in *Proceedings of the 46th AIAA/ASME/ASCE/AHS/ASC Structures, Structural Dynamics and Materials Conference*, pp. 5636–5644, April 2005, AIAA 2005-2253.

[6] H. Mei, Y. Xu, L. Cheng, and L. Zhang, "Nondestructive evaluation and mechanical characterization of a defect-embedded ceramic matrix composite laminate," *International Journal of Applied Ceramic Technology*, vol. 4, no. 4, pp. 378–386, 2007.

[7] J. G. Sun, C. M. Deemer, W. A. Ellingson, and J. Wheeler, "NDT technologies for ceramic matrix composites: oxide and nonoxide," *Materials Evaluation*, vol. 64, no. 1, pp. 52–60, 2006.

[8] F. Levallois, A. Sobeczko, A. Proust, D. Marlot, and J.-C. Lenain, "Non-destructive testing of GAIA frame by means of acoustic emission monitoring during launch simulation tests," in *Proceedings of the 30th European Conference on Acoustic Emission Testing & 7th International Conference on Acoustic Emission*, University of Granada, Granada, Spain, 2012.

[9] P. M. Benson, K. E. Spear, and G. C. Pantano, "Interfacial characterisation of glass matrix/Nicalon SiC fiber composites: a thermodynamic approach," *Ceramic Engineering and Science Proceedings*, vol. 9, pp. 663–670, 1988.

[10] K. G. Dassios, "A review of the pull-out mechanism in the fracture of brittle-matrix fibre-reinforced composites," *Advanced Composites Letters*, vol. 16, no. 1, pp. 17–24, 2007.

[11] M. P. Luong, "Fatigue limit evaluation of metals using an infrared thermographic technique," *Mechanics of Materials*, vol. 28, no. 1–4, pp. 155–163, 1998.

[12] K. G. Dassios, D. G. Aggelis, E. Z. Kordatos, and T. E. Matikas, "Cyclic loading of a SiC-fiber reinforced ceramic matrix composite reveals damage mechanisms and thermal residual stress state," *Composites Part A*, vol. 44, pp. 105–113, 2013.

[13] C. Cady, F. E. Heredia, and A. G. Evans, "In-plane mechanical properties of several ceramic-matrix composites," *Journal of the American Ceramic Society*, vol. 78, no. 8, pp. 2065–2078, 1995.

[14] G. Camus, L. Guillaumat, and S. Baste, "Development of damage in a 2D woven C/SiC composite under mechanical loading: I. Mechanical characterization," *Composites Science and Technology*, vol. 56, no. 12, pp. 1363–1372, 1996.

[15] M. Steen, "Tensile mastercurve of ceramic matrix composites: significance and implications for modelling," *Materials Science and Engineering A*, vol. 250, no. 2, pp. 241–248, 1998.

[16] K. G. M. Dassios and T. E. Matikas, "Residual stress-related common intersection points in the mechanical behavior of ceramic matrix composites undergoing cyclic loading," *Experimental Mechanics*, vol. 53, no. 6, pp. 1033–1038, 2013.

[17] K. G. Dassios and T. E. Matikas, "Large-scale interfacial damage and residual stresses in a glass-ceramic matrix composite," *Composite Interfaces*, vol. 19, no. 8, pp. 523–531, 2012.

[18] C. Koimtzoglou, K. G. Dassios, and C. Galiotis, "Effect of fatigue on the interface integrity of unidirectional Cf-reinforced epoxy resin composites," *Acta Materialia*, vol. 57, no. 9, pp. 2800–2811, 2009.

[19] W. Kaewwaewnoi, A. Prateepasen, and P. Kaewtrakulpong, "Measurement of valve leakage rate using acoustic emission," in

Proceedings of the International Conference on Electrical Engineering/Electronics, Computer, Telecommunications, and Information Technology (ECTI '05), pp. 597–600, Pattaya, Thailand, May 2005.

[20] M. Ohtsu, "Recommendation of RILEM TC 212-ACD: acoustic emission and related NDE techniques for crack detection and damage evaluation in concrete: test method for classification of active cracks in concrete structures by acoustic emission," *Materials and Structures*, vol. 43, no. 9, pp. 1187–1189, 2010.

[21] D. G. Aggelis, S. Verbruggen, E. Tsangouri, T. Tysmans, and D. van Hemelrijck, "Characterization of mechanical performance of concrete beams with external reinforcement by acoustic emission and digital image correlation," *Construction and Building Materials*, vol. 47, pp. 1037–1045, 2013.

13

Investigation of the Effects of Length to Depth Ratio on Open Supersonic Cavities Using CFD and Proper Orthogonal Decomposition

Ibrahim Yilmaz, Ece Ayli, and Selin Aradag

Department of Mechanical Engineering, TOBB University of Economics and Technology, Sogutozu Cad., No. 43, 06560 Ankara, Turkey

Correspondence should be addressed to Selin Aradag; selinaradag@gmail.com

Academic Editors: A. Hadjadj and E. E. Imrak

Simulations of supersonic turbulent flow over an open rectangular cavity are performed to observe the effects of length to depth ratio (L/D) of the cavity on the flow structure. Two-dimensional compressible time-dependent Reynolds-averaged Navier-Stokes equations with k-ω turbulence model are solved. A reduced order modeling approach, Proper Orthogonal Decomposition (POD) method, is used to further analyze the flow. Results are obtained for cavities with several L/D ratios at a Mach number of 1.5. Mostly, sound pressure levels (SPL) are used for comparison. After a reduced order modeling approach, the number of modes necessary to represent the systems is observed for each case. The necessary minimum number of modes to define the system increases as the flow becomes more complex with the increase in the L/D ratio. This study provides a basis for the control of flow over supersonic open cavities by providing a reduced order model for flow control, and it also gives an insight to cavity flow physics by comparing several simulation results with different length to depth ratios.

1. Introduction

In several flow applications, especially for aerospace industry, unsteady, turbulent, and complex flow phenomenon becomes an important part of processes. In aeronautics applications, interior storage carriages, which are used to carry items such as weapons and bombs, are all cavity configurations. As high speed flows pass over cavities, a complex and unsteady flow field emerges in the cavity region. These flow fields lead to pressure fluctuations and relatively high sound pressure levels. Due to the pressure fluctuations and resonant acoustic modes, the flow passing over the cavity can damage the structure of air vehicles and impede successful store release as discussed by Aradag [1].

The importance of the cavity flow mechanism resulted in many studies including some definitions and classifications about cavity flow. Due to the complexity of the flow mechanism in different cavity configurations, cavity flows are categorized based on mainly geometric specifications (L/D ratio, L/W ratio), cavity flow phenomena, and Mach number as discussed by Syed [2]. Geometric specifications of the

cavity region affect the flow propagation in the cavity. When the flow is in the shallow cavity and L/D ratio is greater than 13, it is called closed cavity, whereas if the flow is in deep cavity and L/D ratio is smaller than 10, it is called open cavity configuration as discussed by Aradag [1]. As given in Figure 1, in open cavities, shear layer coming with the free-stream separates at the cavity leading edge and reattaches at the cavity back wall. Due to the shear layer formation, flows inside and outside the cavity are separated. Additionally, the pressure differences between inside and outside of the cavity lead to a single recirculation region inside the cavity. On the other hand, in closed cavity, after the separation of the shear layer, due to containing inadequate energy to pass the cavity, it impinges to cavity base then separates from the base and reattaches at the stagnation point at the trailing edge (Aradag [1]; Lawson and Barakos [3]). As seen in Figure 1, two flow fields are emerged inside the cavity due to the shear layer movements.

Numerous research studies have been performed on supersonic cavity flow to understand the cavity flow mechanism clearly. A pioneering study is performed by Rossiter

Investigation of the Effects of Length to Depth Ratio on Open Supersonic Cavities Using CFD and Proper Orthogonal Decomposition

129

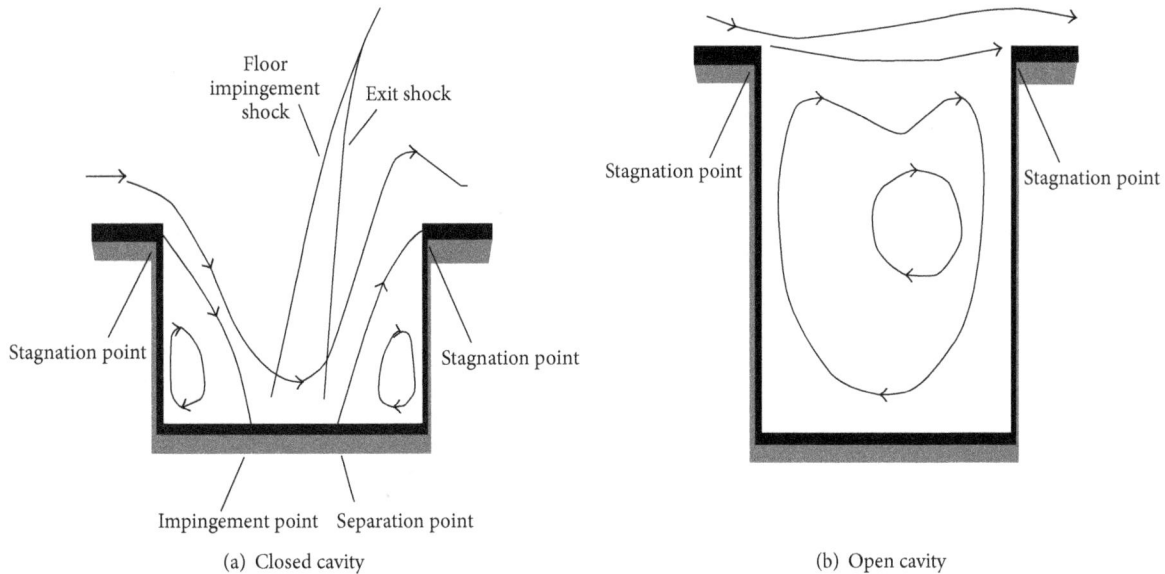

FIGURE 1: Supersonic cavity flow: (a) closed cavity, (b) open cavity.

[4], and an empirical formula to predict the frequencies of the pressure oscillation modes is defined. Heller and Bliss [5] modified this empirical formula to use it in all subsonic, supersonic, and transonic flows conditions. Shieh and Morris [6] used URANS Spalart-Allmaras turbulence model in their 2D and 3D cavity flow simulations. They examined vorticity contours for one period, and a periodic vortex formation in the cavity is observed. In their 3D simulation, the observed vortex is weaker than the one observed in 2D. Shih et al. [7] used k-ε turbulence model at Mach number of 1.5 and L/D ratio of 5.07 open cavity problem. They concluded that the shear layer formation makes the cavity flow more complex due to the mass input and output to the cavity. Rizzetta [8] performed 3D RANS simulations of cavity flow at a Mach number of 1.5. The results of average static pressure and sound pressure levels show good agreement with the experimental ones. In another study, Zhang and Edwards [9] used k-ω turbulence model to solve RANS equations for Mach numbers of 1.5 and 2.5 in different cavity geometries with L/D ratios of 1, 3, and 5. In conclusion, while the cavity flow is more uniform in L/D ratio of 1 cavity, when above the L/D ratio of 3, the flow becomes irregular and oscillations frequencies are increased. These numerical studies include RANS equations. There are also other studies using Large Eddy Simulations (LES) and Detached Eddy Simulation (DES) to perform cavity flow simulations. Hamed et al. [10] used DES method to model supersonic cavity flow with L/D of 5 and Mach number of 1.19. Basu et al. [11], Barakos et al. [12], and Rizzetta [8] also performed simulations on supersonic cavity flow.

Besides the numerical studies, there are experimental studies about supersonic cavity flow. Bueno et al. [13] experimentally investigated the effects of L/D ratio on the cavity flow, and according to their results, when cavity length increases, pressure fluctuations also increase. Ünalmis et al. [14] studied the effects of different L/D ratio on cavity flow

experimentally, and according to them, shock impingement effect is independent of L/D ratio and higher L/D ratios mean higher acoustic oscillations. Perng [15], Lazar et al. [16], and Stallings [17] also studied supersonic cavity flow experimentally.

Proper Orthogonal Decomposition (POD) is a method whose purpose is to obtain reduced order models of the systems by reducing the degree of the data samples collected as results of numerical or experimental studies (Cao et al. [18]). POD was introduced by Karhunen and Loéve, and this method was differently interpreted by researchers mainly with two definitions: Karhunen-Loéve decomposition d and a combination of Karhunen-Loéve decomposition, principal component analysis and singular value decomposition (Holmes et al. [19]; Chatterjee [20]; Feeny and Kappagantu [21]; Ravindra [22]; Kappagantu and Feeny [23]). Due to the compatibility of methods with each other, the second method is usually preferred by researchers (Liang et al. [24]).

Lumley [25] and Aubry et al. [26] performed the pioneering studies using POD for fluid mechanics problems. Other studies including POD applications have been performed by researchers for cavity flows which is an important fluid flow problem. Rowley et al. [27] studied flow pass over an open rectangular cavity for different L/D ratios. They used POD to get reduced order model of the system for control applications. Nagarajan et al. [28] studied POD based modeling of an open cavity with L/D ratio of 2 and Mach number of 0.6 to perform optimal control for cavity flow. In the study of Bortz et al. [29], simulations of an open cavity with a Mach number of 0.85 and L/D ratio of 4.5 are performed for the control of cavity acoustics. Due to high storage requirements of data, also for postprocessing, Proper Orthogonal Decomposition method is used to optimize and eliminate this problem. Colonius [30], Caraballo et al. [31], and Kasnakoglu [32] are other researchers who studied POD-based models of cavity flows. Berkooz et al. [33] and Holmes

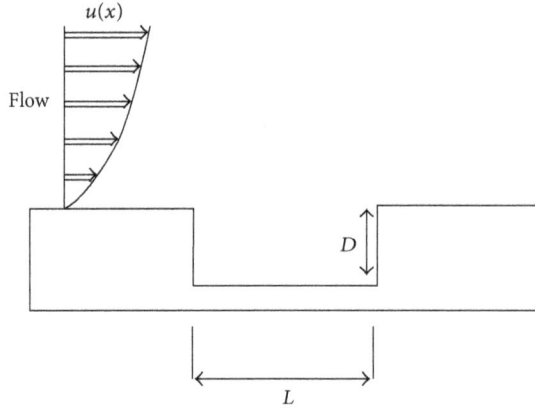

FIGURE 2: Schematic view of the cavity configuration.

TABLE 1: Numerical and experimental parameters for the preliminary CFD study.

Total pressure	66.4 kPa
Total temperature	218 K
Mach number	1.5
Reynolds number	1.09×10^6
Cavity length	0.12065 m
Cavity depth	0.0238 m
Boundary layer thickness	0.0051 m

TABLE 2: L/D ratios of cavities.

	L (m)	D (m)
$L/D = 1$	0.0238	0.0238
$L/D = 3$	0.0714	0.0238
$L/D = 5.07$	0.12065	0.0238
$L/D = 7.6$	0.180975	0.0238
$L/D = 10$	0.2413	0.0238

et al. [19] provide a more general discussion and detailed description of POD.

The main aim of this study is to observe the effects of length to depth ratio (L/D) in an open rectangular cavity simulation for two-dimensional, supersonic turbulent cavity flow. Proper Orthogonal Decomposition (POD) method is used to further analyze the flow. The number of modes necessary to represent the cavity for flow control purposes is determined for each cavity configuration, and the results for each L/D ratio are compared to each other.

2. Methodology

2.1. CFD Methodology. The experimental cavity configuration of Kaufman et al. [34] was used as a basis for the study. Preliminary simulations are performed for L/D ratio of 5.07 to be compared with the experimental results of Kaufman et al. [34]. Parameters used are given in Table 1.

In the numerical study, k-ω turbulence model is utilized. The computations are second order accurate in time and

space. Two-dimensional Reynolds-averaged Navier-Stokes equations are solved. As a result of a mesh independency study, $\Delta x/L$ is 0.00062, $\Delta y/D$ is 0.00252, and average y^+ (average nondimensional cell height) is 3.4. Inflow boundary conditions are obtained from the numerical solution of two-dimensional, steady, turbulent flow over a flat plate by using the program EDDYBL (Wilcox [35]). Schematic view of the cavity geometry is given in Figure 2. The numerical boundary layer thickness for the inflow is matched with the experiment of Kaufman et al. [34]. Boundary conditions are placed far enough to avoid reflection. Pressure far field boundary condition is used for inlet, outlet, and upper wall. For other boundaries, no slip wall boundary condition is given with adiabatic wall temperature 304.8 K. Simulations are performed for 20.000 time steps. For each time step 20 inner iterations are used.

To compare the numerical pressure oscillation frequency values with Rossiter frequency values, given semiempirical equation by Rossiter·[4] is used:

$$f_m = \frac{u_\infty}{L} \left[\frac{m - \alpha}{M_\infty + 1/K_v} \right]. \tag{1}$$

Also the modified Rossiter formulation by Heller and Bliss [5] is used to calculate frequencies:

$$St_m = \frac{f_m L}{U_\infty} = \frac{m - \alpha}{M_\infty \left(1 + \left[(\gamma - 1)/2\right] M_\infty^2\right)^{-1/2} + 1/K}. \tag{2}$$

In (2), K and α are experimental constants. K is a function of Mach number and equals to 0.55 (Aradag [1]). The parameter α is related to cavity geometry and has a value of 0.25 (Syed [2]). U_∞ is free-stream velocity, M_∞ is the free-stream Mach number, St is the Strouhal number, and m is the mode number of the cavity.

Sound pressure levels are used for comparison of performed study with experimental results. SPL levels are obtained with the equations given in the study of Aradag [1].

Using the pressure data obtained as results of simulations, to determine dominant modes frequencies, fast Fourier transform (FFT) is used. Frequency to power spectrum graph is obtained. Additionally, L/D ratios of 1, 3, 7.6, and 10 cavity flow simulations at Mach number of 1.5 are performed. Different simulation cases based on geometrical changes are presented in Table 2.

2.2. POD Methodology. After a CFD simulation is performed for M time steps, snapshots obtained include x-velocity data for M number of time steps. x-velocity data in each snapshot are collected in $U_i(x)$ matrix. As a result, the following equation is obtained:

$$U_i(\vec{x}) = U_1(x), U_2(x), U_3(x), \ldots, U_M(x). \tag{3}$$

To eliminate requirements of scaling in further steps as defined in the studies of Newman [36] and Deane et al. [37], average of all data n is calculated and subtracted from each data matrix:

$$V_i(\vec{x}) = U_i(\vec{x}) - \frac{1}{M} \sum_{i=1}^{M} U_i(\vec{x}) \quad i = 1, 2, \ldots, M. \tag{4}$$

Investigation of the Effects of Length to Depth Ratio on Open Supersonic Cavities Using CFD and Proper Orthogonal Decomposition

131

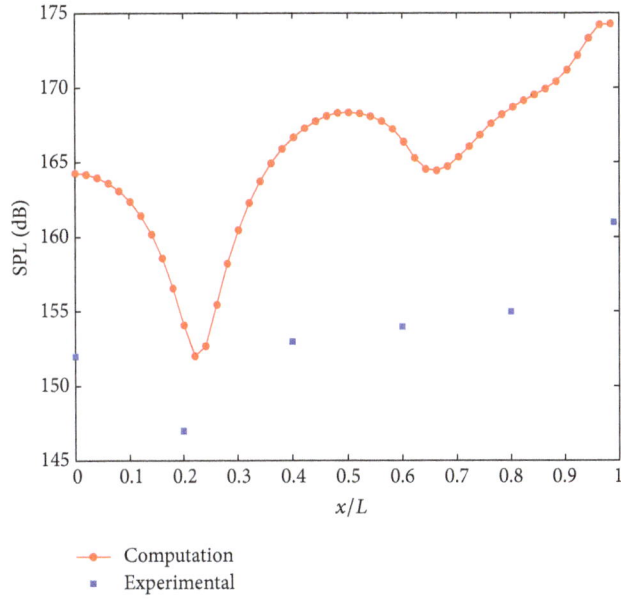

FIGURE 3: The comparison of SPL distribution at cavity floor.

To find basic functions which represent the dominant structures in the system, the function given below is used (Newman [36]):

$$\phi(\vec{x}) = \sum_{i=1}^{M} \alpha_{ik} V_i(\vec{x}) \quad k = 1, 2, \ldots, S \text{ (mode number)} . \quad (5)$$

By using the method of snapshots developed by Sirovich [38], an $M \times M$ dimensional covariance matrix is obtained as (Ly and Tran [39]; Smith et al. [40])

$$C\phi_i = \lambda_i \phi_i \quad i = 1, 2, 3, \ldots, M, \quad (6a)$$

$$(C)_{ij} = \frac{1}{M} \int_{\Omega} V_i(\vec{x}) V_j(\vec{x}) dx \quad i, j = 1, 2 \ldots, M. \quad (6b)$$

This covariance matrix can be solved mathematically. By using singular value decomposition, eigenvalues and eigenvectors are obtained (Volkwein [41]; Chatterjee [20]):

$$C = R \sum P^T. \quad (7)$$

R contains the eigenvectors. After obtaining the eigenvalues, they are sorted starting from the largest as $\lambda_1 > \lambda_2 > \lambda_3 > \cdots > \lambda_M$. By examination of energy information, the number of modes to represent the system can be obtained (Berkooz et al. [33]).

For the reconstruction of the reduced order model, the following equation is used (Cohen et al. [42]):

$$U = \overline{U} + \sum_{k=1}^{S} \alpha_k \phi_k. \quad (8)$$

U is the original data set, $\overline{U}$ is the matrix for the mean values, α_k are time coefficients, ϕ_k are basis functions, and S is total number of modes.

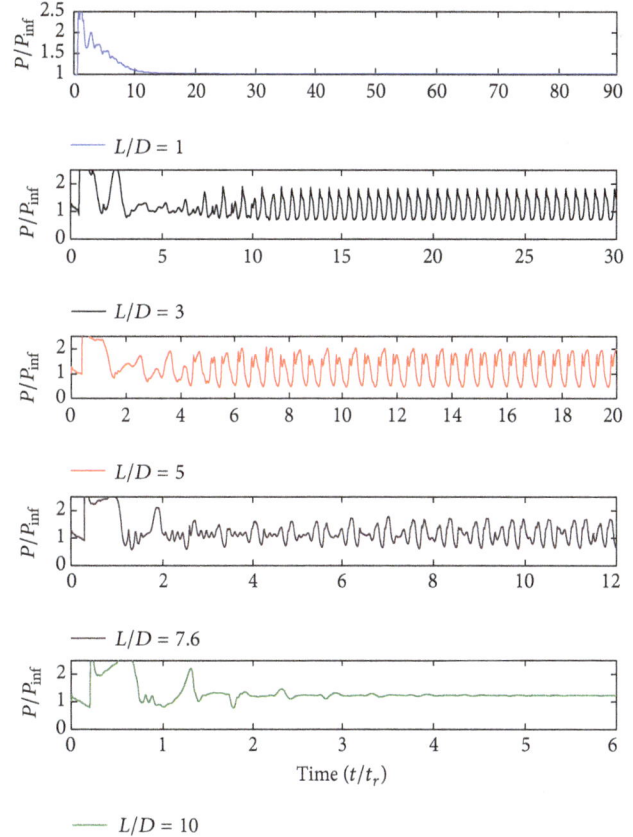

FIGURE 4: Pressure time history of the aft bulkhead at $y/D = 0.6$.

3. Results

3.1. 2D Results for Flow Structure. As it is seen in Figure 3, the SPL values for the preliminary study which utilizes a cavity configuration with an L/D ratio of 5.07 show the same trend with the experimental results.

The flow is composed of vortex-wall, shear layer-wall, and shock waves interactions, and these interactions cause pressure oscillations. According to changes in L/D ratio, pressure oscillation mechanism also changes as it is seen in Figure 4.

Cavity with L/D ratio of 1 is a deep cavity (Garner et al. [43]). In this configuration, due to the inactivity of the shear layer, pressure fluctuations are not observed and cavity region interactions are observed in low levels. Cavities with L/D ratios of 3, 5.07, 7.6, and 10 are all shallow cavity configurations. In cavity configurations with L/D ratios of 3, 5.07, and 7.6, shear layer separates from the leading edge of the cavity and reattaches at the back wall. Shear layer forms two flow zones as cavity flow and free-stream external flow. Since there is a pressure difference between two flow zones, mass inlet and outlet in cavity region are observed. Additionally, pressure fluctuations occur. This flow mechanism represents the open cavity flow as expected. After the flows become fully developed, periodic pressure oscillations are observed as it is seen in Figure 4. The increase in the L/D ratio triggers the pressure oscillations. The highest pressure oscillations

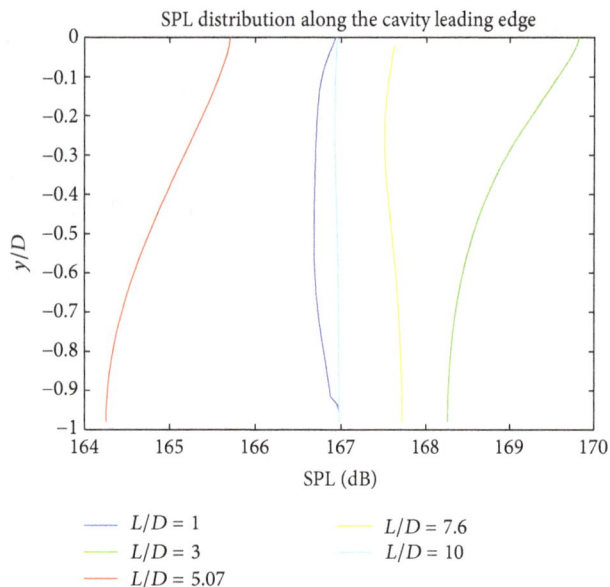

FIGURE 5: Sound pressure level distribution along the cavity leading edge.

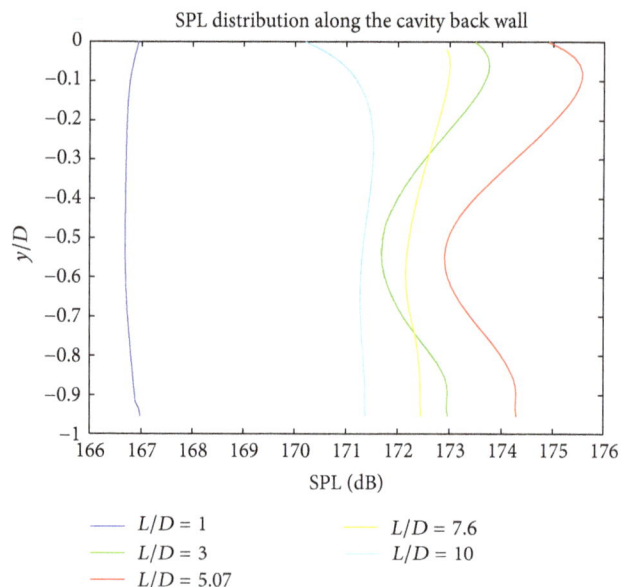

FIGURE 7: SPL distribution along the cavity back wall.

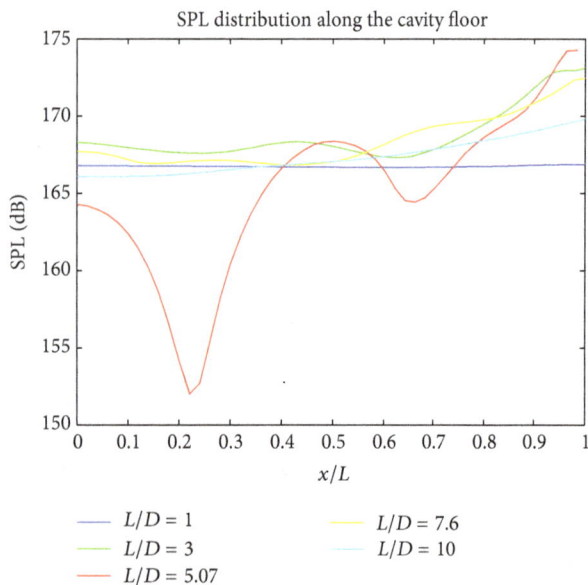

FIGURE 6: SPL distribution along the cavity floor.

layer and cavity wall. The highest SPL values are obtained at the cavity rear wall for all geometries.

For all cavity configurations, frequencies are calculated using the Rossiter and modified Rossiter formulations (Rossiter [4]; Heller and Bliss [5]) to be compared with the results of simulations. Fast Fourier transform is applied to the pressure results of simulations to obtain mode frequencies. As a result, frequency versus power spectrum graph is obtained from pressure history at the experimental measurement point which is located at $y/D = 0.6$ on the aft bulkhead. The results of FFT are given in Figure 8, and the empirical modes frequency formulations are summarized in Table 3.

Peak powers are called dominant modes, and when first mode, is dominant, the flow has a single mode. In flows that include multiple modes, system has more than one mode and dominant frequency is seen in progressive modes. For cavities with L/D ratios of 1 and 10, pressure-time histories prove that flow is dominated by a single mode. For L/D ratio of 1, flow is influenced by a frequency of 4823 Hz, and, for the cavity flow with $L/D = 10$, a single mode exits equal to 388 Hz. For cavities with L/D ratios of 3, 5, and 7.6 dominant modes, which are 3880.1 Hz, 2106 Hz, and 2273 Hz, occur at second mode. When the flow has multiple modes, flow mechanism is more complicated. Results show that with the increase of the mode values, a better agreement between Rossiter formulation and computational Strouhal number values occurs.

For all cases, flow fields include a large trailing-edge vortex and two small corner vortices at rear and trailing edges of the cavity. At the cavities with L/D ratio of 1 and L/D ratio of 10, shear layer deflection is small in size, weak in strength, and no shock waves exist. In other geometries, which have multiple modes, vortex motion, shear layer deflection, and shock wave generation occur. The velocity contours of cavity flow with L/D ratio of 5 is given in Figure 9. These contours

amplitude is observed in cavity flow with L/D ratio of 5. As it is seen clearly in Figure 4, periodicity of pressure oscillations becomes irregular with the increase in L/D ratio. In cavity flow with L/D ratio of 10, the pressure fluctuations are almost damped due to being in transition flow region, between open and closed cavity flow.

In Figures 5, 6, and 7, the acoustic pressure distributions at different regions for all cavity configurations are presented. For the cavities with L/D ratios of 1 and 10, generally lowest SPL levels are obtained. For all L/D ratios at floor edge ($x/L = 1$), SPL levels increase due to the interaction between shear

Investigation of the Effects of Length to Depth Ratio on Open Supersonic Cavities Using CFD and Proper Orthogonal Decomposition

133

TABLE 3: Experimental and calculated mode frequencies.

L/D	Frequency modes	Computational values	Modified Rossiter form.	Rossiter form.	Discrepancy (%)	Discrepancy Rossiter form. (%)
1	$f1$	4823.0	4566.7	5622.2	5.6	14.2
1	$f2$	10034.0	10656.0	11244.4	5.8	10.7
1	$f3$	—	16744.0	16866.6	—	—
3	$f1$	1940.0	1522.2	1663	27.0	16.7
3	$f2$	3880.1	3551.8	3659	9.2	6.0
3	$f3$	5559.2	5581.5	5488	0.4	1.3
5	$f1$	1037.0	900.8	1109.1	15.0	6.5
5	$f2$	2106.0	2102.0	2218.0	0.19	5.0
5	$f3$	3160.0	3303.1	3327.0	4.5	5.0
7.6	$f1$	609.8	600.55	739.4	1.5	17.5
7.6	$f2$	1441.0	1401.3	1478.7	2.8	2.5
7.6	$f3$	2273.0	2202.1	2218.1	3.2	2.5
10	$f1$	388.0	450.41	554.5	13.7	30.0
10	$f2$	1053.0	1051.0	1109.1	0.19	5.1
10	$f3$	1663.0	1651.5	1663.5	0.71	0.1

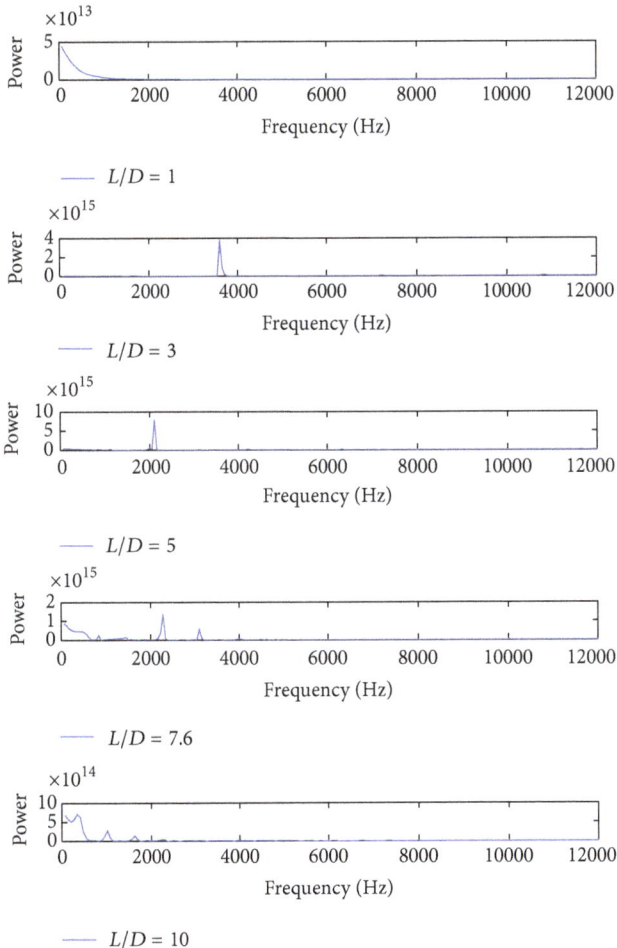

FIGURE 8: Power spectrum on the aft bulkhead $y/D = 0.6$.

represent a clear open cavity flow mechanism using x-velocity data for different times along one Rossiter period.

3.2. POD Results for the Cavity Configuration with L/D Ratio of 5.07. As a result of POD application to the supersonic cavity flow with L/D ratio of 5.07, energy distribution is given in Figure 10 and energy contents of modes are obtained and presented in Table 4. This energy distribution is obtained by using energy contents of each eigenvalue obtained as a result of POD.

As it is seen in Table 4, the first mode includes 70.65% of total energy of the flow after which the energy content values decrease rapidly. 99.18% of the total energy of the system can be represented using 12 modes. There are many small structures which affect the main characteristics of the flow due to turbulent nature of the flow. It is clearly seen in Table 4 that there are many small turbulent structures which contain small energy values that affect the main flow. For future studies for flow control, the small structures are not important since the main idea is to control larger structures from which the smaller structures develop.

In Figure 11, comparison of original cavity contour with reconstructed cavity contours with 4 modes and 12 modes is given. There are small differences between reconstructed contours therefore the system can be represented with 4 modes which contain 96% of the total energy. In Figure 12, modes are given. As it is seen, there are some serrated structures due to the small turbulent structures.

The dominant changes of characteristics of flow are presented with the time coefficient history of modes which is given in Figure 13. The motion of dominant structures in time can be seen clearly. The amplitudes and energy contents of modes are directly proportional to each other. As one mode has higher energy content, it also has higher mode

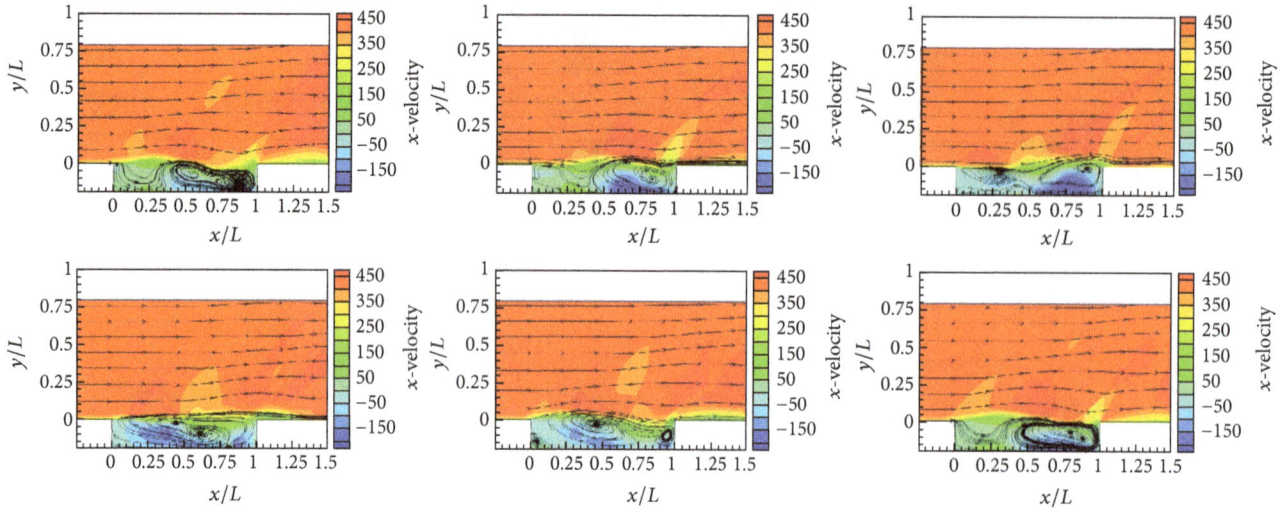

FIGURE 9: x-velocity contours of cavity flow with L/D ratio of 5.07.

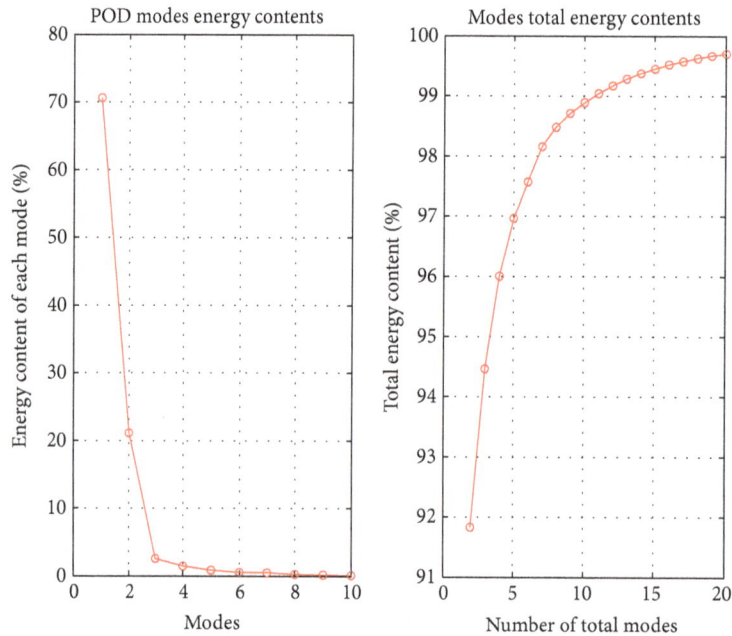

FIGURE 10: Energy distribution of modes.

amplitude. For Mode 1 and Mode 2, the structures show a sinusoidal periodic oscillation. Mode 3 and Mode 4 include periodical but unpredictable oscillations due to the effects of small turbulent structures.

3.3. The Effects of L/D Ratio on Cavity Flow Physics Based on POD Results. POD is applied to the CFD results of several cavity configurations with different L/D ratios. The necessary number of modes to represent the flow without loss of information for all cavity configurations is presented in Table 5. Systems can be represented with number of modes corresponding to 94–97% of the total system energy.

As it is seen in Figure 14, with the increase in L/D ratio, the number of modes increases. Only in the cavity with L/D ratio

of 10, POD results show divergence from the general trend due to being in the transition region. The cavity with L/D ratio of 1 is a deep cavity and flow interactions are low (Ayli [44]). The representation of this cavity configuration using the least number of modes shows that the flow is more uniform than others. Between the range of L/D ratios of 3 and 10, the flows become more complicated with the increase in L/D ratio, and the number of POD modes to define the systems increases with increasing L/D.

Each mode has an energy value. The amount of these energy values is related with how the modes include characteristics of the system at hand. Energy content of each mode is given in Table 6. All cases include multiple modes except for the L/D ratio of 1. Including multiple modes means that

Investigation of the Effects of Length to Depth Ratio on Open Supersonic Cavities Using CFD and Proper
Orthogonal Decomposition

135

TABLE 4: $L/D = 5$ cavity, POD modes energy contents.

					$L/D = 5$, cavity configuration							
Mode number	1	2	3	4	5	6	7	8	9	10	11	12
Energy content % (/96)	70.65	21.29	2.63	1.54	—	—	—	—	—	—	—	—
Energy content % (/99)	70.65	21.29	2.63	1.54	0.96	0.61	0.58	0.32	0.23	0.18	0.15	0.13

TABLE 5: The necessary number of POD modes for all cases.

Cavity configurations	$L/D = 1$		$L/D = 3$		$L/D = 5.07$		$L/D = 7.6$		$L/D = 10$	
Total Energy content %	94.33	99.41	95.58	99.17	96	99.18	95.33	99.19	96.82	99.18
Number of modes	1	4	4	11	4	12	6	14	5	7

TABLE 6: Energy content of each mode for all cavity configurations.

	Energy content of each mode for all cavity configurations %									
Mode number	$L/D = 1$		$L/D = 3$		$L/D = 5.07$		$L/D = 7.6$		$L/D = 10$	
1	94.33	94.33	67.02	67.02	70.65	70.65	60.87	60.87	46.38	46.38
2	—	2.88	20.60	20.60	21.19	21.19	18.15	18.15	31.06	31.06
3	—	1.64	4.86	4.86	2.63	2.63	7.12	7.12	12.15	12.15
4	—	0.56	3.10	3.10	1.54	1.54	5.17	5.17	4.2	4.2
5	—	—	—	0.090	—	0.96	2.16	2.16	3.02	3.02
6	—	—	—	0.078	—	0.61	1.86	1.86	—	1.67
7	—	—	—	0.067	—	0.58	—	1.14	—	0.69
8	—	—	—	0.048	—	0.32	—	0.71	—	—
9	—	—	—	0.030	—	0.23	—	0.55	—	—
10	—	—	—	0.0026	—	0.18	—	0.40	—	—
11	—	—	—	0.0020	—	0.15	—	0.36	—	—
12	—	—	—	—	—	0.13	—	0.30	—	—
13	—	—	—	—	—	—	—	0.23	—	—
14	—	—	—	—	—	—	—	0.17	—	—
Total energy content %	94.33	99.41	95.58	99.17	96.01	99.18	95.33	99.19	96.82	99.18

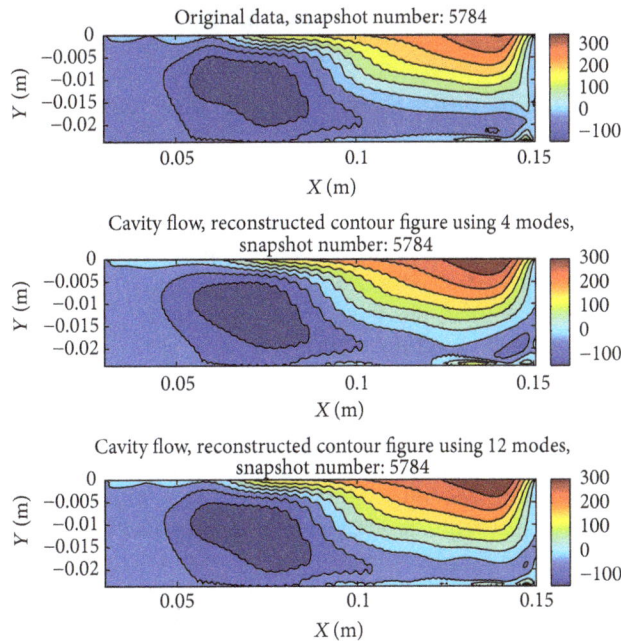

FIGURE 11:Original x-velocity contour and reconstructed x-velocity contours.

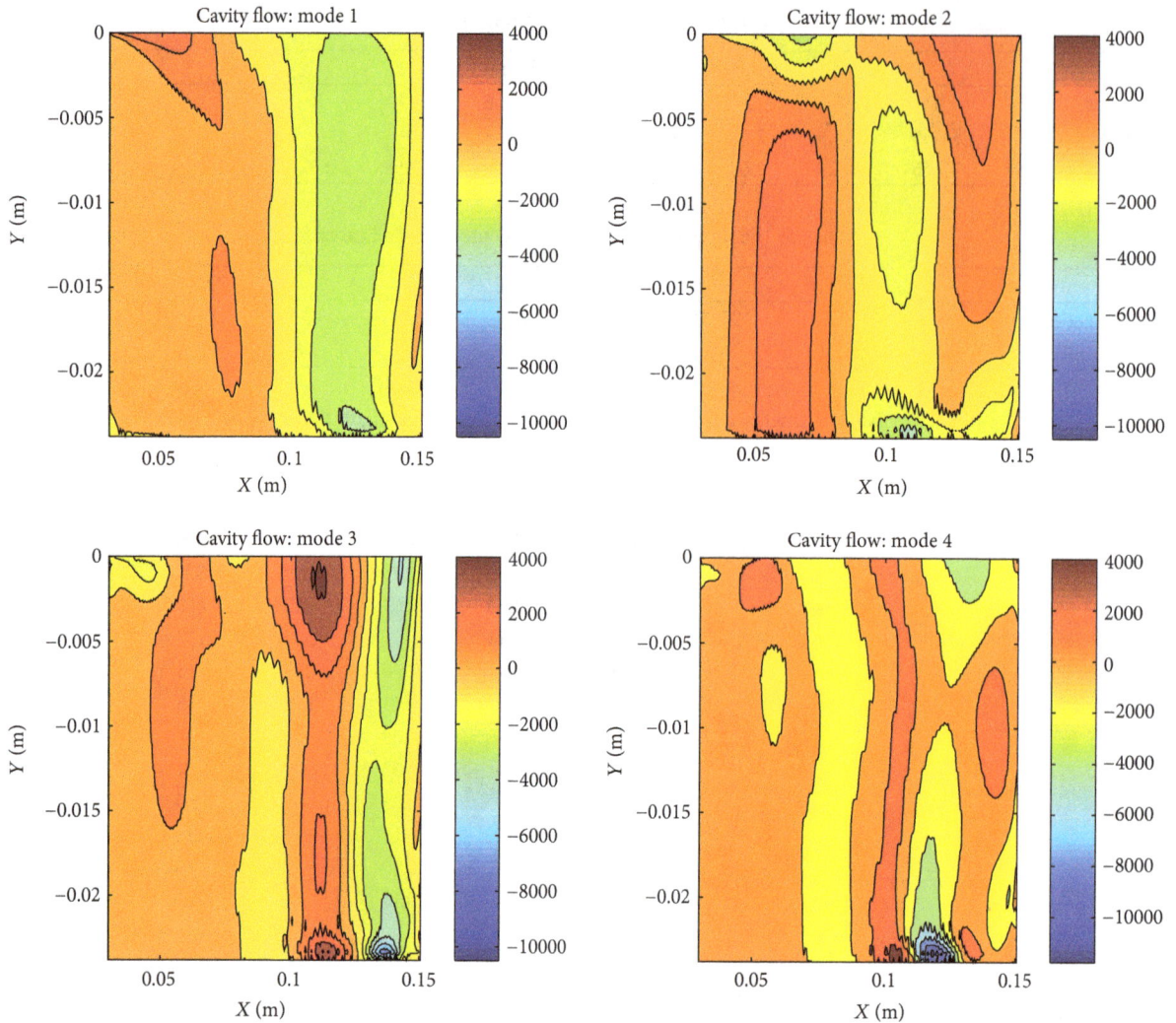

FIGURE 12: POD modes of cavity flow with L/D ratio of 5.07.

4. Discussion and Conclusion

Supersonic cavity flow is examined to present the flow mechanism. CFD simulations are performed. Additionally, Proper Orthogonal Decomposition method is applied to the results of the CFD simulations. The effects of length to depth (L/D) ratio on the flow mechanism are investigated using CFD. Proper Orthogonal Decomposition method is applied to x-velocity component results of simulations, and the cavity configurations are represented with several number of modes which are available for flow control applications. With 2D cavity results, the supersonic cavity flow mechanism is represented for different cavity configurations. POD results and CFD results show good agreement with each other. As a result of Computational Fluid Dynamics simulations and Proper Orthogonal Decomposition,

the flow is affected by small structures. This conclusion shows good agreement with the CFD results of the same cavity configurations by Ayli [44].

(i) in the range of L/D ratios of 3 and 10, the exact open cavity phenomenon is observed. As the L/D ratio increases, regularity of periodic structures are decreased, and flow becomes more complicated. As a result of fast Fourier transform, it is shown that these flows include multiple modes. The number of POD modes to represent these systems also increases as L/D ratio increases.

(ii) In the cavity with L/D ratio of 1, a low level of flow structure interactions is observed and almost no pressure oscillations occur. The flow is not as complex as the others and this is supported by POD results. The least number of POD modes necessary to represent the system is observed in this case. In the cavity with L/D ratio of 10, nearly no pressure oscillations are observed as in the L/D = 1 cavity. However, POD results show that this flow is not as simple as in L/D = 1 cavity flow This L/D ratio is in the range of transition region between open cavity and closed

Investigation of the Effects of Length to Depth Ratio on Open Supersonic Cavities Using CFD and Proper
Orthogonal Decomposition

137

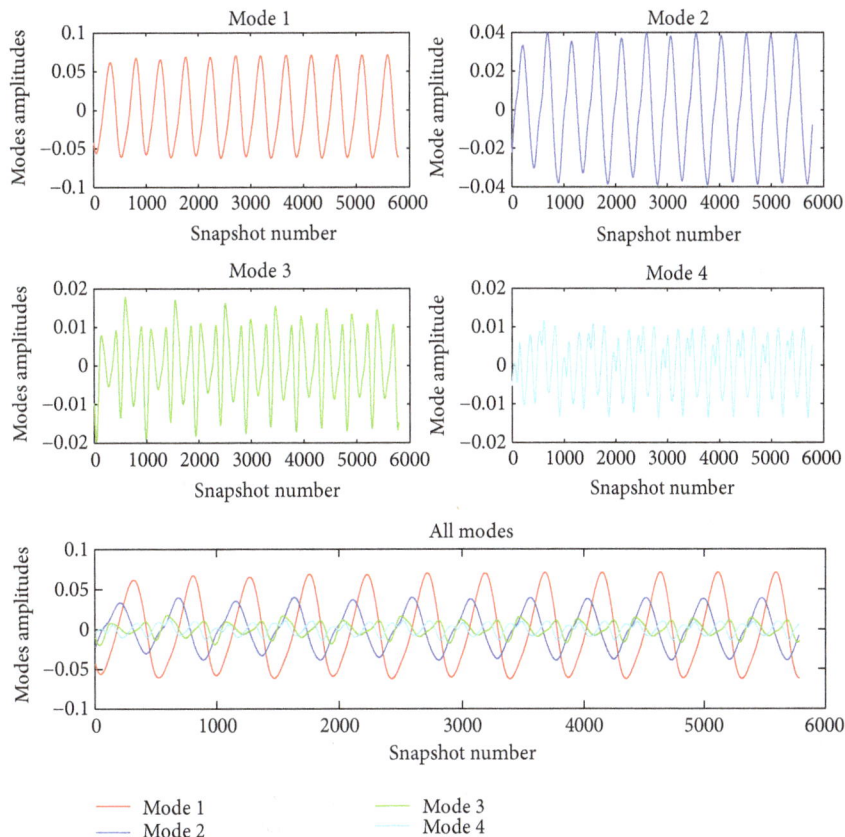

FIGURE 13: Time coefficient history of POD modes of L/D ratio of 5.07 cavity.

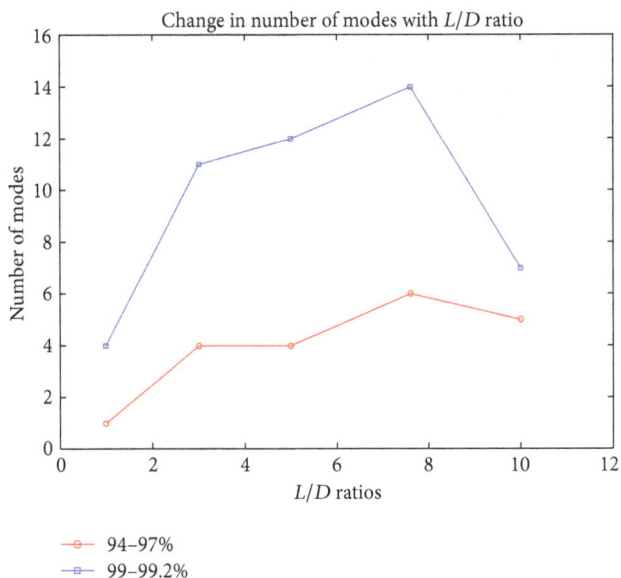

FIGURE 14: Change in the number of POD modes with L/D ratios.

can be expressed with the effects of small turbulent structures on the main flow.

For flow control phenomena, the control of the larger structures is the main objective. Therefore, in this study, it is shown that supersonic cavity configurations can be represented with several POD modes that include 95–97% of the total energy of the flow. For flow control purposes, these large scale structures represented by first several modes can be targeted, and small scale fluctuations may not exist anymore since they develop from the larger ones.

Acknowledgments

This study is supported by TUBITAK (Turkish Scientific and Research Council) under Grant no. 110M539 and Turkish Academy of Sciences Distinguished Young Scientists Awards programme (TUBA-GEBIP).

References

[1] S. Aradag, *CFD for High Speed Flows in Engineering*, VDM Dr. Müller, Saarbrucken, Germany, 2008.

[2] A. Syed, *Detached eddy simulation of turbulent flow over an open cavity with and without cover plates [M.S. thesis]*, Wichita State University, Wichita, Kan, USA, 2010.

[3] S. J. Lawson and G. N. Barakos, "Review of numerical simulations for high-speed, turbulent cavity flows," *Progress in Aerospace Sciences*, vol. 47, no. 3, pp. 186–216, 2011.

cavity; therefore the behavior of the flow is not as clear as others.

(iii) In all cases, POD results show that the modes including small energy values affect the main flow. This

[4] J. E. Rossiter, "Wind tunnel experiments on the flow over rectangular cavities at subsonic and transonic speeds," Tech. Rep. 64037, Royal Aircraft Establishment, Surrey, UK, 1964.

[5] H. H. Heller and D. B. Bliss, "The physical mechanism of flow-induced pressure fluctuations in cavities and concepts for their suppression," in *Proceedings of the 2nd Aeroacoustics Conference*, AIAA Paper No. 75-491, AIAA, Hampton, Va, 1975.

[6] M. C. Shieh and P. Morris, "Comparison of two and three dimensional turbulent cavity flows," in *Proceedings of the 39th Aerospace Sciences Meeting and Exhibit*, AIAA Paper No. 2001-0511, Reno, Nev, USA, 2001.

[7] S. H. Shih, A. Hamed, and J. J. Yeuan, "Unsteady supersonic cavity flow simulations using coupled k-ε and Navier-Stokes equations," *AIAA Journal*, vol. 32, no. 10, pp. 2015–2021, 1994.

[8] D. P. Rizzetta, "Numerical simulation of supersonic flow over a three dimensional cavity," *AIAA Journal*, vol. 26, no. 7, pp. 799–807, 1988.

[9] X. Zhang and J. A. Edwards, "Analysis of unsteady supersonic cavity flow employing an adaptive meshing algorithm," *Computers and Fluids*, vol. 25, no. 4, pp. 373–393, 1996.

[10] A. Hamed, D. Basu, and K. Das, "Effect of Reynolds number on the unsteady flow and acoustic fields of supersonic cavity," in *Proceedings of the 4th ASME/JSME Joint Fluids Engineering Conference*, FEDSM2003-45473, ASME, Honolulu, Hawaii, USA, July 2003.

[11] D. Basu, A. Hamed, and K. Das, "DES, hybrid RANS/LES and PANS models for unsteady separated turbulent flow simulations," in *Proceedings of the ASME Fluids Engineering Division Summer Meeting (FEDSM '05)*, FEDSM, 2005-77421, ASME, Houston, Tex, USA, June 2005.

[12] G. N. Barakos, S. J. Lawson, R. Steijl, and P. Nayyar, "Numerical simulations of high-speed turbulent cavity flows," *Flow, Turbulence and Combustion*, vol. 83, no. 4, pp. 569–585, 2009.

[13] P. C. Bueno, Ö. H. Unalmis, N. T. Clemens, and D. S. Dolling, "The effects of upstream mass injection on a Mach 2 cavity flow," in *Proceedings of the 40th AIAA Aerospace Sciences Meeting & Exhibit*, AIAA Paper No. 2002-0663, AIAA, Reno, Nev, USA, 2002.

[14] Ö. H. Ünalmis, N. T. Clemens, and D. S. Dolling, "Experimental study of shear-layer/acoustics coupling in Mach 5 cavity flow," *AIAA Journal*, vol. 30, no. 2, pp. 242–252, 2001.

[15] S. W. Perng, *Passive Control of Pressure Oscillations in Hypersonic Cavity Flow [Ph.D. dissertation]*, University of Texas at Austin, Austin, Tex, USA, 1996.

[16] E. Lazar, G. Elliotts, and N. Glumac, "Control of the shear layer above a supersonic cavity using energy deposition," *AIAA Journal*, vol. 46, no. 12, pp. 2987–2997, 2008.

[17] R. L. Stallings, "Store separations from cavities at supersonic flight speeds," *Journal of Spacecraft and Rockets*, vol. 20, no. 2, pp. 129–132, 1983.

[18] Y. Cao, J. Zhu, Z. Luo, and I. M. Navon, "Reduced order modeling of the upper tropical pacific ocean model using proper orthogonal decomposition," *Computers and Mathematics with Applications*, vol. 52, no. 8-9, pp. 1373–1386, 2006.

[19] P. Holmes, J. L. Lumley, and G. Berkooz, *Turbulence and Coherent Structures, Dynamical Systems and Symmetry*, Cambridge University Press, New York, NY, USA, 1996.

[20] A. Chatterjee, "An introduction to the proper orthogonal decomposition," *Current Science*, vol. 78, no. 7, pp. 808–817, 2000.

[21] B. F. Feeny and R. Kappagantu, "On the physical interpretation of proper orthogonal modes in vibrations," *Journal of Sound and Vibration*, vol. 211, no. 4, pp. 607–616, 1998.

[22] B. Ravindra, "Comments on 'on the physical interpretation of proper orthogonal modes in vibrations'," *Journal of Sound and Vibration*, vol. 219, no. 1, pp. 189–192, 1999.

[23] R. Kappagantu and B. F. Feeny, "An "optimal" modal reduction of a system with frictional excitation," *Journal of Sound and Vibration*, vol. 224, no. 5, pp. 863–877, 1999.

[24] Y. C. Liang, H. P. Lee, S. P. Lim, W. Z. Lin, K. H. Lee, and C. G. Wu, "Proper orthogonal decomposition and its applications, part I: theory," *Journal of Sound and Vibration*, vol. 252, no. 3, pp. 527–544, 2002.

[25] J. L. Lumley, "The structure of inhomogeneous turbulent flows," in *Atmospheric Turbulence and Radio Propagation*, A. M. Yaglom and V. I. Tatarski, Eds., pp. 166–178, Nauka, Moskow, Russia, 1967.

[26] N. Aubry, P. Holmes, J. L. Lumley, and E. Stone, "The dynamics of coherent structures in the wall region of a turbulent boundary layer," *Journal of Fluid Mechanics*, vol. 192, pp. 115–173, 1988.

[27] C. W. Rowley, T. Colonius, and R. M. Murray, "POD based models of self-sustained oscillations in the flow past an open cavity," in *Proceedings of the 6th Aeroacoustics Conference and Exhibit*, AIAA Paper No. 2000-1969, AIAA, Lahaina, Hawaii, USA, 2000.

[28] K. K. Nagarajan, L. Cordier, C. Airiau, and A. Kourta, "POD based reduced order modelling of a compressible forced cavity flow," in *Le 19ème Congrès Français de Mécanique*, Marseille, France, 2009.

[29] D. M. Bortz, A. D. Rubio, H. T. Banks, A. B. Cain, and R. C. Smith, "Reduced order modeling in control of open cavity acoustics," Tech. Rep. CRSC-TROO-18, Center for Research in Scientific Computation, North Carolina State University, Rayleigh, NC, USA, 2000.

[30] T. Colonius, "An overview of simulation, modeling, and active control of flow/acoustic resonance in open cavities," in *Proceedings of the 39th Aerospace Sciences Meeting and Exhibit*, AIAA Paper No. 2001-0076, AIAA, Reno, Nev, USA, 2001.

[31] E. Caraballo, X. Yuan, J. Little et al., "Further development of feedback control of cavity flow using experimental based reduced order model," in *Proceedings of the 35th AIAA Fluid Dynamics Conference and Exhibit*, AIAA Paper No. 2005-5269, AIAA, Toronto, Canada, 2006.

[32] C. Kasnakoglu, *Reduced order modeling, nonlinear analysis and control methods for flow control problems [Ph.D. thesis]*, The Ohio State University, Columbus, Ohio, USA, 2007.

[33] G. Berkooz, P. Holmes, and J. L. Lumley, "The proper orthogonal decomposition in the analysis of turbulent flows," *Annual Review of Fluid Mechanics*, vol. 25, no. 1, pp. 539–575, 1993.

[34] L. G. Kaufman, A. Maciulaitis, and R. L. Clark, "Mach 0. 6 to 3. 0 flows over rectangular cavities," Tech. Rep. AFWAL-TR-82-3112, Air Force Wright Aeronautical Labs, New York, NY, USA, 1983.

[35] D. C. Wilcox, *Turbulence Modeling for CFD*, DCW Industries, La Canada, Calif, USA, 1993.

[36] A. J. Newman, "Model reduction via the Karhunen-Loéve expansion part II: some elementary examples," Tech. Rep. 9633, Institute for Systems Research, University of Maryland, College Park, Md, USA, 1996, http://hdl.handle.net/1903/5752 .

[37] A. E. Deane, I. G. Kevrekidis, G. E. Karniadakis, and S. A. Orszag, "Low-dimensional models for complex geometry flows:

Investigation of the Effects of Length to Depth Ratio on Open Supersonic Cavities Using CFD and Proper Orthogonal Decomposition

139

application to grooved channels and circular cylinders," *Physics of Fluids A*, vol. 3, no. 10, pp. 2337–2354, 1991.

[38] L. Sirovich, "Turbulence and the dynamics of coherent structures, part 1–3," *Quarterly Applied Mathematics*, vol. 45, no. 3, pp. 561–590, 1987.

[39] H. V. Ly and H. T. Tran, "Modeling and control of physical processes using proper orthogonal decomposition," *Mathematical and Computer Modelling*, vol. 33, no. 1–3, pp. 223–236, 2001.

[40] T. R. Smith, J. Moehlis, and P. Holmes, "Low-dimensional modelling of turbulence using the proper orthogonal decomposition: a tutorial," *Nonlinear Dynamics*, vol. 41, no. 1–3, pp. 275–307, 2005.

[41] S. Volkwein, 1999, Proper Orthogonal Decomposition and Singular Value Decomposition Bericht Nr. 153 (Graz: Spezialforschungsbereich F003 Optimierung und Kontrolle, Projektbereich Kontinuierliche Optimierung und Kontrolle).

[42] K. Cohen, S. Siegel, and T. McLaughlin, "A heuristic approach to effective sensor placement for modeling of a cylinder wake," *Computers and Fluids*, vol. 35, no. 1, pp. 103–120, 2006.

[43] H. C. Garner et al., "Drag of a rectangular planform cavity in a flat plate with a turbulent boundary layer for Mach numbers up to 3, part II: open and transitional flows," Tech. Rep., Engineering Science Data Unit 00007, ESDU, London, UK, 2000.

[44] E. Ayli, *Numerical analysis of supersonic cavity flow [M.S. thesis]*, TOBB University of Economics and Technology, Ankara, Turkey, 2012.

14

Inviscid and Viscous Interactions in Subsonic Corner Flows

Kung-Ming Chung,[1] Po-Hsiung Chang,[2] and Keh-Chin Chang[2]

[1] *Aerospace Science and Technology Research Centre, National Cheng Kung University, Tainan 711, Taiwan*
[2] *Institute of Aeronautics and Astronautics, National Cheng Kung University, Tainan 711, Taiwan*

Correspondence should be addressed to Kung-Ming Chung; kmchung@mail.ncku.edu.tw

Academic Editors: H. Baoyin, A. F. B. A. Prado, and C. Zhang

A flap can be used as a high-lift device, in which a downward deflection results in a gain in lift at a given geometric angle of attack. To characterize the aerodynamic performance of a deflected surface in compressible flows, the present study examines a naturally developed turbulent boundary layer past the convex and concave corners. This investigation involves the analysis of mean and fluctuating pressure distributions. The results obtained indicate strong inviscid-viscous interactions. There are upstream expansion and downstream compression for the convex-corner flows, while the opposite trend is observed for the concave-corner flows. A combined flow similarity parameter, based on the small perturbation theory, is proposed to scale the flow characteristics in both subsonic convex- and concave-corner flows.

1. Introduction

Corner flows occur in a wide variety of internal and external aerodynamic problems. Previous studies have been mainly on supersonic and hypersonic speeds [1–4]. In subsonic flow regime, aircraft designs have employed flaps for take-off and landing performance and ailerons for routine turning maneuver. A study by Bolonkin and Gilyard [5] demonstrated that active modification of control surfaces (variable camber wings) potentially could play a role in performance optimization for fighter aircraft and transport aircraft. While cruising, there could be more than 10 percent in maximizing the lift-to-drag ratio, especially for nonstandard flight conditions [6]. Further, a simplified model of a deflected surface comprised convex-corner and concave-corner flows. For a compressible convex-corner flow (or upper deflected surface), there are strong upstream expansion and downstream compression, caused by viscous-inviscid interactions, near the corner. The displacement thickness near the corner is affected by the overlapping region that lies between the viscous sublayer and the main part of the boundary layer [7]. On the lower deflected surface (or concave corner), the flow decelerates upstream of the corner followed by the downstream

acceleration. Previous studies [8–10] demonstrated that the lift coefficient increases linearly with the deflection angle.

Chung [11] demonstrated that M and η are the two major parameters affecting the type of flow field in compressible convex-corner flows. The hypersonic similarity parameter ($M\eta$) [12] and a similar combined supersonic-hypersonic similarity parameter ($\sqrt{1 - M^2}\eta$) [13] were examined for scaling the expansion flows. However, the parameter $M^2\eta$ appears to be more suitable to characterize the flow characteristics, including peak Mach number, interaction region, and amplitude of peak pressure fluctuations [11, 14]. Further, according to the hypothesis that small streamline deflections produce proportionally small change in Mach number and pressure, a hodograph solution for compressible flow past a corner was given by Verhoff et al. [15]. Another flow similarity parameter, $M^2\eta/\sqrt{1 - M^2}$, was employed by Chung et al. [16] to categorize flow regimes of compressible convex-corner flows. In compressible concave-corner flows, Chung [17] demonstrated that the characteristics of the flow can be scaled with $M\eta$, in which stronger upstream compression and downstream expansion are observed with increasing M and η. Note that there are only slight variations in surface pressure fluctuation coefficients.

Small perturbation theories have been applied to a number of aeronautical problems, in which the flow is characterized by a small deviation of the flow from its original uniform flow. Linearized solutions are useful for explicitly identifying trends and governing parameters. Consider a slender body at hypersonic speeds, the surface pressure coefficient can be written in terms of hypersonic similarity parameter (K) and specific heat ratio (γ), in which [13]

$$C_p = \frac{2\eta^2}{\gamma K^2} \left[\left(1 - \frac{\gamma - 1}{2} K \right)^{2\gamma/(\gamma-1)} - 1 \right] \qquad (1)$$

or

$$\frac{C_p}{\eta^2} = f(K, \gamma). \qquad (2)$$

As mentioned above, the parameters $M^2 \eta / \sqrt{1 - M^2}$ and $M\eta$ were employed to characterize compressible convex- and concave-corner flows, respectively [16]. A common similarity parameter for subsonic corner flows is of interest. Therefore, the present study adopts a combined parameter $K^* (= M^2 / \eta \sqrt{1 - M^2})$ as a governing parameter to scale the flow characteristics in both subsonic convex- and concave-corner flows for rapid estimation of the interaction region, peak Mach number, and peak pressure fluctuations.

2. Experimental Setup

2.1. Transonic Wind Tunnel. The experiments were conducted at a transonic wind tunnel of blowdown type, located at Aerospace Science and Technology Research Center, National Cheng Kung University, Taiwan. Major components of the facility include compressors, air dryers, cooling water system, storage tanks, and the tunnel. The dew point of high-pressure air through the air dryers is maintained at $-40°C$ under normal operation conditions. Operating Mach number ranges from 0.2 to 1.4, and simulated Reynolds number is up to 20 million per meter. The test section is 600×600 mm and 1500 mm long. In the present study, the test section was assembled with solid sidewalls and perforated top/bottom walls. The freestream Mach numbers were 0.33 and 0.64 ± 0.01, and the stagnation pressure and temperature were 172 ± 0.5 kPa and room temperature, respectively. For the data acquisition system, the LeCroy waveform recorders were used. A host computer with CATALYST software controlled the setup of LeCroy waveform recorders through a LeCroy 8901A interface. All input channels were triggered simultaneously.

2.2. Test Model. The test model consisted of a flat plate and an interchangeable instrumentation plate. The test model was 150 mm wide and 600 mm long, which was supported by a single sting mounted on the bottom wall of the test section, as shown in Figure 1. The concave corner with 5-, 7-, 10-, and 15-deg angles or the convex corner with 5-, 10-, 13-, and 15-deg angles was located at 500 mm from the leading edge of the flat plate. Along the centerline of each instrumentation plate, 19 pressure taps (6 mm apart and

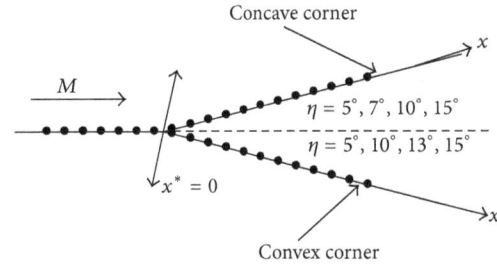

* Solid circles: location of pressure transducers

FIGURE 1: Experimental setup.

2.5 mm in diameter) were drilled perpendicular to the test surface. All the pressure transducers were flush-mounted to the test surface. The side fences of the instrumentation plate were also installed to prevent cross-flow. A study by Miau et al. [18] indicated that the transition of the boundary layer under the present test condition is close to the leading edge of the flat plate, indicating a turbulent boundary layer at the measurement locations. The boundary layer thickness, δ, at 25 mm upstream of the corner was approximately 7.0 mm.

2.3. Experimental Techniques. The Kulite (Model XCS-093-25A, B screen) pressure transducers, which were powered by a TES Model 6102 power supply at 15.0 volts, were employed for pressure measurements. Their outer diameter and sensing element are 2.36 mm and 0.97 mm, respectively. Note that the perforated screen of the pressure transducers might limit the frequency response to only 50 kHz [19]. To improve the signal-to-noise ratio, external amplifiers (Ecreon Model E713) were also employed. With a gain of 20, the roll-off frequency is about 140 kHz. The typical sampling period is 5 μs (200 kHz). Each data record possesses 131,072 data points for statistical analysis. The data were divided into 32 blocks. The mean values of each block (4,096 data points) were calculated. Variations of the blocks are estimated to be 0.43 and 0.13 percent for the mean surface pressure coefficient, C_p, and the fluctuating pressure coefficient, $C_{\sigma p}$, respectively, which were taken as uncertainty of the experimental data.

3. Results and Discussions

3.1. Mean and Fluctuating Surface Pressure Distributions. Distributions of the mean surface pressure coefficient C_p at $M = 0.64$ are shown in Figure 2, where $x^* (= x/\delta)$ is the normalized streamwise distance. The origin of the x coordinate is set at the corner. The solid symbol corresponds to convex-corner flows, while the hollow symbol denotes concave-corner flows. It is known that viscous-inviscid interactions in subsonic corner flows affect displacement thickness (or effective local wall surface) near the corner apex [7]. Thus, as can be seen, convex-corner flows accelerate gradually upstream of the corner followed by stronger expansion and then downstream compression. The minimum pressure coefficient is observed near the corner apex. There are stronger upstream expansion and downstream recompression at $\eta = 15°$. It is also noted that the level of C_p tends to an equilibrium value at further

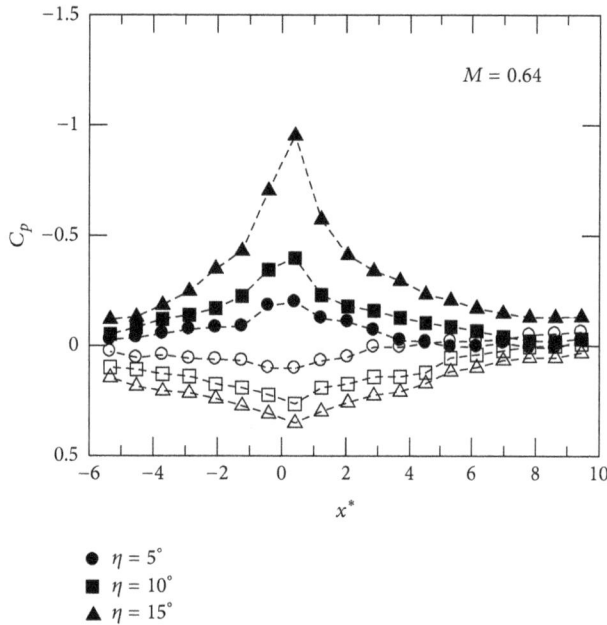

FIGURE 2: Distributions of pressure coefficient, $M = 0.64$. Hallow symbol: concave corner. Solid symbol: convex corner.

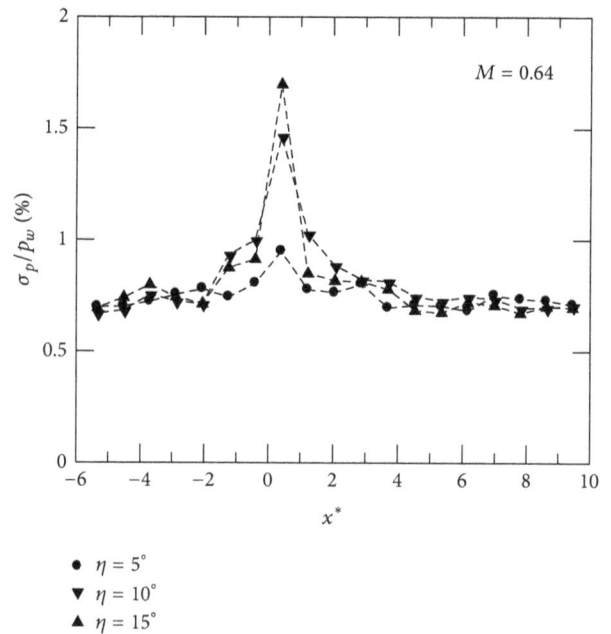

FIGURE 4: Distributions of surface pressure fluctuations, $M = 0.64$.

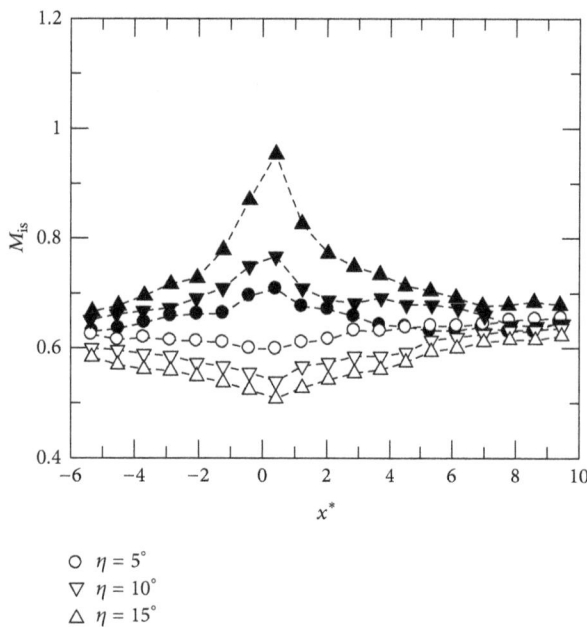

FIGURE 3: Mach number distributions, $M = 0.64$. Hallow symbol: concave corner. Solid symbol: convex corner.

downstream locations and decreases when η increases. For concave-corner flows at $M = 0.64$, the pressure distributions show a similar shape. The flow decelerates upstream of the corner followed by expansion. The amplitude of C_p appears to increase and decrease linearly within the upstream and downstream influence regions. At further downstream locations, there is a more positive C_p at $\eta = 15°$. Moreover, the interaction region tends to expand in both upstream and downstream directions for both test cases, indicating

stronger viscous-viscid interactions when η increases. It is also noted that both upstream and downstream pressure gradients are less significant for convex-corner flows. Further, Mach number distributions at $M = 0.64$ are shown in Figure 3. For concave-corner flows, mild variations in streamwise Mach number are observed. The flow expands suddenly and reaches a peak Mach number, M_p, near the convex corner, approaching sonic condition at $\eta = 15°$.

The distributions of normalized surface pressure fluctuations for convex-corner flows at $M = 0.64$ are shown in Figure 4. σ_p/p_w corresponds to the local variation of surface pressure fluctuations. It can be seen that σ_p/p_w increases upstream of the corner and reaches the maximum at immediately downstream location. The rise in σ_p/p_w corresponds to the initial pressure rise of mean surface pressure or downstream adverse pressure gradient, as shown in Figure 2. The value of peak pressure fluctuations is higher when η increases. At further downstream locations, the amplitude of σ_p/p_w approaches an equilibrium level. Note that there are minor variations in σ_p/p_w for subsonic flow over a concave corner.

3.2. Flow Similarity. The presence of a convex corner in a subsonic uniform flow results in expansion near the corner, corresponding to the increment in displacement thickness by viscous-inviscid interaction. The minimum pressure coefficients $C_{p,\min}$ are associated with freestream Mach number and deflection angle. In Figure 5, $C_{p,\min}/\eta^2$ is plotted against K^* for all test cases. It is noted that η is in radian in the following analyses. Although the data are slightly scattered, it can be seen that $C_{p,\min}/\eta^2$ decreases linearly with K^* in convex-corner flows. For concave-corner flows at $M = 0.64$, there is mild adverse pressure gradient upstream of the

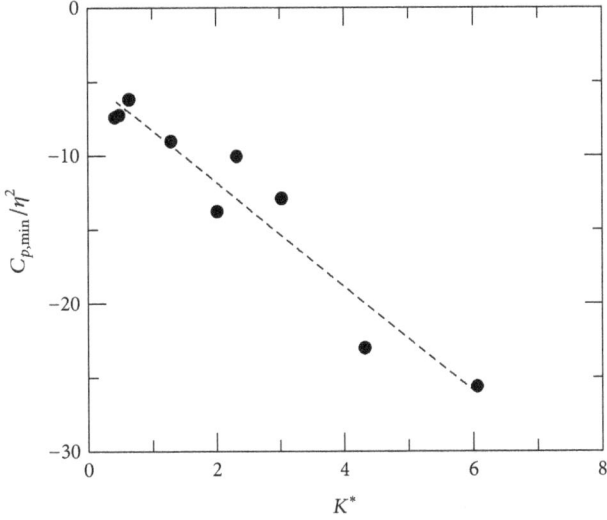

FIGURE 5: Minimum pressure coefficient, convex-corner flows.

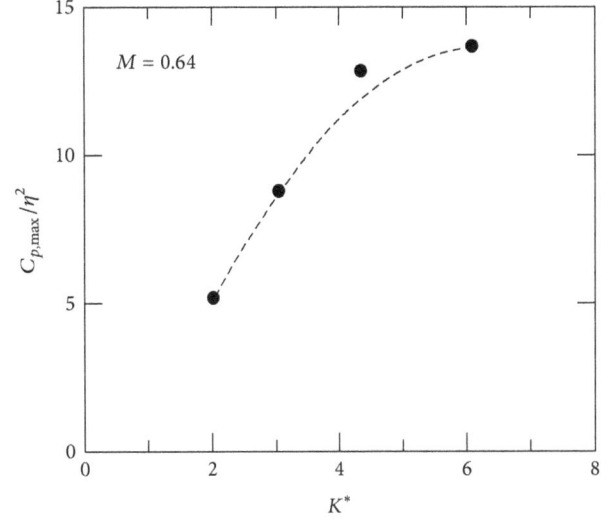

FIGURE 6: Maximum pressure coefficient, concave-corner flows.

corner. $C_{p,\text{max}}/\eta^2$ can also be scaled with K^*, as shown in Figure 6, in which $C_{p,\text{max}}/\eta^2$ increases linearly with K^*, but not for the test case of $K^* = 6.1$. $C_{p,\text{max}}/\eta^2$ appears to be a quadratic function of K^* for concave-corner flows. To elaborate further the correlation of flow expansion/compression near the corner with K^*, M_p^2/η^2 is plotted against K^* for all test cases. As shown in Figure 7, M_p^2/η^2 is a quadratic function of K^* for both test cases. M_p^2/η^2 increases significantly when K^* increases. However, there are only mild variations in M_p^2/η^2 with K^* for concave-corner flows, demonstrating more significant viscous-inviscid interactions in subsonic convex-corner flow than in subsonic concave-corner flows.

Interaction region, including upstream influence X_u and downstream influence X_d, can be employed to highlight viscous-viscid interactions in subsonic convex- and concave-corner flows. Upstream influence, $X_u^*(= X_u/\delta)$, can be determined as the intercept of the tangent to the maximum pressure gradient with the undisturbed surface pressure (or $C_p = 0$). Downstream influence, $X_d^*(= X_d/\delta)$, represents the distance for a disturbed boundary layer back to an equilibrium state and can be estimated from the peak pressure near the corner to the intersection of the tangent through the downstream pressure data with the approximately equilibrium downstream pressure [20]. At $M = 0.64$, variations of X_u^* and X_d^* with K^* are shown in Figure 8. As can be seen, X_u^* and X_d^* decrease linearly with K^* for concave-corner flows. For convex-corner flows, there is a shorter interaction region, corresponding to higher upstream and downstream pressure gradients near the corner. X_u^* and X_d^* appear to be a quadratic function of K^*.

Pressure fluctuations are coupled with global flow unsteadiness. In general, the shear layer structures in the buffer region are responsible for the generation of high-amplitude wall pressure peaks [21]. Laganelli et al. [22] examined wall pressure fluctuations in the attached boundary layer flow. They noted that σ_p/p_w is proportional to M^2.

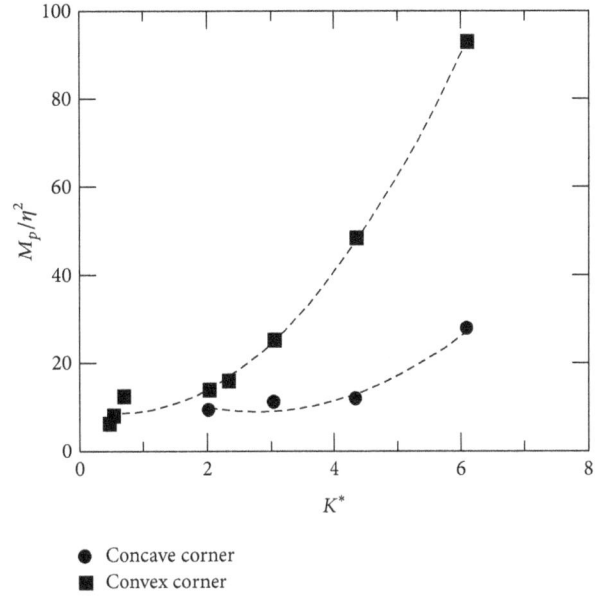

● Concave corner
■ Convex corner

FIGURE 7: Peak Mach number.

In Figure 9, peak pressure fluctuations, $(\sigma_p/p_w)_{\text{max}}/\eta^2$, are plotted against K^* for convex-corner flows. Note that there is only slight variation in surface pressure fluctuations for concave-corner flows. It can be seen that $(\sigma_p/p_w)_{\text{max}}/\eta^2$ increases linearly with K^*. More intense pressure fluctuations are associated with higher peak Mach number or flow expansion near the convex-corner apex.

4. Conclusions

This paper investigates the flow characteristics of subsonic convex- and concave-corner flows. Mean and fluctuating pressures are presented. A combined flow similarity parameter, K^*, based on the small perturbation theory, is employed as a governing parameter to identify the trends in flow

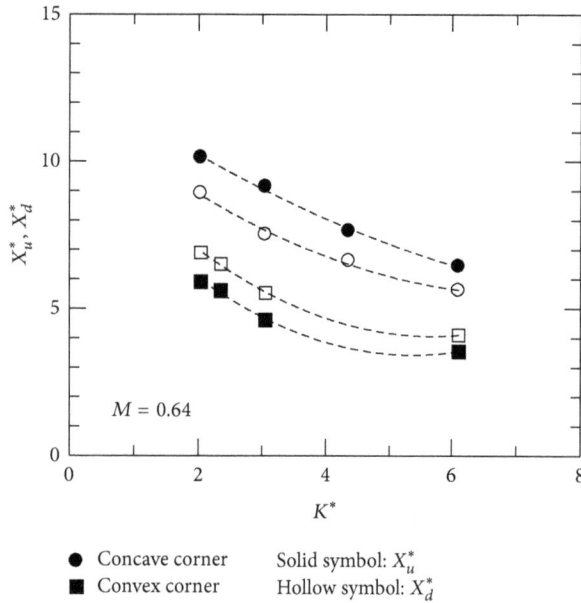

FIGURE 8: Upstream and downstream influences.

● Concave corner   Solid symbol: X_u^*
■ Convex corner   Hollow symbol: X_d^*

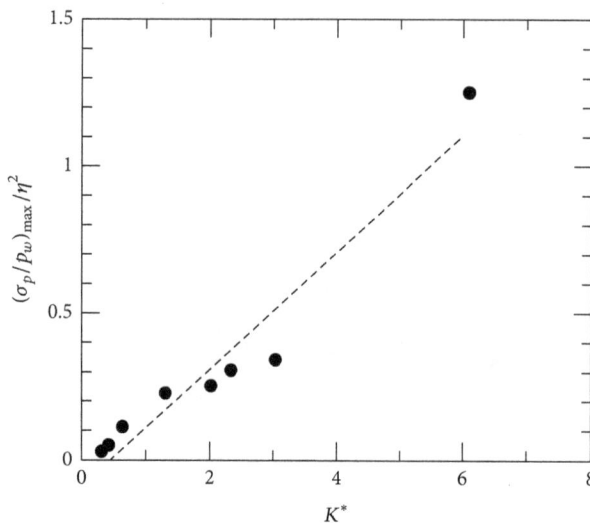

FIGURE 9: Peak pressure fluctuations, convex-corner flows.

M: Freestream Mach number
M_p: Peak Mach number near corner apex
p_w: Mean surface pressure
p_∞: Freestream mean static pressure
q: Freestream dynamic pressure
x: Coordinate along the surface of the corner, cm
x^*: Normalized streamwise distance, x/δ
X_u^*: Upstream influence, X_u/δ
X_d^*: Downstream influence, X_d/δ
K^*: Similarity parameter, $M^2/\eta\sqrt{1-M^2}$
η: Corner angle
δ: Incoming boundary layer thickness, mm
σ_p: Standard deviation of surface pressure.

Acknowledgment

The research has been supported by the National Science Council under Grant NSC 99–2923-E-006-007-MY3.

References

[1] D. S. Dolling and C. T. Or, "Unsteadiness of the shock wave structure in attached and separated compression ramp flows," *Experiments in Fluids*, vol. 3, no. 1, pp. 24–32, 1985.

[2] D. S. Dolling, "Fifty years of shock-wave/boundary-layer interaction research: what next?" *AIAA Journal*, vol. 39, no. 8, pp. 1517–1531, 2001.

[3] T. C. Adamsom Jr., "Effect of transport properties on supersonic expansion around a corner," *Physics of Fluids*, vol. 10, no. 5, pp. 953–962, 1967.

[4] K.-M. Chung and F. K. Lu, "Damping of surface pressure fluctuations in hypersonic turbulent flow past expansion corners," *AIAA Journal*, vol. 31, no. 7, pp. 1229–1234, 1993.

[5] A. Bolonkin and G. B. Gilyard, "Estimated benefits of variable-geometry wing camber control for transport aircraft," NASA TM-1999-206586, National Aeronautics and Space Administration, Dryden Flight Research Center, 1999.

[6] J. Szodruch and R. Hilbig, "Variable wing camber for transport aircraft," *Progress in Aerospace Sciences*, vol. 25, no. 3, pp. 297–328, 1988.

[7] A. I. Ruban, X. Wu, and R. M. S. Pereira, "Viscous-inviscid interaction in transonic Prandtl-Meyer flow," *Journal of Fluid Mechanics*, vol. 568, pp. 387–424, 2006.

[8] F. T. Smith and J. H. Merkin, "Triple-deck solutions for subsonic flow past hymps, steps, concave or convex corners and wedged trailing edges," *Computers and Fluids*, vol. 10, no. 1, pp. 7–25, 1982.

[9] K.-M. Chung, "Aerodynamic characteristics of deflected surfaces in compressible flows," *Journal of Aircraft*, vol. 41, no. 2, pp. 415–418, 2004.

[10] K.-M. Chung, "Investigation on compressible ramp flows," *Journal of Aeronautics, Astronautics and Aviation. Series A*, vol. 38, no. 3, pp. 167–172, 2006.

[11] K.-M. Chung, "Transition of subsonic and transonic expansion-corner flows," *Journal of Aircraft*, vol. 37, no. 6, pp. 1079–1082, 2000.

[12] J. D. Anderson Jr., *Moden Compressible Flow: With Historical Perspective*, Mcgraw-Hill, 1990.

expansion/compression, interaction region, and peak pressure fluctuations. For convex-corner flows, variations in minimum pressure coefficient and peak pressure fluctuations can be scaled linearly with K^*, but not for upstream and downstream influences. Upstream compression in concave-corner flows appears to be a quadratic function of K^*. The results imply that there are small streamline deflections near corner apex, producing proportionally small change in Mach number and pressure. The present results can also be used for a quick estimation of aerodynamic characteristics of a deflected surface in subsonic flows.

Nomenclature

C_p: Pressure coefficient, $(p_w - p_\infty)/q$
C_{σ_p}: Fluctuating pressure coefficient, $(\sigma_p - \sigma_{p\infty})/q$

[13] M. D. van Dyke, "The combined supersonic and hypersonic similarity rule," *Journal of Aeronautical Science*, vol. 18, no. 7, pp. 499–500, 1951.

[14] K.-M. Chung, "Investigation on transonic convex-corner flows," *Journal of Aircraft*, vol. 39, no. 6, pp. 1014–1018, 2002.

[15] A. Verhoff, D. Stockesberry, and T. Michal, "Hodograph solution for compressible flow past a corner and comparison with Euler numerical predictions," AIAA Paper 91-1547, 1991.

[16] K.-M. Chung, P.-H. Chang, and K.-C. Chang, "Flow similarity in compressible convex-corner flows," *AIAA Journal*, vol. 50, no. 4, pp. 985–988, 2012.

[17] K.-M. Chung, "Characteristics of compressible concave-corner flows," *Journal of Aircraft*, vol. 40, no. 4, pp. 797–799, 2003.

[18] J. J. Miau, J. Cheng, K. M. Chung, and J. F. Chou, "The effect of surface roughness on the boundary layer transition," in *Proceeding of the 7th International Symposium in Flow Modeling and Turbulent Measurement*, pp. 609–616, Tainan, Taiwan, 1998.

[19] R. A. Gramann and D. S. Dolling, "Detection of turbulent boundary-layer separation using fluctuating wall pressure signals," *AIAA Journal*, vol. 28, no. 6, pp. 1052–1056, 1990.

[20] K. Chung, "Interaction region of turbulent expansion-corner flow," *AIAA Journal*, vol. 36, no. 6, pp. 1115–1116, 1998.

[21] J. Kim, K. Kim, and H. J. Sung, "Wall pressure fluctuations in a turbulent boundary layer after blowing or suction," *AIAA Journal*, vol. 41, no. 9, pp. 1697–1704, 2003.

[22] A. L. Laganelli, A. Martellucci, and L. L. Shaw, "Wall pressure fluctuations in attached boundary-layer flow," *AIAA Journal*, vol. 21, no. 4, pp. 495–502, 1983.

15

Measurement of Baseline and Orientation between Distributed Aerospace Platforms

Wen-Qin Wang

The School of Communication and Information Engineering, University of Electronic Science and Technology of China (UESTC), Chengdu 611731, China

Correspondence should be addressed to Wen-Qin Wang; wqwang@uestc.edu.cn

Academic Editors: K. M. Isaac and C. Zhong

Distributed platforms play an important role in aerospace remote sensing, radar navigation, and wireless communication applications. However, besides the requirement of high accurate time and frequency synchronization for coherent signal processing, the baseline between the transmitting platform and receiving platform and the orientation of platform towards each other during data recording must be measured in real time. In this paper, we propose an improved pulsed duplex microwave ranging approach, which allows determining the spatial baseline and orientation between distributed aerospace platforms by the proposed high-precision time-interval estimation method. This approach is novel in the sense that it cancels the effect of oscillator frequency synchronization errors due to separate oscillators that are used in the platforms. Several performance specifications are also discussed. The effectiveness of the approach is verified by simulation results.

1. Introduction

Distributed aerospace platforms play a more and more important role in aerospace remote sensing, radar navigation, and wireless communication applications [1–3]. One representative application example is the distributed synthetic aperture radar (SAR) remote sensing [4–6], especially the distributed interferometry SAR (InSAR) remote sensing which is an imaging technique for measuring the topography of surface, along with its changes over time [7]. They are of fundamental importance for a broad range of commercial and scientific applications [8]. For example, many geoscience areas, like hydrology, glaciology, forestry, geology, oceanography, and land environment, require precise and up-to-date information about the earth's surface and its topography.

Conventional repeat-pass InSAR may suffer from temporal decorrelation and atmospheric distortions. Such disadvantages may be avoided by using distributed InSAR, where separate transmitter and receiver are used. Distributed InSAR enables a flexible imaging geometry with large baselines, thereby increasing significantly the interferometric performance [9–11]. Recently, advances in distributed InSAR

techniques have addressed some of the limitations in conventional InSAR systems and subsequently have opened many new remote sensing applications [12–14], because it can avoid time decorrelation between the interferometric signals and improve the topographic measurement sensitivity owing to large baseline formed between the transmitter and receiver. However, besides the requirement of highly accurate time and frequency synchronization for coherent signal processing [15], one has to determine the baseline between transmitter and receiver and the orientation of platform towards each other during data recording. These parameters directly affect the achievable geometrical resolution and the geometry of the radar system.

To determine baseline and orientation for bistatic airborne radars, a method by installing 4×4 dedicated navigation units on the transmitter and receiver was proposed in [16]. The approach is further investigated in [17] by employing only two transmitting antennas and two receiving antennas. However, the influence of oscillator synchronization errors is not considered, and several contributions such as the influence of ionosphere or relativistic effects are also neglected. We are aware that these contributions might become significant.

So this paper extends the approach to include additional factors such as oscillator frequency synchronization errors, Doppler effects, and ionosphere or relativistic effects, which are crucial for ensuring the performance of baseline and orientation estimation.

This paper concentrates on the measurement of baseline and orientation between distributed aerospace platforms for general aerospace applications, not for distributed InSAR remote sensing only. In distributed aerospace platforms, the baseline and orientation will change during the flying time and they should be estimated in real time [18], due to the unstability of the separate platforms. To reach this aim, this paper proposes an improved pulsed duplex microwave ranging technique to determine the spatial baseline and orientation between distributed aerospace platforms.

The remaining sections of this paper are organized as follows. To adaptively estimate the change of spatial baseline and orientation between the transmitting platform and receiving platform, Section 2 proposes a high precision microwave ranging approach. Section 3 presents a high-precision time-interval estimation algorithm. Next, the baseline and orientation estimation performance is analyzed in Section 4. Finally, this paper is concluded in Section 5.

2. Estimation of Baseline and Orientation

In many applications, the baseline between distributed aerospace platforms and the orientation towards each other must be measured in a high precision manner during data recording. Up to now, these parameters are usually determined using independent global navigation satellite system (GNSS)—like the GPS—and differential GPS systems (DGPS) receivers on each platform. The achievable accuracy of a DGPS in real time is about 10 cm and several millimeters after postprocessing. However, much higher accuracy is required for some specific applications; for example, an 0.3 mm accuracy is needed for an X-band distributed InSAR system [16].

To adaptively determine the baseline between the transmitting platform and receiving platform and the orientation towards each other, we propose a pulsed duplex microwave ranging method. In this method, the transmitting navigation unit generates a signal from one stable local oscillator and splits it into three individual navigation signals. Each of these signals is modulated and transmitted over separate antennas. These navigation antennas are pointing towards the receiving platform. In this way, the baseline and orientation between the platforms can be adaptively measured.

Figure 1 shows a configuration of 3×3 two-way ranging technique; the navigation signal is modulated by pseudo-random noise (PN) codes. These PN codes are chosen from a set of gold codes and are unique for each navigation signal; hence, the correlation between any pair of codes is very low. This allows us to use the same carrier frequency for all the navigation signals, so that the effects of oscillator frequency synchronization errors can be canceled out. Moreover, some additional information such as the position of the antennas and the position of the platform is also modulated with these

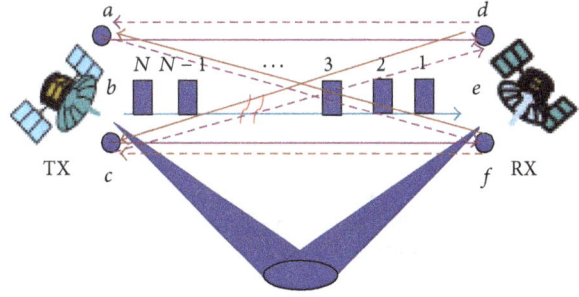

FIGURE 1: Two aerospace platforms and possible combination of navigation signals.

signals. At time T_{TX}, the three navigation signals are transmitted from the TX using separate antennas and therefore travel on different paths towards the RX. Similarly, there are three navigation antennas installed in the RX platform. They are received by the RX antenna d at times T_{RXad}, T_{RXbd}, and T_{RXcd}, respectively. For the other two RX antennas, there are similar results. After frame synchronization, the messages can then be decoded and used for subsequent range calculations. The pseudorange between the TX navigation antennas and each RX navigation antenna can be determined by the time of the signal propagations.

But the measured time may be biased due to the lack of time synchronization between the clocks inside the TX and RX platforms. Additionally, the time may be biased due to the effects of ionosphere and receiver noise. Hence, for the RX antenna d, the measured time equation can be expressed as

$$\Delta T_{ad} = (T_{RXad} - T_{TX}) + \Delta T_{syn} + \Delta t_{ion} + \Delta t_{noi},$$

$$\Delta T_{bd} = (T_{RXbd} - T_{TXab}) + \Delta T_{syn} + \Delta t_{ion} + \Delta t_{noi}, \quad (1)$$

$$\Delta T_{cd} = (T_{RXcd} - T_{TXab}) + \Delta T_{syn} + \Delta t_{ion} + \Delta t_{noi},$$

where ΔT_{syn}, Δt_{ion}, and Δt_{noi} denote the time offsets caused by time synchronization errors, ionosphere effect, and receiver noise, respectively.

As the effect of ionosphere can be compensated to satisfactory accuracy by using experiential radio propagation models, for example, [19]

$$\Delta t_{ion} = \frac{A}{c_0 \cdot f_0^2}, \quad (2)$$

where A is one constant, and c_0 and f_0 are the speed of light and carrier frequency, respectively. Equation (1) can then be further simplified into

$$\Delta T'_{ad} = (T_{RXad} - T_{TX}) + \Delta T_{syn} + \Delta t_{noi},$$

$$\Delta T'_{bd} = (T_{RXbd} - T_{TX}) + \Delta T_{syn} + \Delta t_{noi}, \quad (3)$$

$$\Delta T'_{cd} = (T_{RXcd} - T_{TX}) + \Delta T_{syn} + \Delta t_{noi}.$$

Similarly, for the RX antennas e and f, we also have

$$\Delta T'_{ie} = \left(T_{\text{RX}ie} - T_{\text{TX}}\right) + \Delta T_{\text{syn}} + \Delta t_{\text{noi}}, \tag{4}$$

$$\Delta T'_{if} = \left(T_{\text{RX}if} - T_{\text{TX}}\right) + \Delta T_{\text{syn}} + \Delta t_{\text{noi}}, \tag{5}$$

where $i \in (a, b, c)$.

To cancel out the effect of oscillator frequency synchronization errors, three navigation signals are transmitted from the RX platform to the TX platform at time T_{RX}. In a like manner, we can get

$$\Delta T'_{ji} = \left(T_{\text{TX}ji} - T_{\text{RX}}\right) - \Delta T_{\text{syn}} + \Delta t_{\text{noi}}, \tag{6}$$

with $j \in (d, e, f)$. Note that the $T_{\text{TX}ji}$ are defined as previously. Since

$$T_{\text{RX}ij} - T_{\text{TX}} \simeq T_{\text{TX}ji} - T_{\text{RX}} \tag{7}$$

We can obtain the signal propagation time between the TX antennas and RX antennas:

$$\tau_{ij} = \tau_{ji} = \frac{\Delta T'_{ij} + \Delta T'_{ji}}{2} - \Delta t_{\text{noi}} \approx \frac{\Delta T'_{ij} + \Delta T'_{ji}}{2},$$

$$\Delta T_{\text{syn}} = \frac{\Delta T'_{ij} - \Delta T'_{ji}}{2}. \tag{8}$$

Hence, the pseudo-range between the TX navigation antennas ($i \in (a, b, c)$) and TX navigation antennas ($i \in (d, e, f)$) is the time offset τ_{ij} (or τ_{ji}) multiplied by the speed of light c_0 and is biased by the effect of receiver noise investigated in subsequent section:

$$R_{ij} = R_{ji} = c_0 \cdot \tau_{ij} \approx c_0 \cdot \frac{\Delta T'_{ij} + \Delta T'_{ji}}{2}. \tag{9}$$

Thus, the relative distance between the TX platform and RX platform can be determined from the calculated pseudo-range. As an example, for the RX antenna d we can get

$$\left(x_i - x_d\right)^2 + \left(y_i - y_d\right)^2 + \left(z_i - z_d\right)^2 = R_{id}^2,$$

$$\left(x_i - x_e\right)^2 + \left(y_i - y_e\right)^2 + \left(z_i - z_e\right)^2 = R_{ie}^2, \tag{10}$$

$$\left(x_i - x_f\right)^2 + \left(y_i - y_f\right)^2 + \left(z_i - z_f\right)^2 = R_{if}^2.$$

In this way, using the three determined potions of the navigation antennas on the RX platform (x_j, y_j, z_j), the spatial baseline and orientation between both platforms can be determined. More importantly, this method can cancel the effect of oscillator frequency synchronization errors.

To obtain pseudo-range, frame synchronization is required, which generally involves two steps: PN codes acquisition and tracking. In the first stage, sliding correlator is usually used to reduce the time uncertainty between local replica PN codes and received codes. The timing uncertainty covers a region that is quantized into a finite number of cells. These cells are serially tested until it is determined that a particular cell corresponds to the alignment of the two

sequences to within a fraction of chip. In codes tracking stage, a delay-locked loop (DLL) [20, 21] can be used.

To describe this method, let us consider a rectangular pulse $x(t)$, $0 \leq t \leq T$, the output of a filter matched to $x(t)$ attains its maximum value at time $t = T$, that is, at the peak of the correlation function, as shown in Figure 2. In the presence of noise, the identification of the peak value will be difficult. Instead of sampling the signal at peak, suppose we sample early at $t = T - \delta$ and late at $t = T + \delta$. The absolute value of the early samples $|S[n(T - \delta)]|$ and the late samples $|S[n(T+\delta)]|$ will be smaller than the samples of the peak value $|S[n(T)]|$. As the autocorrelation function is even with respect to the optimum sampling time $t = T$, the absolute values of the correlation function at $t = T - \delta$ and $t = T + \delta$ are equal. Under this condition, the proper sampling time is the midpoint between $t = T - \delta$ and $t = T + \delta$.

This condition forms the basis for the tracking technique, as shown in Figure 3. The two correlators integrate over the symbol interval T, but one correlator starts integrating δ seconds early relative to the estimated optimum sampling time, and the other integrator starts integrating δ seconds late relative to the estimated optimum sampling time. An error signal is formed by taking the difference between the absolute values of the two correlator outputs. If the timing is off relative to the optimum sampling time, the average error signal at the output of the low-pass filter is nonzero, and the clock signal is either retarded or advanced, depending on the sign of the error. In this way, the smoothed error signal is used to drive a voltage compensation controller (VCC), whose output is the desired clock signal.

3. High-Precision Time-Interval Estimation

To obtain a high-precision baseline estimation, the time difference between transmission and reception must be measured. We use the high-precision time-to-phase conversion measurement technique shown in Figure 4. This technique uses the two measured signals T_1 and T_2 to trigger an analog-to-digital converter (DAC) to sample one sinewave signal generated from the local oscillator directly.

We start to describe this approach from a sinewave signal expressed as

$$s(t) = a_0 \cos\left(2\pi f_{lo} t + \phi(t) + \phi_0\right), \tag{11}$$

where a_0 is the amplitude, f_{lo} is the center frequency, $\phi(t)$ is the phase fluctuation, and ϕ_0 is the starting phase. We then have

$$s_a = a_0 \cos\left(2\pi f_{lo} t_a + \phi(t_a) + \phi_0\right),$$

$$s_b = a_0 \cos\left(2\pi f_{lo} (t_a + T_1) + \phi(t_a + T_1) + \phi_0\right),$$

$$s_c = a_0 \cos\left(2\pi f_{lo} t_c + \phi(t_c) + \phi_0\right),$$

$$s_d = a_0 \cos\left(2\pi f_{lo} (t_c + T_2) + \phi(t_c + T_2) + \phi_0\right), \tag{12}$$

where t_a and t_c are the time at a and c, respectively, T_1 is the time interval between a and b, and T_2 is the time interval between c and d.

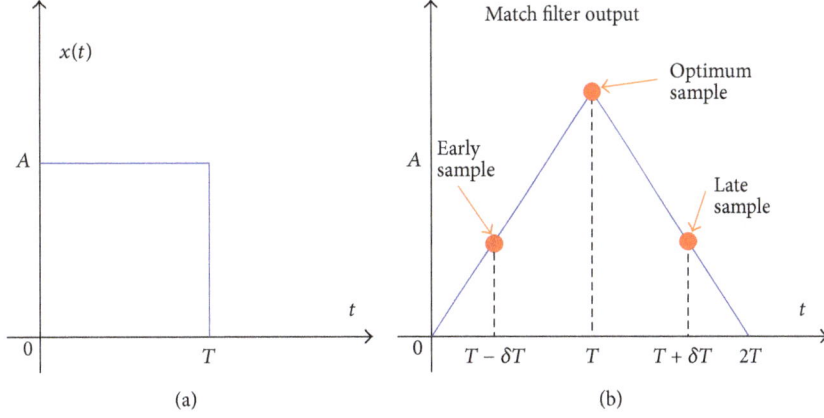

FIGURE 2: Rectangular signal pulse (a) and its matched filter output (b).

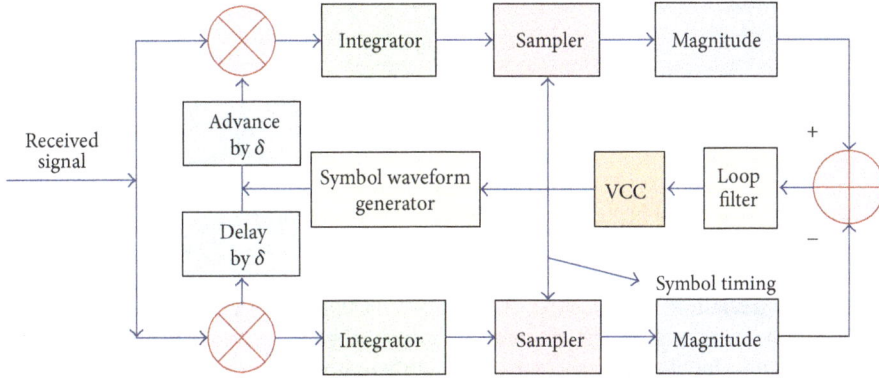

FIGURE 3: Block diagram of synchronizing tracker.

The phase terms can be further expressed as

$$\Phi_a = 2\pi f_{lo} t_a + \phi(t_a) + \phi_0 = \arccos\left(\frac{s_a}{a_0}\right) + 2k_1\pi,$$

$$\Phi_b = 2\pi f_{lo}(t_a + T_1) + \phi(t_a + T_1) + \phi_0$$

$$= \arccos\left(\frac{s_b}{a_0}\right) + 2k_1\pi,$$

$$\Phi_c = 2\pi f_{lo} t_c + \phi(t_c) + \phi_0 = \arccos\left(\frac{s_c}{a_0}\right) + 2k_2\pi,$$

$$\Phi_d = 2\pi f_{lo}(t_c + T_2) + \phi(t_c + T_2) + \phi_0$$

$$= \arccos\left(\frac{s_d}{a_0}\right) + 2k_2\pi. \tag{13}$$

After the 2π-phase ambiguities in Φ_a, Φ_b, Φ_c, and Φ_d being removed in a like manner that is performed in the GPS processing [22], we can get

$$T_1 = \frac{(\Phi_b - \Phi_a) - (\phi(t_a + T_1) - \phi(t_a))}{2\pi f_{lo}},$$

$$T_2 = \frac{(\Phi_d - \Phi_c) - (\phi(t_c + T_2) - \phi(t_c))}{2\pi f_{lo}}. \tag{14}$$

Hence, the time interval between T_A and T_B signals is

$$T_{int} = mT_{clk} - T_1 + T_2$$

$$= mT_{cl} - \frac{(\Phi_b - \Phi_a) - (\phi(t_a + T_1) - \phi(t_a))}{2\pi f_{lo}}$$

$$+ \frac{(\Phi_d - \Phi_c) - (\phi(t_c + T_2) - \phi(t_c))}{2\pi f_{lo}} \tag{15}$$

As the parameters T_{clk}, Φ_a, Φ_b, Φ_c, Φ_d, m, and f_{lo} are all measurable or calculable and the parameters $\phi(t_a + T_1) - \phi(t_a)$ and $\phi(t_c + T_2) - \phi(t_c)$ are neglectable, the time difference between T_1 and T_2 signals can be calculated from (15).

4. Design Specifications

Several factors will influence the phase of the microwave measurement link. The receiver noise determined by the signal-to-noise ratio (SNR) is of special interest; its influence on the signal phase is described by the receiver phase noise spectral density function $S_{\theta SNR}(f)$. The hardware system error is represented by $\theta_{sys}(t)$. Furthermore, the phase is sampled, which requires a later interpolation. We may choose to filter this phase with an arbitrary transfer function $H_{pf}(f)$. The link error, that is, the residual error contribution, is

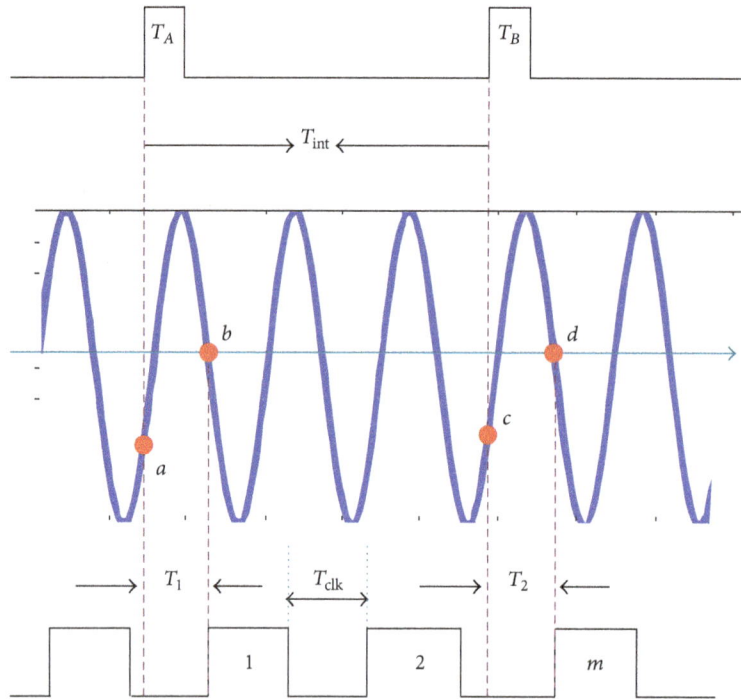

FIGURE 4: One time-interval measurement approach.

represented by the standard deviation (STD) σ_{link}. This section makes an investigation on these residual influence factors such as receiver noise and Doppler effects as well as contributions known from the sampling theory.

4.1. Receiver Noise. The receiver noise consists of thermal noise, and the noise collected by antenna will introduce both amplitude and phase variations to the navigation signal. Here, only the phase variations described by their spectral density function are of interest. For band-limited white noise, the spectral density function $S_{\varphi\text{SNR}}(f)$ is related to the SNR by [23]

$$S_{\varphi\text{SNR}}(f) = \frac{1}{2B_w \cdot \text{SNR}}, \tag{16}$$

where B_w is the receiver noise bandwidth. Correspondingly, the receiver noise variance then is

$$\sigma_{\text{SNR}}^2 = \frac{1}{2f_{\text{sys}} \cdot \text{SNR}} \int_{-f_{\text{sys}}/2}^{f_{\text{sys}}/2} \left| H_{\text{pf}(f)} \right|^2 df, \tag{17}$$

where f_{sys} is the measurement rate. Note that here uncorrelated noise and equal SNR value are assumed for both receivers.

As an example, assuming a bistatic spaceborne InSAR system with the following parameters: B_w = 100 MHz, PRF = 5000 Hz, pulse duration τ_p = 15 μs, and synthetic aperture time T_s = 1 s, then $S_{\varphi\text{SNR}}(f)$ is found to be smaller than −120 dBc/Hz. This noise may degrade ranging precision consequently. Generally, it is about 0.2 ~ 0.5 mm.

4.2. Doppler Effects. As the transmit instance of RX is delayed by τ_{sys} with respect to TX, the navigation signals are inherently decoupled. This offers the possibility of using a single carrier frequency. The Doppler phenomenon due to the relative velocity v_{sat} between the TX and RX platforms may be a problem. However, the Doppler phase contribution is constant for constant relative satellite velocity v_{sat}. Only a relative satellite acceleration, that is, a time-dependent relative satellite velocity $v_{\text{sat}}(t)$, will cause a measurement error because the Doppler contribution of constant relative satellite velocity v_{sat} can be canceled out in range calculation. For severe intersatellite acceleration, a Doppler phase compensation that requires the satellite separation to be known is necessary. Fortunately, severe intersatellite acceleration is usually not existing in bistatic spaceborne InSAR systems.

4.3. Analog Digital Converter (ADC). As range estimation is processed in digital signal, analog digital converter (ADC) quantization errors may degrade the estimation precision consequently. The quantization errors result in what appears to be a white noise floor. The amplitude quantization errors e_A can be assumed to be totally uncorrelated and uniformly distributed within each quantization step, that is,

$$-\frac{\Delta_A}{2} \le e_A \le \frac{\Delta_A}{2}. \tag{18}$$

For a D-bit ADC, the quantization step size is

$$\Delta_A = \frac{1}{2^{D-1}}. \tag{19}$$

TABLE 1: Possible phase errors caused by ADC.

D	4	8	10	12	14	16
e_A	$3.13E-2$	$2.0E-3$	$4.9E-3$	$1.2E-3$	$3.1E-5$	$7.6E-6$
$\delta\varphi$	$3.8°$	$0.22°$	$0.056°$	$0.014°$	$0.0035°$	$0.00087°$

Then the amplitude error power is [24]

$$E\left(e_A^2\right) = \frac{1}{\Delta_A}\int_{-\Delta_A/2}^{\Delta_A/2} e_A^2 de_A = \frac{\Delta_A^2}{12}. \tag{20}$$

Accordingly, the noise variance is

$$\sigma = \sqrt{\frac{\Delta_A^2}{12}} = \frac{\Delta_A}{2\sqrt{3}}. \tag{21}$$

As the signal power is

$$S = \left(2^D \Delta_A\right)^2, \tag{22}$$

hence the noise-signal-ratio is

$$\frac{N}{S} = 10\log\frac{\sigma^2}{S} = 20\log\frac{\Delta_A/2\sqrt{3}}{2^D\Delta_A}. \tag{23}$$

Denoting the sampling noise bandwidth as B_n, we then have

$$S_{\varphi_{AD}}(f) = \frac{N}{S}\cdot\frac{1}{B_n} = 20\log\frac{\Delta_A/2\sqrt{3}}{2^D\Delta_A} - 20\log(B_n). \tag{24}$$

As an example, assuming the quantization bits are 12 bits and sampling rate is 300 MHz, then $S_{\varphi_{AD}}(f)$ is found to be -167.77 dBc/Hz. Correspondingly, the phase errors caused by ADC can be modeled as [25]:

$$\delta\varphi_{\max} = \tan^{-1}\left[\frac{1}{2^D-1}\right]. \tag{25}$$

From Table 1, we can conclude that ADC has neglectable effects on the performance of range estimation.

4.4. Other Possible Factors. Practically, the transmitted navigation signal (from the signal generator through frequency conversion, amplification, and transmission by the navigation antenna) has some unwanted phase characteristics. Sources of unwanted errors include nonlinearity in amplifiers, antenna, link path, frequency dependent phase effects in filters and waveguide dispersion. The phase of the hardware system, dominated by active and passive radar radio frequency (RF) components, will change within the duration of data collection. Concerning the performance of the navigation link, all the contributions from components common to the TX and RX path will cancel out, due to two-day operation. In [26], an exemplary measured phase variation in a two-way synchronization link is about $0.39°$. This effect is small and can be ignored.

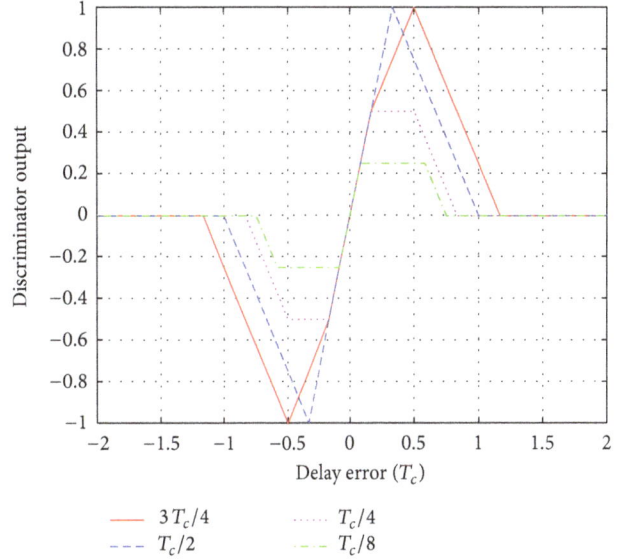

FIGURE 5: Characteristics of the discriminator.

4.5. DLL Ranging Estimation. An important performance criterion for DLL ranging estimation is the tracking jitter, that is, the variation of the delay error around the origin due to input noise, and this will ultimately affect the ranging accuracy. Consider the discriminator characteristics of the DLL as shown in Figure 5; it can be noticed that smaller delay offset gives a higher accuracy but a slightly smaller threshold acquisition range and a substantially smaller quasilinear region. There is another advantage of a smaller early-late gate delay spacing Δ_ξ. For example, if $\Delta_\xi = T_c/2$ (T_c is one chip width of the PN codes), the squared autocorrelation is $R^2(T_c/2) = 1/4$, whereas, for $\Delta_\xi = T_c/8$, it increases to $R^2(T_c/2) = 0.766$, a 5.4 dB improvement. Note also that there are certain disadvantages if the early-late spacing becomes too small such as a slightly smaller threshold acquisition range and a substantially smaller quasilinear region. As a good compromise, $\Delta_\xi = T_c/4$ is used in the following simulations. Under the assumption that early-late gate behaves like a linear filter and the error is in the linear tracking area, the tracking jitter can be evaluated by [27]

$$\sigma_\delta = T_c\sqrt{\frac{\Delta_\xi\omega_L}{2\cdot\text{SNR}}}, \tag{26}$$

where ω_L is the loop filter bandwidth. Figure 6 shows the tracking performance of the DLL versus SNR. From Figure 7 we can conclude that this method can achieve satisfied results.

5. Conclusion

In distributed aerospace platforms, the baseline between the transmitting platform and receiving platform and the orientation of platforms towards each other are the most important parameters, and they are crucial for ensuring the system performance. To adaptively resolve the baseline and orientation estimation between distributed platforms

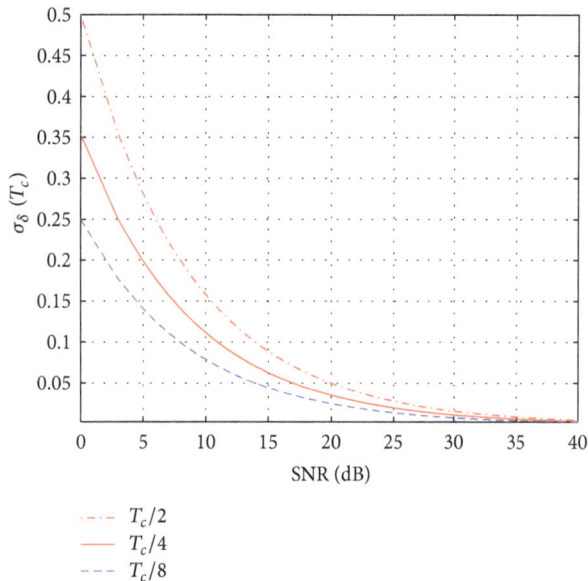

FIGURE 6: Tracking performance of the DLL versus SNR.

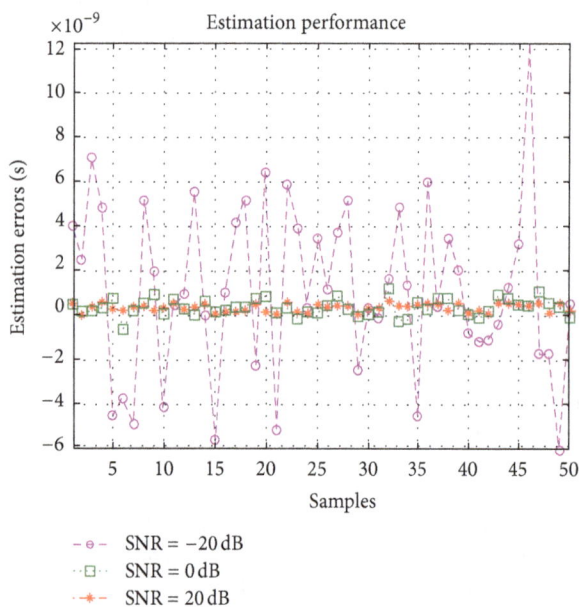

FIGURE 7: Statistical performance of the baseline estimation errors.

for high-precision applications, an improved pulsed duplex microwave ranging method is presented in this paper, which allows determining the spatial baseline and orientation of two spaceborne platforms to each other in real time. Simulation results show that satisfactory measurement accuracy can be obtained for the approach. This approach is novel in the sense that it cancels the effect of oscillator synchronization errors due to that separate oscillators are used in the TX and RX platforms. Note that the influence of Doppler effect is neglected in this paper; we are aware that this contribution might become significant when there is severe intersatellite acceleration between the platforms; however, we rather

choose to maintain a general overview character of our approach.

Acknowledgments

This work was supported in part by the National Natural Science Foundation of China under Grant no. 41101317 and the Program for New Century Excellent Talents in University under Grant no. NCET-12-0095.

References

[1] M. Berioli, A. Molinaro, S. Morosi, and S. Scalise, "Aerospace communications for emergency applications," *Proceedings of the IEEE*, vol. 99, no. 11, pp. 1922–1938, 2011.

[2] R. J. Terrile, "Pathways and challenges to innovation in aerospace," *IEEE Aerospace and Electronic Systems Magazine*, vol. 26, no. 12, pp. 4–9, 2011.

[3] C. H. Yang, J. H. Everitt, Q. Du, B. Luo, and J. Chanussot, "Using high-resolution airborne and satellite imagery to assess crop growth and yield variability for precision agriculture," *Proceedings of the IEEE*, vol. 101, no. 3, pp. 582–592, 2013.

[4] I. Walterscheid, T. Espeter, A. R. Brenner et al., "Bistatic SAR experiments with PAMIR and TerraSAR-X-setup, processing, and image results," *IEEE Transactions on Geoscience and Remote Sensing*, vol. 48, no. 8, pp. 3268–3279, 2010.

[5] M. Rodriguez-Cassola, P. Prats, D. Schulze et al., "First bistatic spaceborne SAR experiments with TanDEM-X," *IEEE Geoscience and Remote Sensing Letters*, vol. 9, no. 1, pp. 33–37, 2012.

[6] W. Q. Wang and H. Z. Shao, "Azimuth-variant signal processing in highaltitude platform passive SAR with spaceborne/airborne transmitter," *Remote Sensing*, vol. 5, no. 3, pp. 1292–1310, 2013.

[7] S. Duque, P. Lopéz-Dekker, and J. J. Mallorqui, "Single-pass bistatic sar interferometry using fixed-receiver configurations: theory and experimental validation," *IEEE Transactions on Geoscience and Remote Sensing*, vol. 48, no. 6, pp. 2740–2749, 2010.

[8] P. A. Rosen, S. Hensley, I. R. Joughin, and S. N. Madsen, "Synthetic aperture radar interferometry," *Proceedings of IEEE*, vol. 88, no. 2, pp. 333–382, 2000.

[9] A. Moccia and G. Rufino, "Spaceborne along-track SAR interferometry: performance analysis and mission scenarios," *IEEE Transactions on Aerospace and Electronic Systems*, vol. 37, no. 1, pp. 199–213, 2001.

[10] D. Massonnet, "Capabilities and limitations of the interferometric cartwheel," *IEEE Transactions on Geoscience and Remote Sensing*, vol. 39, no. 3, pp. 506–520, 2001.

[11] M. Stangl, R. Werninghaus, B. Schweizer et al., "TerraSAR-X technologies and first results," *IEE Proceedings: Radar, Sonar and Navigation*, vol. 153, no. 2, pp. 86–95, 2006.

[12] G. Krieger, A. Moreira, H. Fiedler et al., "TanDEM-X: a satellite formation for high-resolution SAR interferometry," *IEEE Transactions on Geoscience and Remote Sensing*, vol. 45, no. 11, pp. 3317–3340, 2007.

[13] D. Cerutti-Maori and J. H. G. Ender, "Performance analysis of multistatic configurations for spaceborne GMTI based on the auxiliary beam approach," *IEE Proceedings: Radar, Sonar and Navigation*, vol. 153, no. 2, pp. 96–103, 2006.

[14] R. Romeiser and H. Runge, "Theoretical evaluation of several possible along-track InSAR modes of TerraSAR-X for ocean

current measurements," *IEEE Transactions on Geoscience and Remote Sensing*, vol. 45, no. 1, pp. 21–35, 2007.

[15] W. Q. Wang, C. B. Ding, and X. D. Liang, "Time and phase synchronisation via direct-path signal for bistatic synthetic aperture radar systems," *IET Radar, Sonar and Navigation*, vol. 2, no. 1, pp. 1–11, 2008.

[16] M. Weiß, "Determination of baseline and orientation of platforms for airborne bistatic radars," in *Proceedings of the IEEE International Geoscience and Remote Sensing Symposium (IGARSS '05)*, pp. 1967–1970, Seoul, Korea, July 2005.

[17] W. Wang, "Baseline estimation in distributed spaceborne interferometry SAR systems," in *Proceedings of the IEEE Aerospace Conference, AC*, Big Sky, Mont, USA, March 2008.

[18] W. Q. Wang, "Optimal baseline design and error compensation for bistatic spaceborne InSAR," in *Proceedings of the ESA Workshop Advances in SAR Interferometry from Envisat and ERS Missions*, pp. 1–4, Italy, May 2005.

[19] A. Ishimaru, Y. Kuga, J. Liu, and T. Freeman, "Ionospheric effects on synthetic aperture radar at 100 MHz to 2 GHz," *Radio Science*, vol. 34, no. 1, pp. 257–268, 1999.

[20] J. Louveaux, L. Vandendorpe, and T. Sartenaer, "Early-late timing recovery with decision-feedback equalizers," *IEEE Communications Letters*, vol. 7, no. 7, pp. 332–334, 2003.

[21] S. M. Simon, "Nonlinear analysis of an absolute value type of an early-late bit synchronization," *IEEE Transactions on Communications*, vol. 18, no. 5, pp. 589–596, 1970.

[22] B. H. Wellenhof, H. Lichtenegger, and J. Collins, *Global Positioning System: Theory and Practice*, Springer, New York, NY, USA, 4th edition, 1997.

[23] M. Younis, R. Metzig, and G. Krieger, "Performance prediction of a phase synchronization link for bistatic SAR," *IEEE Geoscience and Remote Sensing Letters*, vol. 3, no. 3, pp. 429–433, 2006.

[24] W. R. Bennett, "Spectra of quantized signals," *Bell System Technology Journal*, vol. 27, pp. 467–472, 1948.

[25] M. M. Abousetta, "Noise analysis of digitised FMCW radar waveforms," *IEEE Proceedings on Radar, Sonar and Navigation*, vol. 145, no. 3, pp. 209–215, 1998.

[26] M. Younis, R. Metzig, and G. Krieger, "Performance prediction and verification for bistatic SAR synchronization link," in *Proceedings of the Europe Synthetic Aperture Radar Conference*, pp. 1–4, Dresden, Germany, May 2006.

[27] E. A. Y. Gadallah, *Global position system receiver design for multipath mitigation [Ph.D. dissertation]*, Air Force Institute of Technology, Ohio, USA, 1998.

16

Response of a Hypersonic Boundary Layer to Freestream Pulse Acoustic Disturbance

Zhenqing Wang, Xiaojun Tang, and Hongqing Lv

College of Aerospace and Civil Engineering, Harbin Engineering University, Harbin 150001, China

Correspondence should be addressed to Xiaojun Tang; tangxiaojun2214@163.com

Academic Editors: G. Pascazio and S. Torii

The response of hypersonic boundary layer over a blunt wedge to freestream pulse acoustic disturbance was investigated. The stability characteristics of boundary layer for freestream pulse wave and continuous wave were analyzed comparatively. Results show that freestream pulse disturbance changes the thermal conductivity characteristics of boundary layer. For pulse wave, the number of main disturbance clusters decreases and the frequency band narrows along streamwise. There are competition and disturbance energy transfer among different modes in boundary layer. The dominant mode of boundary layer has an inhibitory action on other modes. Under continuous wave, the disturbance modes are mainly distributed near fundamental and harmonic frequencies, while under pulse wave, the disturbance modes are widely distributed in different modes. For both pulse and continuous waves, most of disturbance modes slide into a lower-growth or decay state in downstream, which is tending towards stability. The amplitude of disturbance modes in boundary layer under continuous wave is considerably larger than pulse wave. The growth rate for the former is also considerably larger than the later the disturbance modes with higher growth are mainly distributed near fundamental and harmonic frequencies for the former, while the disturbance modes are widely distributed in different frequencies for the latter.

1. Introduction

The accurate predictions about the aerodynamic, drag, and heat transfer rate of hypersonic vehicle surface can provide important basis to develop more efficient hypersonic vehicle in aerodynamic, thermal protection, and flight control designs, which is a challenge for hypersonic vehicle development [1]. These aerothermodynamics parameters depend considerably on boundary layer flow characteristics [2]. Therefore, investigations on boundary layer stability characteristics have very practical significance. Boundary layer is the regions near wall surface, where viscous force plays a critical role. Under certain circumstances, boundary layer flow can move from an orderly laminar state to a disorderly turbulent state, which is called transition. Aerodynamic frictional forces and heat transfer rate of hypersonic vehicle surface become elevated significantly once boundary layer flow changes from laminar to turbulent. Turbulence can make the friction drag on vehicle surface ten times larger. Therefore, the accurate predictions about the flow state of hypersonic vehicle surface have implications for new hypersonic vehicle development. In recent years, hypersonic boundary layer stability has caused wide public concern. A series of experimental and numerical investigations on the subject are performed [3–7], and a variety of theory interpretations and experiment analysis were conducted on the influential factor of boundary layer stability.

The process of boundary layer flow moving from laminar state to turbulent state is affected by many factors, such as freestream instability, wall surface roughness, wall temperature, and Reynolds number. Though the stability characteristics of hypersonic boundary layer had been investigated by many scholars, the transition mechanism of boundary layer, especially hypersonic boundary layer, is still not fully understood. The accurate prediction methods about transition position are still not fully reliable [8]. The hypersonic boundary layer stability characteristic is rather different from the boundary layer stability characteristic of incompressible flow, subsonic flow, and low Mach number supersonic flow. Under hypersonic condition, many new complex problems occur [9], which do not appear for low velocity flow conditions and should be understood, for instance, the appearance of second unstable disturbance mode, the effects of wall temperature on boundary layer stability, the sensitivity of flow factors to Mach number, and so forth. Thus, hypersonic boundary layer was not wholly explicable in terms of subsonic or low Mach number supersonic flow. The emergence of such problems makes the accurate transition prediction more difficult. For incompressible flow, subsonic flow, and low Mach number supersonic flow, boundary layer only contained unstable vorticity wave with low frequency, which is called the first disturbance mode or Tollmien-Schlichting (T-S) wave [10]. However, as the Mach number increases, the disturbance wave with high frequency occurs; apart from unstable vorticity wave with low frequency, there are a series of acoustic waves in boundary layer, which is unstable Mack2 mode. Mack2 mode becomes the least stable mode when Mach number is large enough [11]. Wang et al. [1] numerically studied the response of a Mach 8 flow over a 5.3° half-angle sharp wedge to wall blowing-suction and investigated the spatial development of boundary layer waves. They found that mode F, mode S, acoustic waves, and entropy/vorticity waves are simultaneously excited by wall blowing-suction. Maslov et al. [12] investigated the stability of a hypersonic shock layer on a flat plate. A new experimental technique is introduced for the investigation of artificially generated disturbances in planar laminar hypersonic boundary layers in [13]. Jiang et al. investigated [14] the instability wave propagation in boundary layer flows at subsonic through hypersonic Mach numbers, and three separate flow configurations are investigated. The linear and nonlinear developments of instability waves in a range of boundary layer flows are discussed. Fedorov and Khokhlov [15] investigated the prehistory of instability in a hypersonic boundary layer and presented a detailed analysis about how the forcing environmental disturbances enter into boundary layer and produce unstable wave that further develops and induces typical unstable wave in boundary layer. Based on direct numerical simulation (DNS) and linear stability theory (LST) analysis, Liang et al. studied the effects of wall temperature on stabilities of hypersonic boundary layer over a 7° half-cone-angle blunt cone under freestream small disturbance in [8] and found that the growth of disturbance waves is significantly affected by wall temperature; cooling the surface can accelerate unstable Mack II mode waves and decelerate Tollmien-Schlichting mode. As shown in previous studies, many investigations on the stability characteristic of hypersonic boundary layer have been presented, and most of these researches focused on the receptivity to freestream disturbance wave, response of hypersonic boundary layer to wall blowing-suction, the development of disturbance wave in boundary layer, and the effects of some flow parameters on boundary layer stability as well as laminar-turbulent transition. However, very few works were conducted on the effects of freestream pulse wave on hypersonic flow and boundary layer stability characteristic, whereas the interactions between freestream pulse wave and hypersonic flow as well as boundary layer are rather different from freestream continuous wave. Thus, the evolution mechanism of boundary layer disturbance wave and stability characteristic for the action of freestream continuous disturbance is rather different from pulse wave.

Investigations on the area will help to understand the stability characteristic of hypersonic boundary layer under freestream pulse wave, which also can provide a different perspective for investigations on the stability of hypersonic boundary layer and is helpful to elucidate the underlying mechanisms of hypersonic boundary layer laminar-turbulent transition. Therefore, investigations on the receptivity and boundary layer stability for freestream pulse wave have very practical significance, and it is necessary to complete more systematic investigations.

In the present paper, hypersonic unsteady flows over a blunt wedge under the action of freestream pulse/continuous disturbance wave are computed. The interactions between freestream disturbance wave and hypersonic flow as well as boundary layer are analyzed. The response of hypersonic boundary layer to freestream disturbance wave and the evolution of boundary layer disturbance wave are investigated. The stability characteristics of hypersonic boundary layer over an 8° half-wedge-angle blunt wedge for freestream pulse wave and continuous wave were compared.

2. Governing Equations and Numerical Methods

2.1. Governing Equations. For steady and unsteady hypersonic flow computation, the Navier-Stokes equations in conservation form at Cartesian coordinates (x, y) are transformed into equations at general curvilinear coordinates (ψ, ϕ) and can be expressed as follows:

$$\frac{\partial \left(\mathbf{J}^{-1} \mathbf{U} \right)}{\partial t} + \frac{\partial \mathbf{F}^*}{\partial \psi} + \frac{\partial \mathbf{G}^*}{\partial \phi} + \frac{\partial \mathbf{F}_v^*}{\partial \psi} + \frac{\partial \mathbf{G}_v^*}{\partial \phi} = 0, \qquad (1)$$

where the variables $\mathbf{U}$, $\mathbf{J}$, t are the state vector, the Jacobin matrix, and time, respectively, $\mathbf{F}^*$ and $\mathbf{G}^*$ are nonviscid terms at general curvilinear coordinates, and $\mathbf{F}_v^*$ and $\mathbf{G}_v^*$ are viscid terms at general curvilinear coordinates, while

$$\mathbf{F}^* = \frac{\mathbf{F}\psi_x + \mathbf{G}\psi_y + \mathbf{U}\psi_t}{\mathbf{J}},$$

$$\mathbf{G}^* = \frac{\mathbf{F}\phi_x + \mathbf{G}\phi_y + \mathbf{U}\phi_t}{\mathbf{J}},$$

$$\mathbf{F}_v^* = \frac{\mathbf{F}_v\psi_x + \mathbf{G}_v\psi_y}{\mathbf{J}}, \qquad (2)$$

$$\mathbf{G}_v^* = \frac{\mathbf{F}_v\phi_x + \mathbf{G}_v\phi_y}{\mathbf{J}},$$

where the variables $\mathbf{F}$ and $\mathbf{G}$ are nonviscid terms at Cartesian coordinates and $\mathbf{F}_v$ and $\mathbf{G}_v$ are viscid terms at Cartesian coordinates.

2.2. Numerical Method. To accurately simulate hypersonic unsteady flow filed under disturbances wave in freestream, a high-order direct numerical simulation method will be established and is used to solve compressible Navier-Stokes equations. The splitting of convection terms is conducted

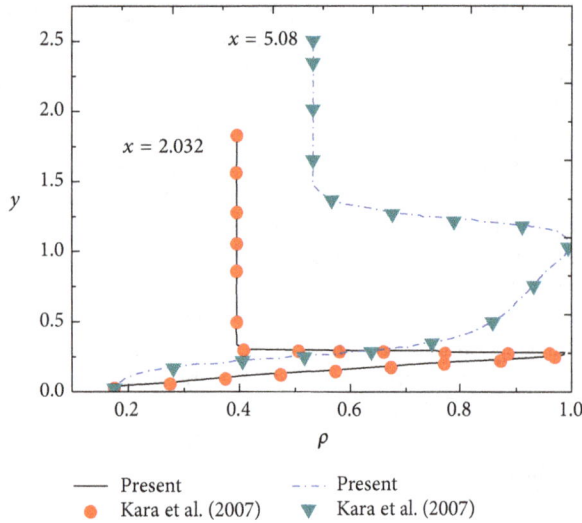

FIGURE 1: Comparison of density profile with results of Kara et al.

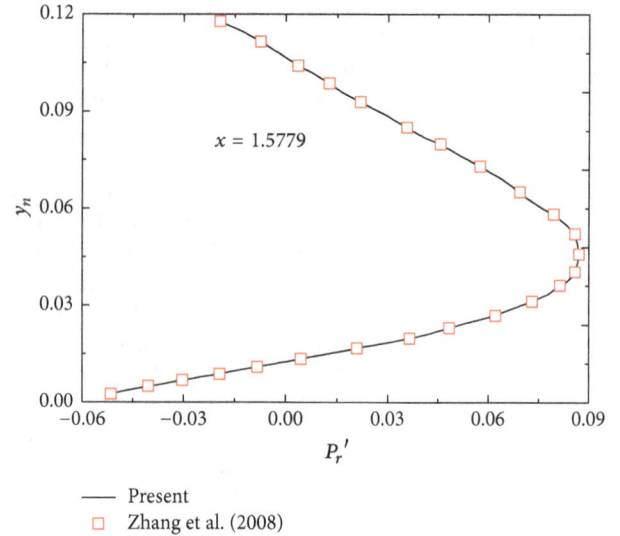

FIGURE 2: Comparison of the real part of Fourier transform for pressure disturbance with Zhang et al.'s result.

by Steger-Warming splitting method [16], as is shown in (3). Since the high-order weighted essentially nonoscillatory (WENO) methods have been widely implemented in the DNS of compressible turbulent flows, show strong robustness and high resolution, and are able to keep the higher-order approximations in smooth regions and to eliminate or suppress the oscillatory behavior near the discontinuities [17], positive convection and negative convection are discretized with 5th order upwind WENO scheme [18], as shown in (4) and (5), respectively. Viscous terms are discretized with 6th order center difference scheme [19], as is shown in (6); third order Runge-Kutta TVD type method [20] is employed for time advancing. To avoid the fact that accuracy of interior grid node is polluted, the 5th order WENO weighted scheme is used in the boundary points of positive convection terms, negative convection terms, and viscous terms.

Steger-Warming splitting method for convection terms splitting can be expressed as follows:

$$W = W^+ + W^-, \qquad (3)$$

where the variables W^- and W^+ are positive convection terms and negative convection terms, respectively.

The positive and negative convection terms are discretized by the 5th upwind scheme and can be expressed as follows in (4) and (5), respectively:

$$W^{+\prime} = \frac{1}{\Delta}\left(\sum_{N=1}^{6} m_N W^-_{j+3-N}\right), \qquad (4)$$

$$W^{-\prime} = \frac{1}{\Delta}\left(\sum_{N=1}^{6} n_N W^{+\prime}_{j+4-N}\right), \qquad (5)$$

where Δ is the grid spacing, $W^{+\prime}$ and $S^{-\prime}$ are the difference approximation of the derivative of W^+ and W^-, respectively, and m_i and n_i are weighting coefficients.

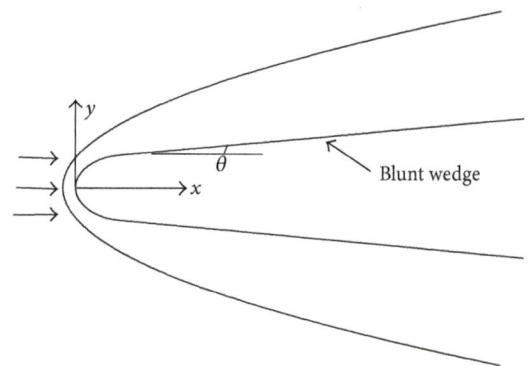

FIGURE 3: Computational mode.

Center difference scheme is employed for viscous terms discretion and can be expressed as follows:

$$H' = \frac{1}{\Delta}\left(\sum_{N=1}^{3} K_N\left(H_{j+N} - H_{j-N}\right)\right), \qquad (6)$$

where the variables H, H', and K_i are viscous terms, the difference approximation of the derivative of viscous terms, and weighting coefficient, respectively.

The numerical simulations of two similar modes are conducted as follows: (1) Mach number of 6 over a blunt wedge with a half-wedge-angle $\theta = 5°$ under freestream acoustic disturbances and (2) Mach number of 15 over a blunt body with $\theta = 5°$ under freestream acoustic disturbances, entropy disturbances, and vorticity disturbances. Figures 1 and 2 show the comparison of density profile with results of Kara et al. [17] and the comparison of Fourier transform for pressure disturbance with Zhang et al.'s result [19]. From Figures 1 and 2, it can be seen that the effectiveness of the numerical method is demonstrated.

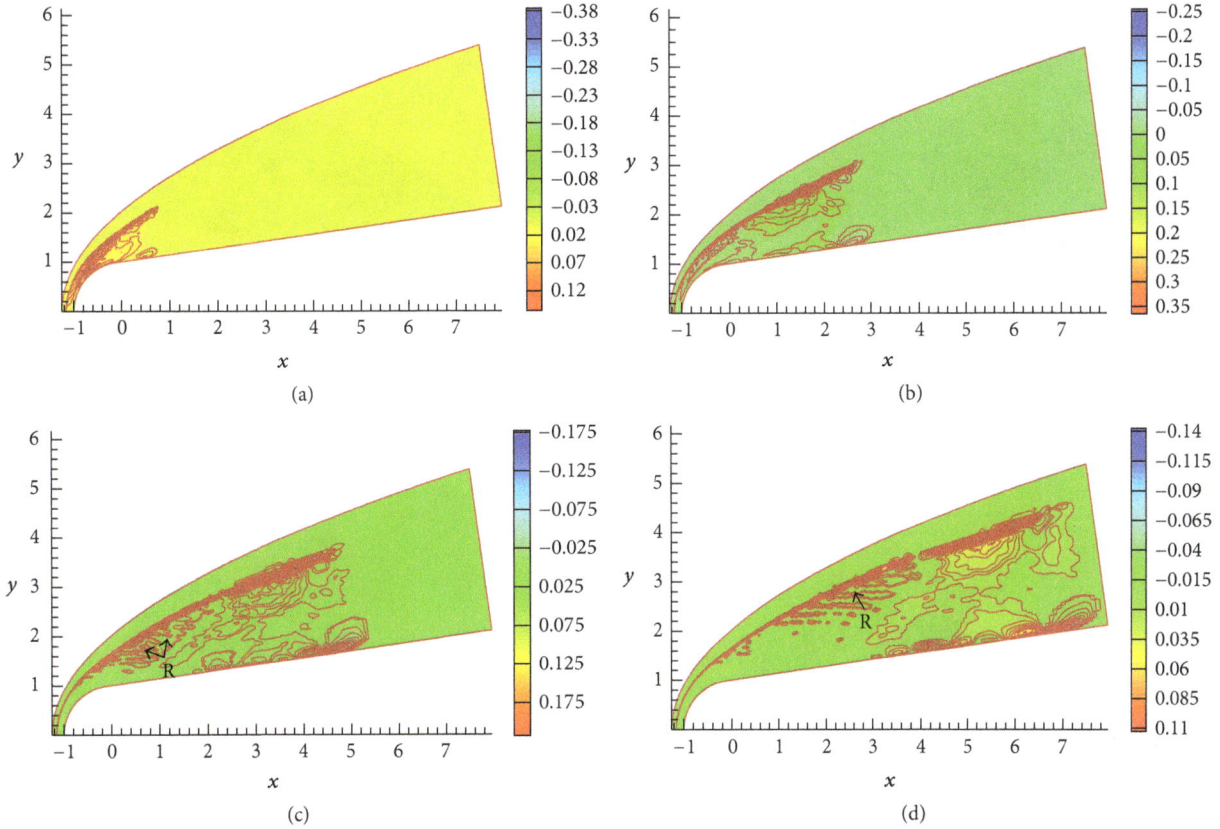

FIGURE 4: Contours of velocity disturbances along y-axis $v'(x, y, t)$ at different times.

3. Computational Conditions

Based on a hypersonic flow over a wedge with blunt noses, the numerical simulations of hypersonic unsteady flow under freestream fast acoustic wave are performed. The computational mode and schematic diagram are shown in Figure 3. Freestream condition and extrapolation are employed at computational field upstream and the exit of hypersonic flow fields, respectively. No-slip and isothermal wall conditions are enforced at the wall surface. The solutions presented in this paper are resolved by 300×120 grids. The mesh grid density in this paper matches that in Zhang et al.'s and Prakash's investigations [19, 21]. The stretching function is used to cluster more points in noses area and boundary layer. The effectiveness of the stretching method is demonstrated in Figures 1 and 2. An asymmetry condition is introduced at $y = 0$. The parameters used in the paper are dimensionless. The details are expressed as follows: the velocity, the length scales, the density ρ, the pressure P, the temperature T, and the time t are nondimensionalized by the freestream velocity, the nose radius R, the freestream density ρ_∞, $\rho_\infty u_\infty u_\infty$, the freestream temperature T_∞, and R/u_∞, respectively. Based on the dimensionless method above, nondimensional computation is conducted in this paper; the results are nondimensional. For example, the nondimensional frequency f is obtained by u_∞/R. The Reynolds number, $\mathrm{Re}_n = \rho_\infty R u_\infty/\mu_\infty$, based on freestream parameters

and the cone's nose radius, is equal to 10000, where μ_∞ is viscosity coefficient. The wave fields are represented by disturbances of instantaneous flow variables with respect to the local steady base flow variables at the same location. Subscripts "∞" and "$'$" denote freestream condition and the disturbances of instantaneous flow variables, respectively. The flow parameters are shown in Table 1. The variables Re, Ma_∞, T_∞, T_w, α_n, and θ are Reynolds number, freestream Mach number, freestream temperature, wall temperature, angle of attack, and half-wedge-angle, respectively.

Freestream fast acoustic continuous and pulse waves are separately introduced to the upstream end of the steady flow field without disturbance to do direction numerical simulation of hypersonic unsteady flow at $t1 = 52$. The form of the disturbance wave employed can be expressed as follows:

TABLE 1: Computational conditions.

Re	Ma_∞	T_∞ (K)	T_w (K)	α_n ($^\circ$)	R (mm)	θ ($^\circ$)
10000	6	169	200	0	1	8

$$\mathbf{L} = \begin{cases} \mathbf{0} & t < t1 \\ \mathbf{L_A} e^{i(kx - ((F \cdot \mathrm{Re})/10^6)t + (\pi/2))} & t1 \leq t < t2 \\ \mathbf{0} & t \geq t2, \end{cases} \quad (7)$$

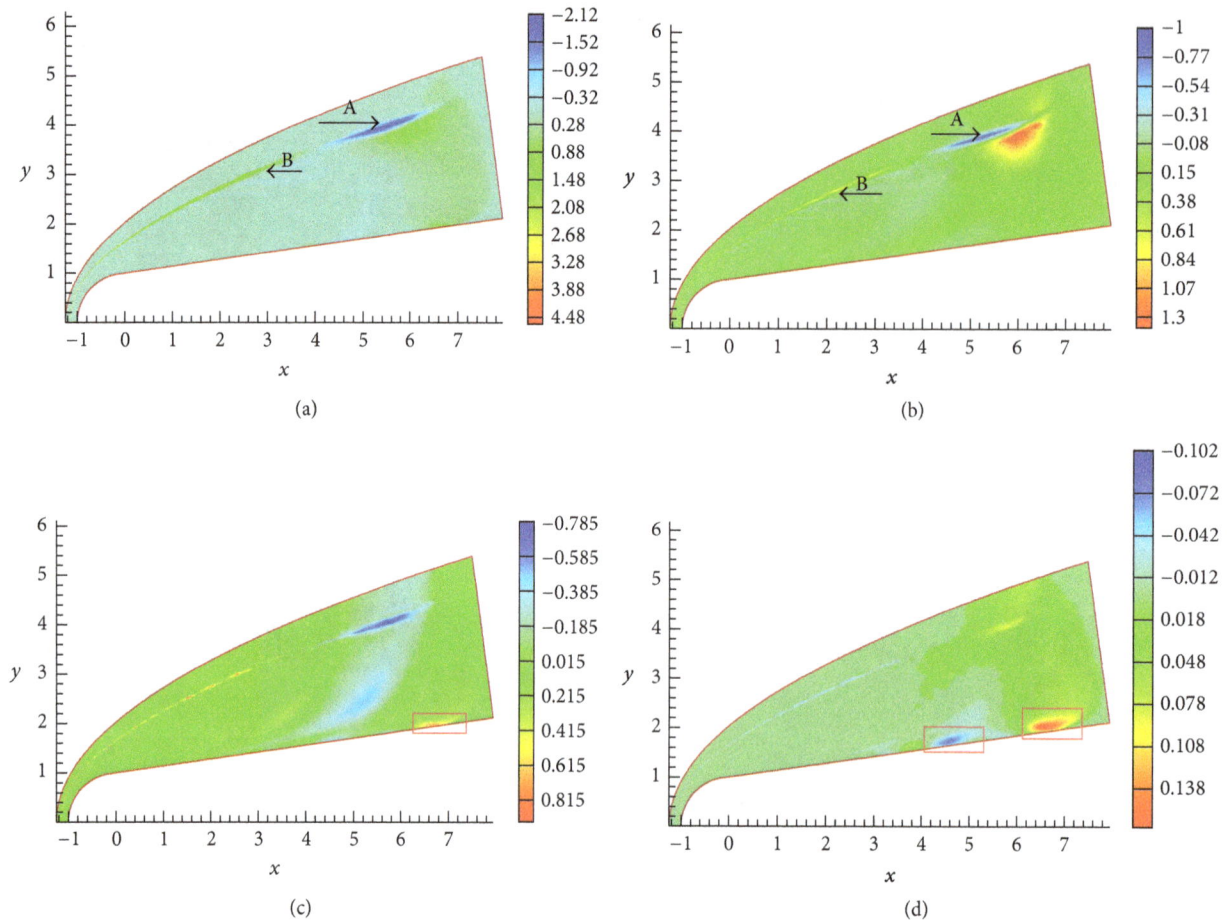

FIGURE 5: Contours of pressure $P'(x, y, t)$, density $\rho'(x, y, t)$, temperature $T'(x, y, t)$, and velocity disturbance.

where the matrix $\mathbf{L} = [u'\ v'\ p'\ \rho']^T$, $\mathbf{L_A} = [A\ 0\ A/Ma\ AMa]^T$, the variables u', v', p', and ρ' are the disturbance values of the velocity along x-axis, velocity along y-axis, pressure, and density, respectively, the amplitude $A = 0.06$, wave number $k = 3.1446 \times 10^{-4}$, generalized frequency $F = 50\pi$, and freestream Mach $Ma = 6.0$. When continuous wave in freestream is simulated, $t2$ should be taken to infinity theoretically; however, $t2$ should only be large enough to enable the flow field to reach period state in practice. When pulse wave in freestream is simulated, the length of time is $2.0(u_\infty/R)$, namely, 1/2 cycle; $t1$ and $t2$ should be taken to 52.0 and 54.0, respectively.

4. Numerical Results and Discussion

4.1. Response of Hypersonic Flow Field to Pulse Fast Acoustic Wave. The contours of velocity disturbances along y-axis at $t = 54.0$, 56.0, 58.0, and 60.0 are shown in Figures 4(a), 4(b), 4(c), and 4(d), respectively. The instantaneous velocity disturbances along y-axis on the areas of interactions of disturbance wave with the bow shock and shock layer and inside the boundary layer are significantly affected. In particular, the flow velocity gradient near wall surface is

larger. There exist significant differences between velocity disturbance modes outside and inside the boundary layer. The forcing disturbance wave interacts with bow shock and then enters the boundary layer; a part of disturbance wave propagated along streamwise, and the other repeat moved between shock and nose [8]. The reciprocating motion of disturbance wave is continual for freestream continuous wave, while the number of reflection is limited for freestream pulse wave due to viscous dissipation. It should be mentioned that the velocity disturbances behind bow shock still exist when pulse wave leaves away, as the mark R shown in Figure 4. This is the result of the residual reflection wave being enlarged by shock wave, and then the residual reflection wave will be dissipated due to viscous dissipation. Therefore, the residual reflection wave only appears near the bow shock. Disturbance wave with different modes will be generated after the forcing disturbance wave interacts with boundary layer; the generated waves further interact with boundary layer and induce more unstable modes [8]. The interactions between boundary layer and unstable wave become more complicated due to the reflection between bow shock wave and wall.

The contours of pressure $P'(x, y, t)$, density $\rho'(x, y, t)$, temperature $T'(x, y, t)$, and velocity disturbance are shown in Figures 5(a), 5(b), 5(c), and 5(d), respectively. As seen,

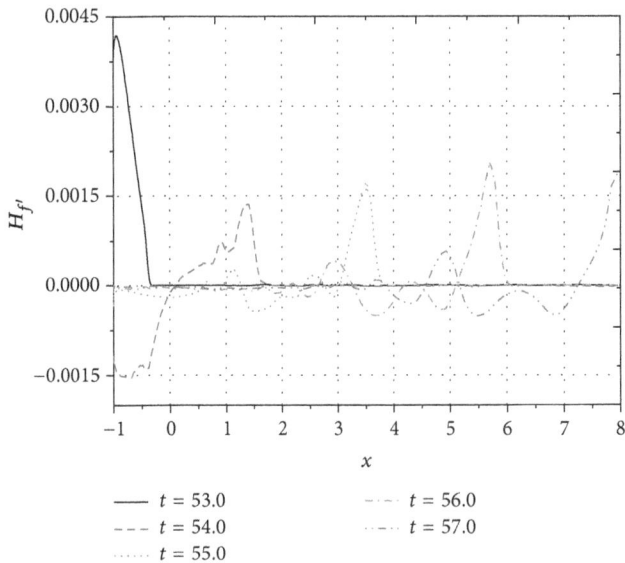

FIGURE 6: Distribution of wall heat flux disturbance $H'_f(x, y, t)$ at different times.

bow shock is bent obviously inwards when being subjected to pulse wave. The pressure $P'(x, y, t)$, density $\rho'(x, y, t)$, and temperature $T'(x, y, t)$ become smaller sharply in the deformation area of shock wave, as the mark A shown in the figure. The amplitude of flow variables change is significantly larger than that of initial disturbance in freestream. As discussed above, a part of disturbance wave will be reflected between shock and nose [8]. It should be noted that the bow shock wave slightly deforms when being subjected to the refection wave, as the mark B shown in the figure. Liang et al. [8] studied the evolution of continuous small disturbance waves in hypersonic flow by using direct numerical simulation (DNS) and found that, because of the normal shock wave, the forcing disturbance in freestream is enlarged. Figure 5 shows that the disturbance amplitude of flow variables is enlarged sharply relative to that of initial disturbance in freestream. The numerical results agree with Liang et al.'s results. From the figure, the velocity and temperature near the wall surface are changed sharply as confronts freestream acoustic disturbances, as the rectangular mark shown in the figure. Obviously, the structure of strong shear flow and the thermodynamics properties in boundary layer are changed significantly under freestream pulse wave; the structure of strong shear flow and the thermal state near wall surface will affect the behavior of hydrodynamic stability [2]. The evolvement of unstable disturbance waves in boundary layer and the boundary layer stability characteristic will be discussed in the following.

4.2. Evolvement of Unstable Disturbance Waves in Boundary Layer. Figure 6 shows the variation of heat flux disturbance on the wall surface at different times (t = 53.0, 54.0, 55.0, 56.0, 57.0, 58.0, 59.0, 60.0, and 61.0). It shows that when the flow field was subjected to the semisinusoidal shape of the pulse disturbance wave, the heat flux disturbance

firstly increases and then decreases, the trend of which is analogous to freestream disturbance wave. However, when the pulse disturbance wave passes through entirely, the effect remains. Furthermore, the disturbance amplitude will undergo a damped oscillation until the amplitude decreases to 0. It is due to the reflection of the disturbance wave between bow shock and wall. From Figure 6, it is clearly indicated that the heat flux on wall changed sharply under freestream pulse acoustic disturbance with $A = 6 \times 10^{-2}$. Since heat flux is a characteristic variable of heat conductivity in boundary layer, the thermal conductivity of boundary layer flow is significantly affected by freestream pulse wave. The influences of the thermal conduct mechanism of boundary layer flow on the evolution of disturbance wave modes and boundary layer stability state were confirmed in [2, 22, 23]. Meanwhile, thermal surface state will be changed by the thermal conduct on the wall surface. Hirschel [2] found that the thermal surface state has a strong back-coupling to the state of the boundary layer. It is reasonable to believe that the behaviors of hydrodynamic stability, even laminar-turbulent transition, will be affected under freestream pulse disturbance.

Before broaching the following discussions, it should be mentioned that a fitted coordinate system s is introduced to represent the distance from the points on the generatrix of blunt wedge to the stagnation point. The relationship between x and s is expressed in [8]. In order to study the disturbance wave evolvement in boundary layer, the comparison of variation of pressure disturbance locations with time between freestream continuous wave and pulse wave at different surface locations is given in Figure 7. It can be seen that when the flow field is subjected to freestream pulse disturbance wave, the pressure disturbance firstly increases and then decreases, the trend of which is similar to freestream pulse wave. However, when the pulse disturbance wave passes through entirely, the effect remains. The disturbance amplitude will undergo a damped oscillation until the amplitude decreases to 0. The damped oscillation is due to the reflection of disturbance waves between wall and bow shock, as the rounded mark shown in the figure. From the variation of pressure disturbance in periodic states under the action of freestream continuous wave as shown in Figure 7, the oscillation also can be seen. There exist the disturbance wave propagation and the complex mutual interferes of the reflection wave and boundary layer, which complicate the boundary layer flow. This leads to the fact that the pressure disturbance is changed from relatively unitary shape in upstream to complex shape in downstream. From Figure 7, for both pulse wave and continuous wave, the amplitudes of pressure disturbance are enlarged significantly relative to initial freestream disturbance. In general, due to the fact that the amplification caused by normal shock wave (approximates) in nose is larger than that caused by oblique shock wave in nonnose, the amplitude of pressure disturbance in nose is significantly larger than that in nonnose. This agrees with Liang et al.'s results [8]. It should be mentioned that, for continuous wave, the pressure disturbance is periodic; however, for pulse wave, the pressure disturbance changed with the time. That is, there only exists spatial evolution (along streetwise) in boundary layer for

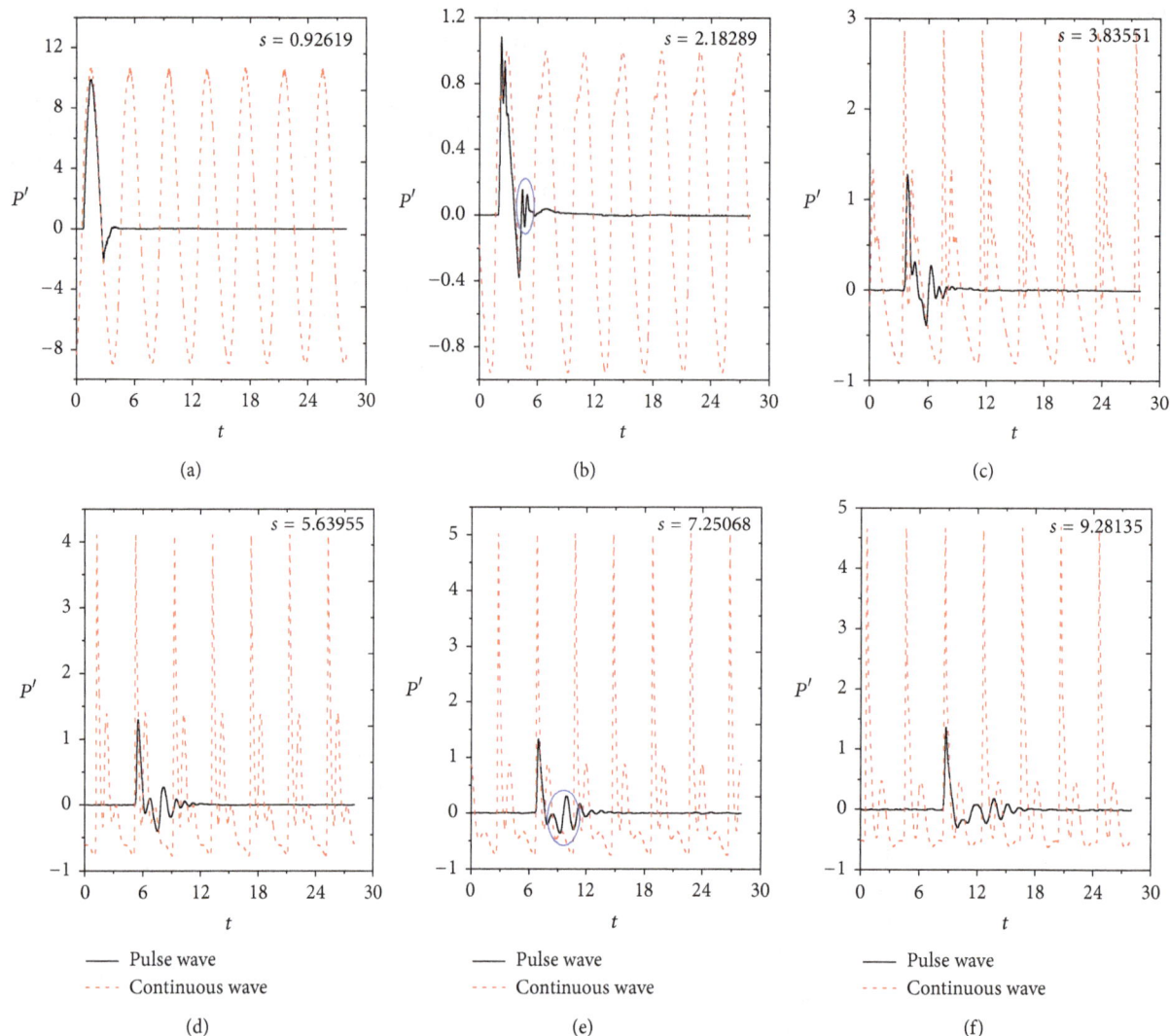

FIGURE 7: Pressure disturbance at different surface locations varying with time.

continuous wave, while there exist not only spatial evolution but also temporal evolution in boundary layer for pulse wave.

The temporal signals of pressure disturbance are decomposed into frequency signals using Fourier transform. Figure 8 shows the Fourier frequency analysis of pressure disturbance at different surface locations under freestream pulse fast acoustic wave. It can be seen that (a) the frequencies of most disturbance are less than 0.5 in nose boundary layer. There are 5 main disturbance mode clusters at $s = 2.18289$; however, the number is reduced from 5 at $s = 2.18289$ to 2 at $s = 5.63955$, and the two clusters are mainly distributed near $f = 0.25$ and $f = 0.75$. When $5.63955 < s < 7.25078$, the amplitude of $f = 0.75$ is once larger than that of $f = 0.25$, and this makes the dominant mode transfer in boundary layer. It indicates that the development of dominant mode in boundary layer restrains the growth of other modes. (b) The Fourier amplitude of pressure disturbance at $s = 0.92616$ ($s < \pi/2$, the nose region) is greater than that at other locations. It is mainly owing to the bow wave going

from normal shock on the nose region to oblique shock wave on the non-nose region. (c) With the development of disturbance waves in boundary layer along streamwise from upstream to downstream, various tends are revealed for the different frequency disturbance modes, such as continuous growth, decay, and first growth and then decay. It indicates that there is competition among different modes in boundary layer. In general, when $s > \pi/2$, the Fourier amplitude of pressure disturbance of low frequency mode ($f < 0.25$) decreases and the Fourier amplitude of high frequency ($f > 0.5$) increases. The ratio between low frequency component and high frequency component transformed quickly, and the high frequency components increase quickly. With the development of disturbance waves along streamwise from upstream to downstream, the frequency band narrows.

Figure 9 shows the Fourier frequency analysis of pressure disturbance at different surface location under freestream continuous fast acoustic wave. It indicates that, under freestream continuous wave with single frequency, the

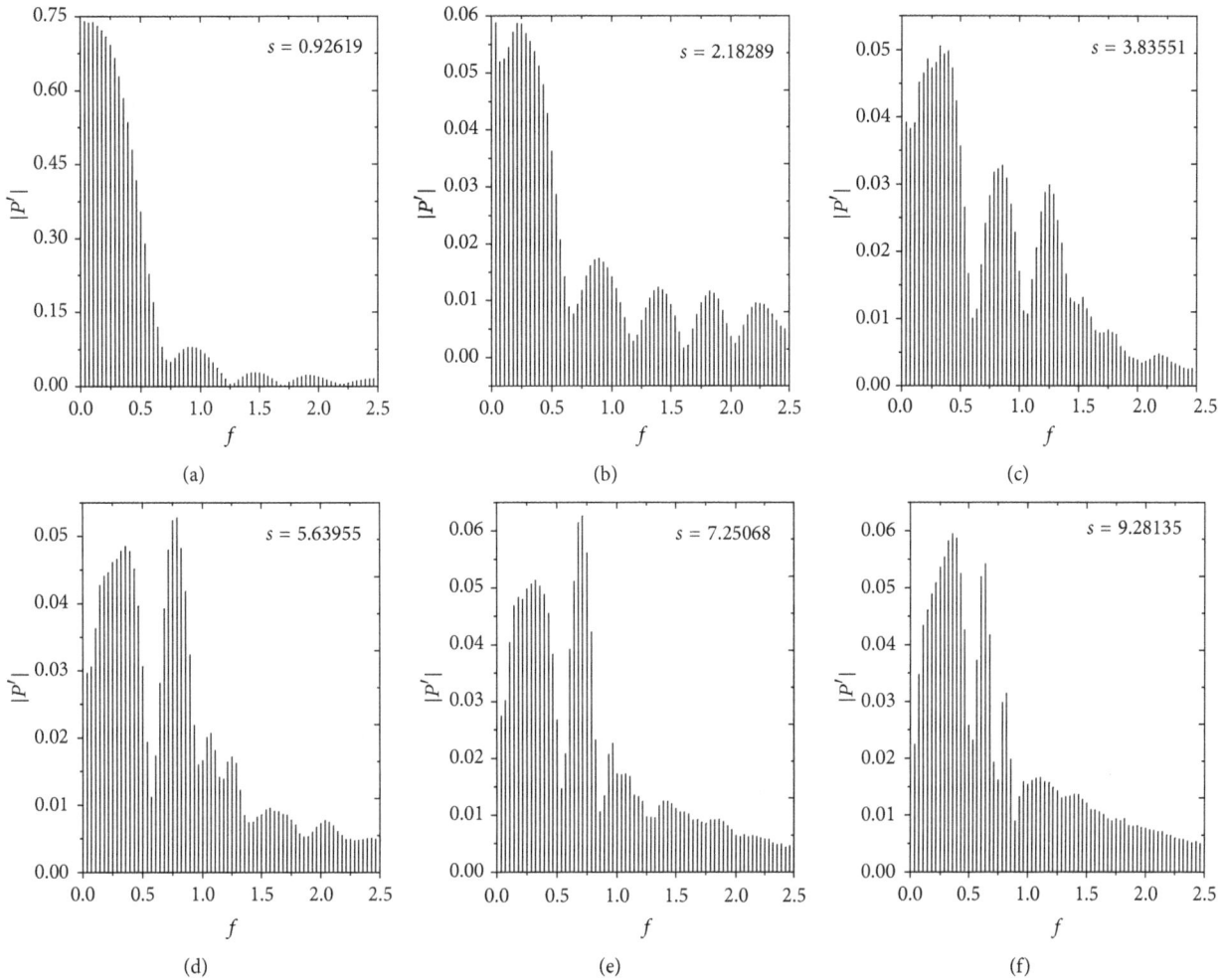

FIGURE 8: Fourier frequency analysis of pressure disturbance at different surface locations under freestream pulse fast acoustic wave.

disturbance modes in boundary layer are mainly distributed near fundamental frequency and harmonic frequencies (i.e., P_n, where n is positive integer), and the amplitudes of the other modes (n is not integer) are tiny. Comparing Figures 8 and 9, it can be seen that, (1) for both freestream continuous wave and pulse wave, the amplitudes of fundamental frequency (for freestream continuous wave) near fundamental frequency (freestream pulse) decrease sharply in the nose boundary layer. With the development of disturbance waves along streamwise from upstream to downstream, the components of high frequency disturbance modes (the frequency is larger than the second harmonic; that is, $f > 0.5$) increase quickly. (2) The Fourier amplitudes of pressure disturbance in boundary layer under freestream continuous wave are considerably larger than those under freestream pulse wave; there is an order of magnitude difference between the two conditions. Obviously, this is due to the fact that the disturbance energy in boundary layer under freestream continuous wave is persistent, while that under freestream pulse wave is temporary. (3) Under freestream continuous wave, the disturbance modes in boundary layer are mainly distributed near fundamental frequency and harmonic frequencies, and

the amplitudes of the other modes are very tiny, while, under freestream pulse wave, the disturbance modes in boundary layer are widely distributed in different modes. Namely, the frequency band for the former is narrower than the latter. This means that the disturbance energy is transferred from both fundamental frequency and harmonic frequencies to other modes, which increases the amplitudes of other modes (P_n, where n is not integer). Obviously, it has also become another important reason why the Fourier amplitudes of pressure disturbance under continuous wave are considerably larger than those under pulse wave. (4) For pulse wave, there are only 2 main disturbance mode clusters in boundary layer when $s \geq 5.63955$. The two main mode clusters are distributed near fundamental frequency and the third harmonic frequency, and the other modes remain basically stable, even to attenuate. However, for continuous wave, there are still 3 main disturbance modes in boundary layer when $s = 9.28315$. The 3 main modes are fundamental frequency, the second harmonic frequency, and the third harmonic frequency.

Figure 10 shows the comparison of pressure disturbances amplitudes for fundamental frequency (P_1) and harmonic

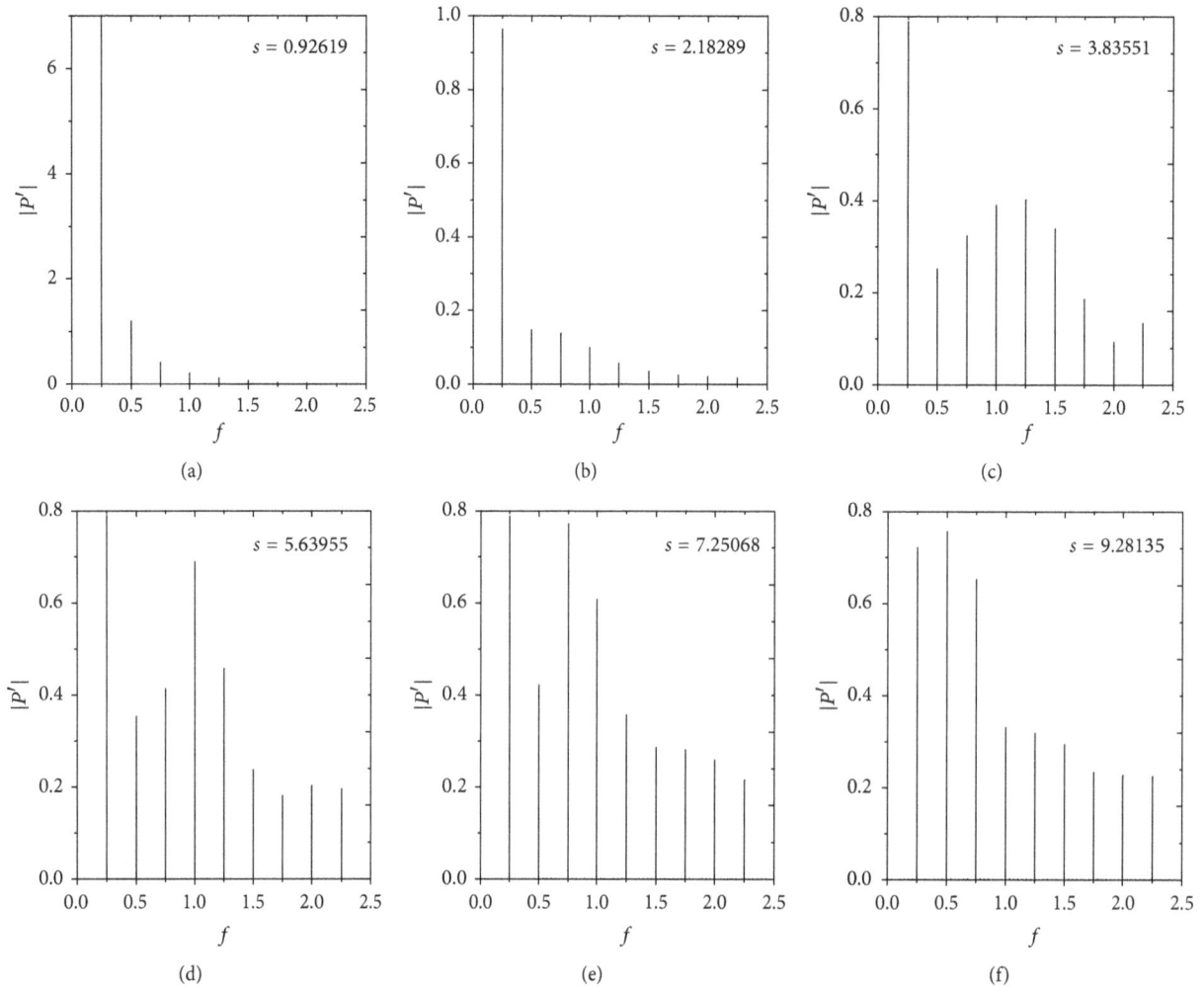

FIGURE 9: Fourier frequency analysis of pressure disturbance at different surface locations under freestream continuous fast acoustic wave.

frequencies (P_{2-4}) in boundary layer under both pulse wave and continuous wave. It can be seen that, for both pulse wave and continuous wave, the pressure disturbance amplitudes for fundamental frequency (P_1) and harmonic frequencies (P_{2-4}) in nose boundary layer are larger than those in the non-nose boundary layer; the amplitudes of pressure disturbance for fundamental frequency (P_1) are significantly larger than those for harmonic frequencies (P_{2-4}) in the nose boundary layer. It indicates that the nose boundary layer is dominated by fundamental frequency (for freestream continuous wave) or the modes near fundamental frequency (for freestream pulse wave). The results are consistent with investigations by Zhang et al. [19]. It also can be seen that, for both fundamental frequency and harmonic frequencies, the disturbance amplitudes for freestream pulse wave are significantly less than those for continuous wave. It indicates that the growth of boundary layer disturbance wave for pulse wave is much later than that for continuous wave. For both pulse wave and continuous wave, fundamental frequency remained basically stable in non-nose boundary layer. However, the variation of harmonic frequencies in non-nose boundary layer along streamwise for continuous wave differs from that for pulse

wave. For the former, there exists a significant growth for the second harmonic frequency P_2, the third harmonic frequency P_3, and the fourth harmonic frequency P_4, when $s > \pi/2$. For the latter, only the third harmonic frequency P_3 increases significantly when $s > \pi/2$; the second harmonic frequency P_2 and the fourth harmonic frequency P_4 remain basically stable, even to attenuate when $s > \pi/2$.

Figure 11 shows the growth rate α for different frequency disturbances in boundary layer along streamwise under freestream pulse wave. It can be seen that, when $s = 2.18289$, the growth of pressure disturbance mode with low frequency $(f < 0.5)$ is negative, and the disturbance amplitudes tend to decline; the decay of low frequency modes becomes slow with the frequency increasing; the growth of pressure disturbance modes for $f > 0.5$ generally increases; there exist 4 disturbance mode clusters with high growth and 4 growth peaks. When $s = 5.63955$, the decay of pressure disturbance modes for $f < 0.25$ becomes slow, even to grow; the pressure disturbance modes for $0.5 < f < 0.8$ continue to grow. When $f > 0.8$, the growth of different frequency disturbances in boundary layer is relatively larger at $s = 2.18289$, while the value becomes small, even to decay. When $s = 9.28135$, the

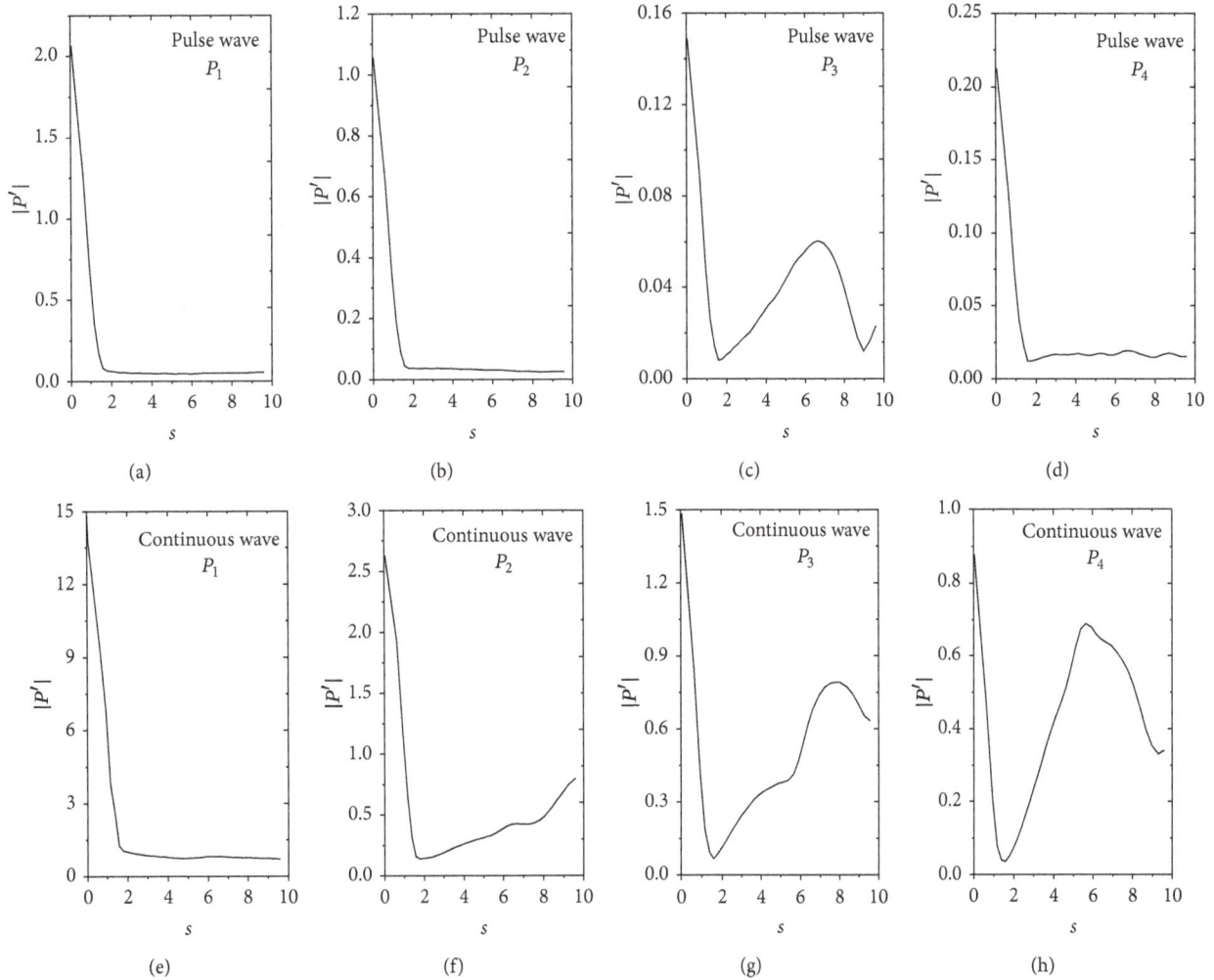

FIGURE 10: Comparison of pressure disturbances amplitudes in boundary layer for both pulse and continuous fast acoustic waves.

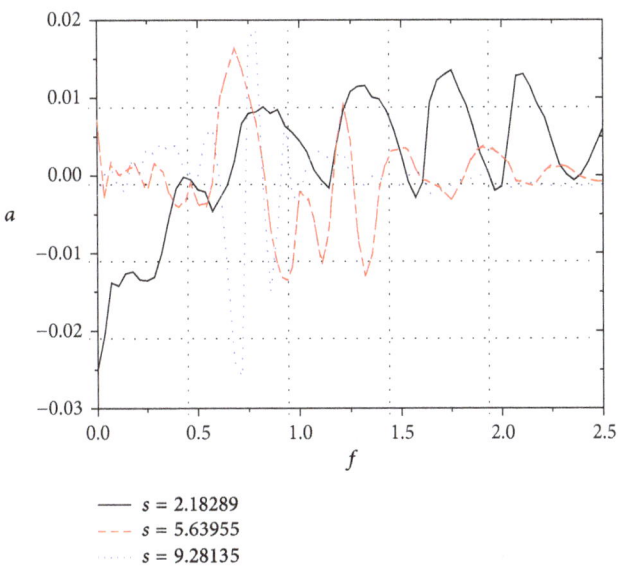

FIGURE 11: Growth of different frequency disturbances in boundary layer along streamwise under freestream pulse fast acoustic wave.

amplitude of growth rate for $f > 1.0$ and $f < 0.5$ is tiny, which maintains total stability; the modes for $0.7 < f < 0.8$, namely, the third harmonic frequency, increase sharply; these indicate that the development of dominant mode inhibited the other modes in boundary layer.

Figure 12 shows the growth for different frequency disturbances in boundary layer along streamwise under freestream continuous wave. It shows that the growth rate of the disturbance modes for fundamental frequency and harmonic frequencies (i.e., P_n, where n is positive integer) is significantly larger than for other modes (n is not integer). When $s = 2.18289$, the fundamental frequency decays and all harmonic frequencies increase. When $s = 5.63955$, the fundamental frequency increases slowly; all harmonic frequencies increase except the disturbance modes for $f = 1.25$ (P_5). When $s = 9.28135$, P_2 continued to grow, P_3 began to decay, and the other modes slide into a lower-growth or no-growth state, which presents a total stability. From Figures 11 and 12, (1) for both pulse wave and continuous wave, with the development of disturbance waves in boundary layer from upstream to downstream, most of disturbance modes slide

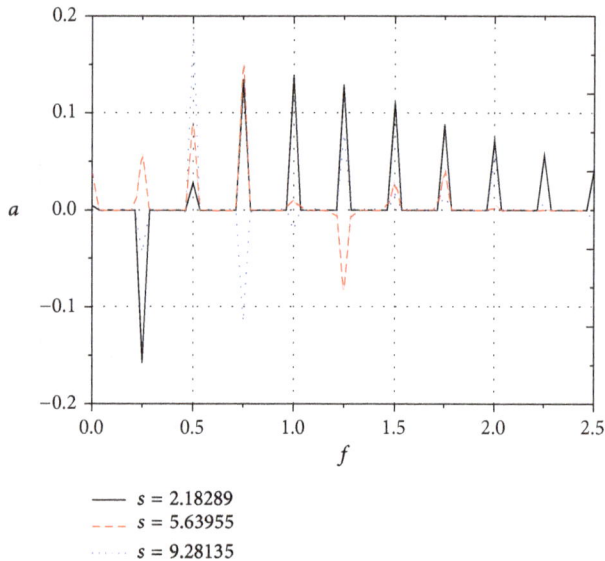

FIGURE 12: Growth of different frequency disturbances in boundary layer along streamwise under freestream continuous fast acoustic wave.

into a lower-growth or decay state, which is tending towards stability. (2) The amplitudes of the growth rate of disturbance modes in boundary layer under freestream continuous wave are considerably larger than those under freestream pulse wave. (3) The disturbance modes with higher growth are mainly distributed near fundamental frequency and harmonic frequencies under freestream continuous wave, while the disturbance modes with higher growth are widely distributed in different modes under freestream pulse wave.

5. Conclusions

The response of hypersonic boundary layer to freestream pulse acoustic wave is analyzed, and the stability characteristics of boundary layer under freestream pulse and continuous fast acoustic wave are compared. We draw some conclusions.

(1) Under freestream pulse acoustic disturbance, the bow shock is bent obviously, and the thermal conductivity of strong shear flow in boundary layer is significantly changed. Various tends are revealed for the different frequency disturbance modes along streamwise, such as continuous growth, decay, and first growth and then decay. With the development of disturbance waves along streamwise, the number of main disturbance mode clusters reduced quickly, and the frequency band narrows. The dominant mode in boundary layer restrains the growth of other modes. There is competition among different modes in boundary layer. There also is disturbance energy transfer between fundamental frequency and harmonic frequencies (P_n, where n is positive integer) and other modes (n is not integer).

(2) For both pulse wave and continuous wave, the nose boundary layer is dominated by fundamental frequency or the modes near fundamental frequency. The amplitudes of fundamental frequency or near fundamental frequency decrease sharply in the nose boundary layer. However, the fundamental frequency remains basically stable, and the components of high frequency disturbance modes ($f > 0.5$) increase quickly along streamwise in non-nose boundary layer. The Fourier amplitudes of disturbance in boundary layer under freestream continuous wave are considerably larger than those under freestream pulse wave; there is an order of magnitude difference between the two conditions. The disturbance modes in boundary layer are mainly distributed near fundamental frequency and harmonic frequencies, and the amplitudes of the other modes are tiny under freestream continuous wave, while the disturbance modes in boundary layer are widely distributed in different modes under freestream pulse wave. Namely, the frequency band for the former is narrower than the latter.

(3) For both pulse wave and continuous wave, with the development of disturbance waves along streamwise, most of disturbance modes slide into a lower-growth or decay state, which is tending towards stability. The amplitudes of the growth rate of disturbance modes in boundary layer under freestream continuous wave are considerably larger than those under freestream pulse wave. The disturbance modes with higher growth are mainly distributed near fundamental frequency and harmonic frequencies for the former, while the disturbance modes with higher growth are widely distributed in different modes for the latter.

Conflict of Interests

The authors declare that there is no conflict of interests regarding the publication of this paper.

Acknowledgments

This paper is funded by the National Science Foundation of China (11272096), the Ph.D. Programs Foundation of Ministry of Education of China (20112304110015), and the Fundamental Research Funds for the Central Universities (HEUCF130216). The authors acknowledge Harbin Engineering University of China.

References

[1] X. Wang, X. Zhong, and Y. Ma, "Response of a hypersonic boundary layer to wall blowing-suction," *AIAA Journal*, vol. 49, no. 7, pp. 1336–1353, 2011.

[2] E. M. Hirschel, *Basics of Aerothermodynamics*, Springer, Berlin, Germany, 2010.

[3] D. Park and O. S. Park, "Linear and non-linear stability analysis of incompressible boundary layer over a two-dimensional hump," *Computers & Fluids*, vol. 73, pp. 80–96, 2013.

[4] I. A. Halatchev and J. P. Denier, "The stability of boundary-layer flows under conditions of intense interfacial mass transfer: the effect of interfacial coupling," *International Journal of Heat and Mass Transfer*, vol. 46, no. 20, pp. 3881–3895, 2003.

[5] I. H. Herron, "Floquet theory for the stability of boundary layer flows," *Journal of Approximation Theory*, vol. 42, no. 4, pp. 387–406, 1984.

[6] S. Armfield and R. Janssen, "A direct boundary-layer stability analysis of steady-state cavity convection flow," *International Journal of Heat and Fluid Flow*, vol. 17, no. 6, pp. 539–546, 1996.

[7] A. N. Bogdanov and V. N. Diyesperov, "The stability of a transonic boundary layer on an elastic surface," *Journal of Applied Mathematics and Mechanics*, vol. 75, no. 3, pp. 357–362, 2011.

[8] X. Liang, X. Li, D. Fu, and Y. Ma, "Effects of wall temperature on boundary layer stability over a blunt cone at Mach 7.99," *Computers and Fluids*, vol. 39, no. 2, pp. 359–371, 2010.

[9] F. Dexun, M. Yanwen, and L. Xinliang, *Direct Numerical Simulation of Compressible Turbulence*, Science Press, Beijing, China, 2010.

[10] L. M. Mack, "Boundary layer linear stability theory," AGARD 709-3-1C3-81, California Institute of Tech, 1984.

[11] X. Li, D. Fu, and Y. Ma, "Direct numerical simulation of hypersonic boundary-layer transition over a blunt cone," *AIAA Journal*, vol. 46, no. 11, pp. 2899–2913, 2008.

[12] A. A. Maslov, T. V. Poplavskaya, and B. V. Smorodsky, "Stability of a hypersonic shock layer on a flat plate," *Comptes Rendus Mécanique*, vol. 332, no. 11, pp. 875–880, 2004.

[13] D. Kastell and A. N. Shiplyuk, "Experimental technique for the investigation of artificially generated disturbances in planar laminar hypersonic boundary layers," *Aerospace Science and Technology*, vol. 3, no. 6, pp. 345–354, 1999.

[14] L. Jiang, C.-L. Chang, M. Choudhari, and C. Liu, "Instability-wave propagation in boundary-layer flows at subsonic through hypersonic Mach numbers," *Mathematics and Computers in Simulation*, vol. 65, no. 4-5, pp. 469–487, 2004.

[15] A. V. Fedorov and A. P. Khokhlov, "Prehistory of instability in a hypersonic boundary layer," *Theoretical and Computational Fluid Dynamics*, vol. 14, no. 6, pp. 359–375, 2001.

[16] J. L. Steger and R. F. Warming, "Flux vector splitting of the inviscid gasdynamic equations with application to finite-difference methods," *Journal of Computational Physics*, vol. 40, no. 2, pp. 263–293, 1981.

[17] K. Kara, P. Balakumar, and O. A. Kandil, "Receptivity of hypersonic boundary layers due to acoustic disturbances over blunt cones," in *Proceedings of the 37th AIAA Fluid Dynamics Conference and Exhibit*, pp. 11464–11481, NASA Langley Research Center, Miami, Fla, USA, January 2007.

[18] X.-D. Liu, "Weighted essentially non-oscillatory schemes," *Journal of Computational Physics*, vol. 115, no. 1, pp. 200–212, 1994.

[19] Y. Zhang, D. Fu, Y. Ma, and X. Li, "Receptivity to free-stream disturbance waves for hypersonic flow over a blunt cone," *Science in China G*, vol. 51, no. 11, pp. 1682–1690, 2008.

[20] C.-W. Shu, "Essentially non-oscillatory and weighted essentially non-oscillatory schemes for hyperbolic conservation laws," NASA/CR-97-206253 and ICASE Report 97-6, 1997.

[21] A. Prakash, N. Parsons, X. Wang, and X. Zhong, "High-order shock-fitting methods for direct numerical simulation of hypersonic flow with chemical and thermal nonequilibrium," *Journal of Computational Physics*, vol. 230, no. 23, pp. 8474–8507, 2011.

[22] A. J. Laderman, "Effect of wall temperature on a super-sonic turbulent boundary layer," *AIAA Journal*, vol. 16, no. 7, pp. 723–729, 1978.

[23] S. Özgen, "Effect of heat transfer on stability and transition characteristics of boundary-layers," *International Journal of Heat and Mass Transfer*, vol. 47, no. 22, pp. 4697–4712, 2004.

17

Robust Stabilization Control Based on Guardian Maps Theory for a Longitudinal Model of Hypersonic Vehicle

Yanbin Liu, Mengying Liu, and Peihua Sun

College of Astronautics, Nanjing University of Aeronautics and Astronautics, Nanjing 210016, China

Correspondence should be addressed to Yanbin Liu; nuaa_liuyanbin@139.com

Academic Editor: Chin-Chia Wu

A typical model of hypersonic vehicle has the complicated dynamics such as the unstable states, the nonminimum phases, and the strong coupling input-output relations. As a result, designing a robust stabilization controller is essential to implement the anticipated tasks. This paper presents a robust stabilization controller based on the guardian maps theory for hypersonic vehicle. First, the guardian maps theories are provided to explain the constraint relations between the open subsets of complex plane and the eigenvalues of the state matrix of closed-loop control system. Then, a general control structure in relation to the guardian maps theories is proposed to achieve the respected design demands. Furthermore, the robust stabilization control law depending on the given general control structure is designed for the longitudinal model of hypersonic vehicle. Finally, a simulation example is provided to verify the effectiveness of the proposed methods.

1. Introduction

With the development of advanced flight control technologies, the resulting demands become higher due to the complicated and diverse flight tasks. Also, some new design challenges in combination with the uncertain flight conditions and the unknown internal dynamics will have a tremendous impact on the conventional design means of flight control [1]. As a result, the design ideas of flight control are required to constantly update to satisfy the complex performance requirements. In the past decades, the flight control technologies rapidly develop with the help of the mathematical tools, the manufacturing means, and the computer simulations [2]. Typically, some traditional design ways such as the root locus method, the eigenvalue assignment method, and the Bode analysis method are used to obtain the control parameters related to the according tasks [3]. However, if the control structure becomes more and more complicated and associated with the large flight range, these control design means depending on the given trim point may become insufficient because the realization of the multiple demands of flight stability is difficult in consideration of the different trim states, especially the presence of some bizarre flight points due to the design errors corresponding to the

control parameters. These may make overall stability unable to be satisfied throughout the whole flight envelope [4–6].

In order to enhance flight stability and manipulating capability, some advanced control strategies are introduced in the design process of flight control. These design methods consist of the robust adaptive control [7], the sliding mode control [8], the dynamic inversion control [9], the fuzzy control [10], the neural control [11], and so on. Nevertheless, the flight control laws in terms of these modern control methods need to acquire the feedback of the complete flight states which are difficult to measure in the practical application [12–15]. On the other hand, these proposed controllers are limited to use due to some unrealistic assumptions and harsh prerequisites [16]. For example, the dynamic inversion control can relieve the coupling relations corresponding to the coupling states by building the affine linearization model [17]. Besides that, such resulting model is more accurate for the control design in comparison to the linear model acquired by Taylor's series expansion; the control robustness is improved accordingly [18]. Unfortunately, the dynamic inversion control relies too much on the airplane model, and in some case the affine linearization model is difficult to obtain due to the strong coupling dynamics between the inputs and the outputs. In addition, the dynamic inversion control is necessary to connect with the other control methods so as to improve system robustness and control adaptability [19]. In general, the conventional flight control approaches cannot guarantee overall stability; the control structures and parameters need to be changed with regard to the different trim conditions. The advanced flight control methods depend on the established model, providing the entire feedback states in the control design process. For this reason, new control methods should be introduced to design the satisfactory control law for the modern aerospace vehicle.

Theoretically, the realization of the flight control system is based on the vehicle models gotten by small perturbation linearization or linear variable parameter transformation [20]. As soon as the vehicle model is built, the control law can be designed by means of the linear interpolation or the application of linear parameter varying (LPV) control structure [21]. Although the linear interpolation is commonly used in practice, the stability issues in the control boundaries are difficult to be addressed. Alternatively, the LPV control is too conservative for implementing the stabilization control over the large flight range. On the other hand, the LPV control needs to solve the matrix inequalities which might be singular in some flight points [22]. As a result, the new control ideas need to be introduced to guarantee generalized stability, whereas the robust stabilization control based on the guardian maps can establish the constraint relations between the open subsets of complex plane and the eigenvalues of the state matrix of closed-loop control system [23]. Accordingly, the control parameters corresponding to the fixed control structure can be automatically regulated by assigning the poles of the closed-loop control system in accordance with the anticipated stability region [24].

As for hypersonic vehicle, the overall structure is special to implement the challenging tasks under the complicated flight conditions corresponding to the high altitude and speed [25]. To this end, the vehicle model exhibits the unconventional dynamic characteristics such as the unstable poles and the rigid-elastic coupling mode [26]. In addition, some new challenges need to be faced for hypersonic flight, including the trajectory optimization [27], the scramjet propulsion design [28], and the multidisciplinary integrated iteration [29]. More importantly, the hypersonic vehicle is easily affected by the external disturbances, the unknown model dynamics, and the uncertain coupling actions, leading to worsening of the anticipated performances due to the deviation of the waveriding design points. As a result, the robust stabilization controller designed is critical for hypersonic vehicle to maintain the continuous and stable flight states. Based on that, this paper puts forward the robust stabilization control methods using the guardian maps theories for a longitudinal model of hypersonic vehicle. Also, a scheduling algorithm of the control gains is applied for this hypersonic vehicle model, while adopting the state matrix in relation to the needed parameters determined by the guardian map restrains. Furthermore, a simulation example is performed to verify the effectiveness of the proposed control methods.

2. Matrix Polynomial Family and Guardian Maps Theory

In principle, the generalized stability with regard to the matrix polynomial is characterized by all system poles located in the left half-plane. However, if some design goals such as the bandwidth, damping, and response time are considered, the design area must be specifically defined to meet the respected handling qualities. According to these stability demands, the resulting stability region can be set using the guardian maps defined as follows.

Definition 1 (see [30]). Let χ be the set of all $n \times n$ square matrices, or the set of all polynomials of degree at most n, and let S be an open subset of χ. Let v map χ into the complex field $\mathscr{C}$. One thinks that v guards S if for all $x \in \bar{S}$, the equivalence

$$v(x) = 0 \Longleftrightarrow x \in \partial S \qquad (1)$$

holds. In this situation, one says that v is a guardian map for S when ∂S denotes the border of S. Based on (1), one further assumes that $x(r_0) \in S$ for some $r_0 \in U$ then has [30]

$$\forall r \in U, \quad x(r) \in S \Longleftrightarrow \forall r \in U, \quad v(x(r)) \neq 0. \qquad (2)$$

Equation (2) demonstrates that the stability issue can be replaced by guarded stability sets using the guardian map. Also, we find that this guardian map $v(x(r))$ is the function of r, and in this case $x(r)$ is stable with regard to the given region Ω only if (1) $x(r_0)$ is stable corresponding to Ω; (2) $v(x(r))$ has no zeros along with $[r_-, r_+]$, where r_-, r_+ are, respectively, the minimum and maximum of r. In other words, if v constrains $x(r)$ and $x(r_0) \in \Omega$, $x(r)$ is stable with regard to Ω only if $v(x(r))$ is not equal to zero in the interval of $[r_-, r_+]$ [24]. As a result, we need to identify whether $x(r)$ maintains stable along with Ω, so the related work is to check whether $v[A(r)]$ exists zeros for the interval $[r_-, r_+]$ such that testing stability problem is convenient for the control design. Accordingly, the guardian maps are provided for some typical regions as follows [23].

(1) The left half-plane region χ_1 with regard to $\text{Re}(z) < \alpha$ is guarded by

$$v_\alpha(\chi_1) = \det\left(\chi_1^2 \odot I - \alpha I \odot I\right) \det\left(\chi_1 - \alpha I\right), \qquad (3)$$

where $\odot$ represents the bialternate product.

(2) The conic area χ_2 with inner angle 2θ is guarded by

$$v_\xi(\chi_2) = \det\left(\chi_2^2 \odot I + \left(1 - 2\xi^2\right) \chi_2 \odot \chi_2\right) \det\left(\chi_2\right) \xi$$
$$= \cos\theta. \qquad (4)$$

(3) The circle sector χ_3 of radius $\omega > 0$ is guarded by

$$v_\omega(\chi_3) = \det\left(\chi_3 \odot \chi_3 - \omega^2 I \odot I\right)$$
$$\cdot \det\left(\chi_3 - \omega I\right) \det\left(\chi_3 + \omega I\right). \qquad (5)$$

Furthermore, the resulting sector $\chi_1 \cap \chi_2 \cap \chi_3$ is guarded by [12]

$$v(\chi_1 \cap \chi_2 \cap \chi_3) = v_\alpha(\chi_1) v_\xi(\chi_2) v_\omega(\chi_3). \qquad (6)$$

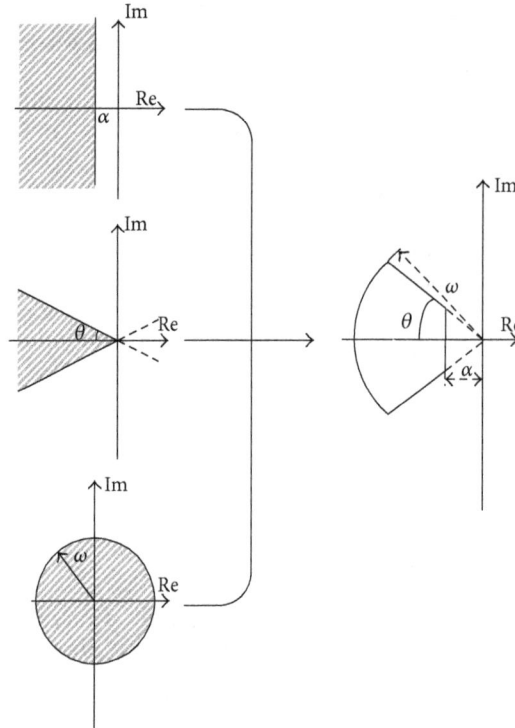

FIGURE 1: Construction sector in terms of classic stability section.

According to (6), we can construct new domains from the classic region to satisfy the real design demands, and this stability area is demonstrated in Figure 1. In addition, the guardian maps have other intrinsic transformation properties with regard to symmetry, translation, and scaling [30]. Using these characteristics, the guardian maps equations corresponding to some new sections can be obtained accordingly, such that other design requirements with consideration of the anticipated goals can be realized.

3. General Control Law Design Based on Guardian Maps Theory

For the given system model, this paper applies the classical control structure to realize the stability and track control. It is expressed in Figure 2.

In Figure 2, as soon as the initial control gains K_a, K_p, K_i, K_d are given, the according poles of the closed system lie in the region $\Omega = \Omega(\alpha, \xi, \omega)$ shown in Figure 1. Correspondingly, this closed system can be written in the following form:

$$A(r) = A_0 + rA_1 + \cdots + r^k A_k, \tag{7}$$

where A denotes the state matrix of this closed system. Therefore, the resulting guardian map $v_\Omega[A(r)]$ is only relevant to the control gain matrix r, composed of K_a, K_p, K_i, K_d, and the anticipated region can be designed as $\Omega_t = \Omega(\alpha_t, \xi_t, \omega_t)$ where in α_t, ξ_t, ω_t are required to meet the performance indices, so the design goal corresponding to the control action is to make that $\Omega = \Omega(\alpha, \xi, \omega)$ can

approach to $\Omega_t = \Omega(\alpha_t, \xi_t, \omega_t)$ by constantly iterating the control gains K_a, K_p, K_i, K_d. In each iterative process, the according region in Figure 1 is used if all eigenvalues defined as $\Lambda = \{\lambda_1, \lambda_2, \ldots, \lambda_n\}$ lie in the left half plane. In this case, the corresponding region $\Omega_\Lambda = \Omega(\alpha_\Lambda, \xi_\Lambda, \omega_\Lambda)$ determined by $\Lambda = \{\lambda_1, \lambda_2, \ldots, \lambda_n\}$ will be compared with the respected goal region. Therefore, the integrated region is obtained as follows [24]:

$$\alpha_U = \max\{\alpha_t, \alpha_\Lambda\}$$
$$\xi_U = \min\{\xi_t, \xi_\Lambda\} \tag{8}$$
$$\omega_U = \max\{\omega_t, \omega_\Lambda\},$$

where

$$\alpha_\Lambda = \max\{\mathrm{Re}\,(\lambda_i)\}$$
$$\xi_\Lambda = \min\{\xi\,(\lambda_i)\} \tag{9}$$
$$\omega_\Lambda = \max\{|\lambda_i|\}.$$

In particular, if any pole among $\Lambda = \{\lambda_1, \lambda_2, \ldots, \lambda_n\}$ is located in the right half plane, the resulting region is redefined as in Figure 3.

According to Figure 3, the resulting sector can be depicted by

$$\alpha_U = \alpha_\Lambda$$
$$\omega_U = \max\{\omega_t, \omega_\Lambda\}, \tag{10}$$

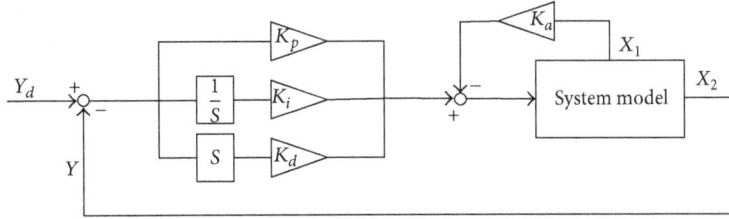

FIGURE 2: General structure of control system.

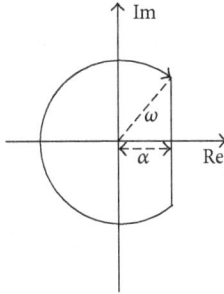

FIGURE 3: Construction sector in combination with instability region.

where

$$\alpha_\Lambda = \max\{\text{Re}(\lambda_i)\}$$
$$\omega_\Lambda = \max\{|\lambda_i|\}. \tag{11}$$

Based on (8) and (10), the according sector is gotten with regard to each iteration process that does not stop until this resulting area completely matches with the anticipated region. In addition to the selection of the iterative region, how to change the control gain matrix is equally important. Following that, we assume the control parameters r_0 can make that the eigenvalues belong to the iterative region decided by (8) and (10). At this time, we can obtain the interval range corresponding to r by solving $\nu_\Omega[A(r)]$. They are expressed by [23]

$$r^- \approx \sup\{r < r_0 : \nu_\Omega[A(r)] = 0\} \ (\text{or} -\infty \text{ if none exists})$$

$$r^+ \approx \inf\{r < r_0 : \nu_\Omega[A(r)] = 0\} \ (\text{or} -\infty \text{ if none exists}). \tag{12}$$

To regulate the control gain matrix $r = [K_a, K_p, K_i, K_d]$, first we select K_a as the variable, whereas other parameters are fixed, and in this case calculate r^- and r^+ for $\Omega_\Lambda = \Omega(\alpha_\Lambda, \xi_\Lambda, \omega_\Lambda)$ to get $K_{a\,\max}$ and $K_{a\,\min}$. Following that, K_a is assigned to $(K_{a\,\max} + K_{a\,\min})/2$, and then the similar works are performed for each control gain such that the control gain matrix is continually updated as $r' = [K_a', K_p', K_i', K_d']$. Thus, new design region Ω_Λ is identified with $r' = [K_a', K_p', K_i', K_d']$. Furthermore, this iterative process does not stop until $\Omega_\Lambda \subset \Omega_t$.

4. Modeling and Flight Control Design for Hypersonic Vehicle

The dynamic models of hypersonic vehicle tend to be highly complex and multidisciplinary due to strong interaction between aerodynamics, propulsion, structure, and controls [31]. Not only that, the special configuration applied to design hypersonic vehicle makes that the established model is unstable, nonminimum phase, and strong coupling relations. These will bring new challenges for the control system design, as well as the model establishment. Based on the Lagrange equation, the longitudinal model of hypersonic vehicle is built by [31]

$$\dot{V} = \frac{1}{m}(T\cos\alpha - D) - g\sin(\theta - \alpha)$$

$$\dot{\alpha} = \frac{1}{mV}(-T\sin\alpha - L) + q + \frac{g}{V}\cos(\theta - \alpha)$$

$$\dot{q} = \frac{M}{I_{yy}} \tag{13}$$

$$\dot{h} = V\sin(\theta - \alpha)$$

$$\dot{\theta} = q,$$

where V, h, α, θ, q represent the flight velocity, the flight altitude, the angle of attack, the pitch angle, and the pitch angle change rate of hypersonic vehicle, respectively; L, D, M denote, respectively, the lift, the drag, and the pitch moment; m, I_{yy} indicate the vehicle mass and the moment of inertia, respectively. The aerodynamic forces and moments of the nonlinear model in (13) need to be acquired to build the complete relations between the systems inputs and outputs. They are expressed by

$$L \approx \frac{1}{2}\rho V^2 S C_L(\alpha, \delta_e)$$

$$D \approx \frac{1}{2}\rho V^2 S C_D(\alpha, \delta_e) \tag{14}$$

$$M \approx \frac{1}{2}\rho V^2 S\bar{c}[C_{M,\alpha}(\alpha) + C_{M,\delta e}(\delta_e)],$$

where ρ represents air density; S, $\bar{c}$ denotes the reference area and the mean aerodynamic chord, respectively. Additionally, the oriented control model established is critical to design the flight control system. In this paper, the aerodynamic coefficients with regard to the oriented control model are selected as the following polynomials [18]:

$$C_L(\alpha, \delta_e) = C_L^\alpha \alpha + C_L^{\delta_e} \delta_e + C_L^0$$

$$C_D(\alpha, \delta_e) = C_D^{\alpha^2} \alpha^2 + C_D^\alpha \alpha + C_D^{\delta_e^2} \delta_e^2 + C_D^{\delta_e} \delta_e + C_D^0$$

$$C_{M,\alpha} = C_{M,\alpha}^{\alpha^2} \alpha^2 + C_{M,\alpha}^\alpha \alpha + C_{M,\alpha}^0 \tag{15}$$

$$C_{M,\delta_e} = c_e \delta_e.$$

Besides these aerodynamic forces and moments, this paper considers the thrust expression as [18]

$$C_T = c_\Phi \Phi$$

$$\ddot\Phi = -2\xi\omega\dot\Phi - \omega^2\Phi + \omega^2\Phi_c, \tag{16}$$

where Φ denotes the fuel to air ratio, whereas Φ_c indicates the propulsive ratio command. In this study, the elevator deflection angles δ_e and Φ_c are selected as the control inputs to stabilize and follow the according commands. In particular, applying the approximate polynomials to depict the aerodynamic coefficients may bring some modeling errors, but these expressions are more suitable for designing the control law, as well as analyzing the dynamic characteristics. Therefore, in order to obtain these model polynomials of hypersonic vehicle, the corresponding identification methods need to be used to estimate these polynomial coefficients. For example, the trust region method can be adopted to identify these polynomial coefficients [32]. The core idea of this method is to first define an adjacent area with regard to the current iteration point x_k, and this area is called the trust region written by

$$\Pi_k = \{x \mid \|x - x_k\| \le h_k\}, \tag{17}$$

where h_k denotes the upper bound of step. Furthermore, we adopt the trust region method to get the according coefficients in (15). For the lift expression, the criterion function is provided by

$$G(K) = \frac{1}{2}\left|C_L\left(K, \overline{X}\right) - \overline{Y}\right|^2, \tag{18}$$

where K is the coefficients matrix $K_L = [C_L^\alpha, C_L^{\delta_e}, C_L^0]$; $\overline{X}$ is the matrix constituted of the given flight state datum, whereas $\overline{Y}$ represents the given lift coefficients. $\overline{X}$ and $\overline{Y}$ can be acquired by means of the parametric modeling and computational fluid dynamics methods. Moreover, the tentative mean is used to implement the iteration process, and then we have [32]

$$s = x_{i+1} - x_i$$

$$\min_x F = \min_x \left\{ \frac{1}{2} s^T H_f s + s^T g_i \right\} \tag{19}$$

$$\|D_i d\| \le h_i,$$

where s indicates the iterative step; H_f is the selected symmetric matrix that is obtained by

$$H_f = \begin{bmatrix} \dfrac{\partial^2 F}{\partial x_1^2} & \cdots & \dfrac{\partial^2 F}{\partial x_1 \partial x_n} \\ \vdots & \ddots & \vdots \\ \dfrac{\partial^2 F}{\partial x_n \partial x_1} & \cdots & \dfrac{\partial^2 F}{\partial x_n^2} \end{bmatrix} \tag{20}$$

$$g_i = \mathrm{grad}(F) = \nabla(F) = \left(\frac{\partial F}{\partial x_1}, \cdots \frac{\partial F}{\partial x_n} \right).$$

Once the expression $H_f + \lambda I$ is positive definite matrix when $\lambda \ge 0$, (20) is solvable. In this case, we have

$$\left\|(H_f + \lambda I)^{-1} s\right\| = h. \tag{21}$$

Further, we can set the next step to perform the iterative process, and lastly the optimal results can be obtained for the lift coefficients matrix K_L. Similarly, the polynomial coefficients with regard to C_D, C_M can be gotten accordingly; thus these aerodynamic expressions are introduced to construct the nonlinear model. For the further application in the control design, this model needs to be simplified. Commonly, there are two means to complete the model simplification. One is to obtain the linear model by the small perturbation linearization, and the other is to transform the nonlinear model to the equivalent model, such as the feedback linearized model or the linear varying parameter model. This paper utilizes the linear model to design the control system for hypersonic vehicle. It is expressed by [33]

$$\Delta \dot X = A \cdot \Delta X + B \cdot \Delta U$$

$$\Delta X = [\Delta V, \Delta \alpha, \Delta q, \Delta h, \Delta \theta]^T \tag{22}$$

$$\Delta U = [\Delta \Phi, \Delta \delta_e]^T,$$

where

$$A = \begin{bmatrix} X_v & X_\alpha & 0 & X_h & -g \\ \dfrac{Z_v}{V_{T_0}} & \dfrac{Z_\alpha}{V_{T_0}} & \dfrac{1 - Z_q}{V_{T_0}} & \dfrac{Z_h}{V_{T_0}} & 0 \\ M_v & M_\alpha & M_q & M_h & 0 \\ 0 & -V_0 & 0 & 0 & V_0 \\ 0 & 0 & 1 & 0 & 0 \end{bmatrix} \tag{23}$$

$$B = \begin{bmatrix} X_{\delta_e} & X_\Phi \\ \dfrac{Z_{\delta_e}}{V_{T_0}} & \dfrac{Z_\Phi}{V_{T_0}} \\ M_{\delta_e} & M_\Phi \\ 0 & 0 \\ 0 & 0 \end{bmatrix},$$

where X, Z, M indicate the coefficients of the linear model. For (22), the characteristic polynomial regarding the short-term mode is written by

$$\Delta_{sp} = V_t s^2 - (Z_\alpha + V_t M_q) s + M_q Z_\alpha - V_t M_\alpha. \tag{24}$$

The according solutions of (24) are gotten by

$$s = \frac{Z_\alpha}{2V_t} \pm \frac{\sqrt{\left(Z_\alpha - V_t M_q\right)^2 + 4V_t^2 M_\alpha}}{2V_t}. \tag{25}$$

Under normal circumstances, hypersonic vehicle is designed as the special slim construction such that the aerodynamic forces suffered from the forebody section are large enough in comparison to the other parts. Such configuration makes that the aerodynamic focus is prior to the gravity centre, namely, $M_\alpha > 0$. For (25), there exists the positive root, and this means the short-term motion is unstable for hypersonic vehicle. Furthermore, when removing the long term parts in (23), we have

$$\begin{bmatrix} s - \dfrac{Z_\alpha}{V_0} & -1 + \dfrac{Z_q}{V_0} \\ -M_\alpha & s - M_q \end{bmatrix} \begin{bmatrix} \Delta\alpha(s) \\ \Delta q(s) \end{bmatrix}$$

$$= \begin{bmatrix} \dfrac{Z_\phi}{V_0} & M_\phi \\ \dfrac{Z_{\delta_e}}{V_0} & M_{\delta_e} \end{bmatrix}^T \begin{bmatrix} \Delta\phi(s) \\ \Delta\delta_e(s) \end{bmatrix}. \tag{26}$$

According to (26), the frequency and damping corresponding to the short-term mode are estimated by

$$\omega_{sp} \approx \sqrt{\frac{Z_\alpha}{V_0} M_q - \left(1 + \frac{Z_q}{V_0}\right) M_\alpha}$$

$$\zeta_{sp} \approx \frac{\left(-Z_\alpha/V_0 - M_q\right)}{\left(2\omega_{sp}\right)}. \tag{27}$$

Based on that, we see that the angle of attack and pitch angle change rate will affect the short-term mode. Thus, if these flight states can be introduced into the control inputs, the system performances will be ameliorated accordingly. Additionally, we can design the augmentation control loop in line with the structure in Figure 2. In this case, only $K_{a\alpha}$, K_{aq} need to be adjusted and tuned. Alternatively, the control structure can be adopted for the pitch angle control loop, the altitude control loop, and the velocity control loop by applying the general control structure in Figure 2. Especially, the according control laws based on Figure 2 can be provided as follows:

$$\Delta\delta_{ea} = K_{a\alpha}\Delta\alpha + K_{aq}\Delta q$$

$$\Delta\delta_e = \Delta\delta_{ea} + \left[K_{p\theta}\left(\Delta\theta_d - \Delta\theta\right) + K_{i\theta}\int\left(\Delta\theta_d - \Delta\theta\right)\right]$$

$$\Delta\theta_d = K_{ah}\Delta\dot{h} + K_{ph}\left(\Delta h_d - \Delta h\right) + K_{ih}\int\left(\Delta h_d - \Delta h\right) \tag{28}$$

$$\Delta\Phi = K_{pV}\left(\Delta V_d - \Delta V\right) + K_{iV}\int\left(\Delta V_d - \Delta V\right),$$

where $K_{a\alpha}$, K_{aq} denote the augmentation control parameters; $K_{p\theta}$, $K_{i\theta}$ are the pitch angle control gains; K_{ah}, K_{ph}, K_{ih}

represent the altitude control gains; K_{pV}, K_{iV} indicate the speed control gains. The control block diagram of hypersonic vehicle is demonstrated in Figure 5.

From Figure 5, we note that the control block diagram is constituted of the general control structure in Figure 2, and the according control parameters can be obtained based on the iteration process in Figure 4. With the combination of the inner loop and the outer loop, the complete control configuration is formed to achieve the track control responses with regard to the altitude and the velocity.

5. Simulation Example and Analysis

To verify the effectiveness of the control system based on the guardian maps theories, this paper applies the hypersonic vehicle model in [26], and the aerodynamic coefficients of this model are provided by

$$C_L = 0.6203\alpha$$

$$C_D = 0.6450\alpha^2 + 0.0043378\alpha + 0.003772$$

$$C_M = -0.035\alpha^2 + 0.036617\alpha \tag{29}$$

$$+ 5.3261 \times 10^{-6} + 0.0292\left(\delta_e - \alpha\right)$$

$$C_T = 0.02576\left[1 - 164(\alpha - \alpha_{\text{trim}})^2\right]\beta,$$

where α_{trim} is the trim angle of attack of this hypersonic vehicle model. Also, the model properties are used in [34]. By introducing these aerodynamic parameters to the nonlinear model, the trim states with regard to $V_0 = 4500\,\text{m/s}$, $h_0 = 33500\,\text{m}$ include $\alpha_{\text{trim}} = 2.845°$, $\delta_{e\text{trim}} = -0.564°$, $\beta_{c\text{trim}} = 0.217$. The resulting characteristic roots of this model are gotten as -0.737, 0.623, $-0.0000663 \pm 0.0366102j$ wherein there exists one positive root that demonstrates that this model is unstable. Additionally, the characteristic roots with regard to the long-term mode are close to the imaginary axis, and this shows the long-term motion is underdamped. As a result, the control action is crucial to guarantee flight stability as well as to realize the command track.

Because this model has the unstable and nonminimum phase characteristics, first we need to design the stability augmentation system to improve system performances. Based on the control structure in Figure 2, the guardian maps theories are applied to acquire the stability augmentation gain $K_{a\alpha}$. Beyond this, we select the design region in consideration of the system instability in Figure 3, including $\alpha_{ta} = 0.01$, $\omega_{ta} = 10$. By the continuous iteration, the augmentation parameter is gotten as $K_{a\alpha} = 4.0368$ such that the characteristic roots lie in the left-half plane or the imaginary axis. Furthermore, the performance qualities with regard to the short-term mode are considered, and they consist of $\alpha_{ts} = -1$, $\omega_{ts} = 10$, $\xi_{ts} = 0.35$ which are satisfied with the flight qualities demands [35]. In this situation, the control gain of the pitch angle rate is gotten as $K_{aq} = 2.0687$, and the resulting characteristic roots of the short-term mode are $-3.37 \pm 1.2i$. After that, the quality requirements with respect to the long-term mode are given as $\alpha_{tl} = -0.02$, $\omega_{tl} = 5$, $\xi_{tl} = 0.1$,

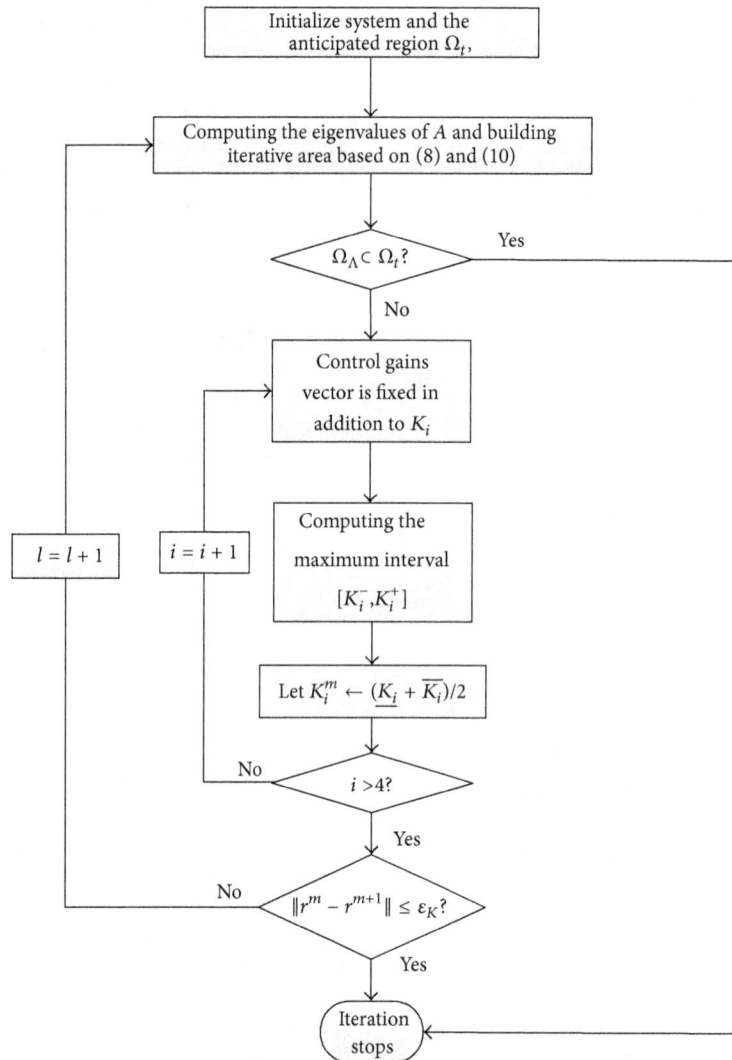

FIGURE 4: Iteration flowchart with regard to control gains.

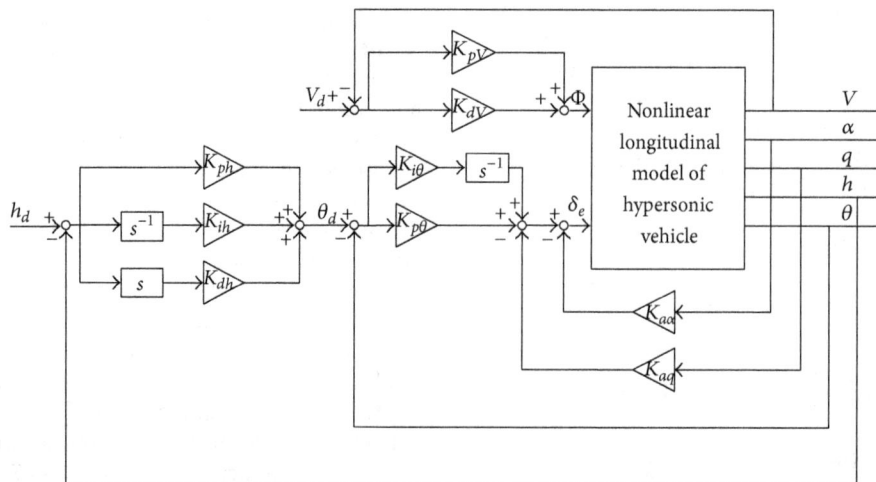

FIGURE 5: Control block diagram of hypersonic vehicle.

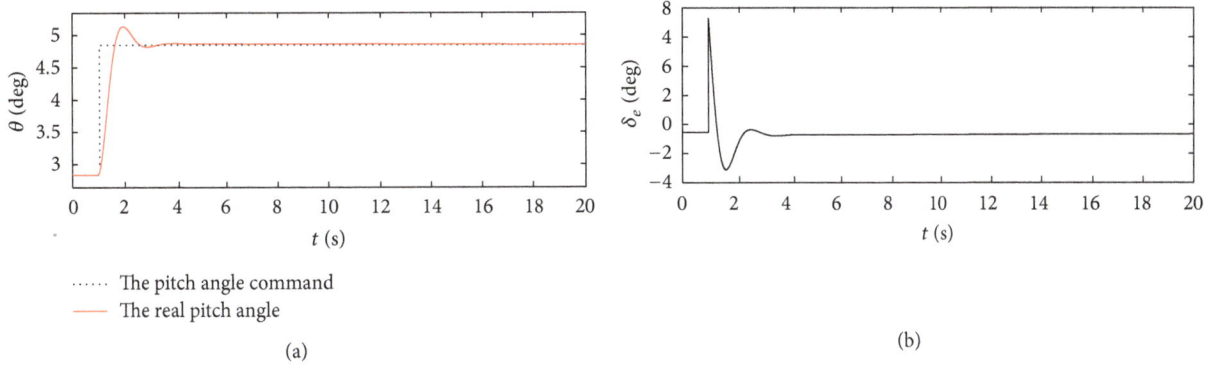

FIGURE 6: Pitch angle response and elevator deflection angle.

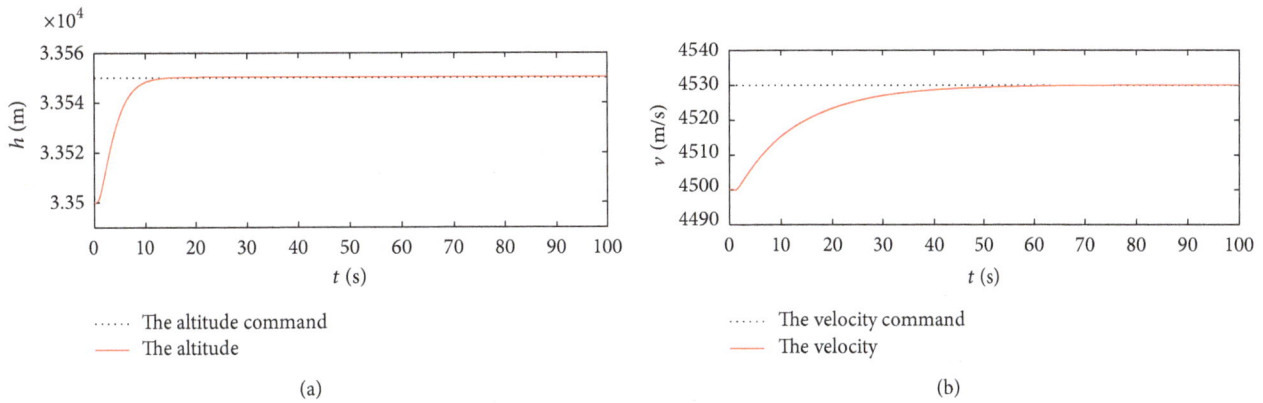

FIGURE 7: Response curves to the altitude and velocity commands.

and accordingly the control gains are obtained as $K_{p\theta}$ = 3.9245, $K_{i\theta}$ = 13.1620 such that the resulting characteristic roots of the long-term mode are $-0.0200 \pm 0.0300i$. Based on these acquired control parameters, the track simulation can be done to test the validity of the attitude controller. Correspondingly, the response curves to the step signal $\Delta\theta_d$ = 2 deg at the 2 second are shown in Figure 6.

According to Figure 6, we find that the pitch angle can rapidly follow the attitude command signal, and the elevator deflection angle reaches the resulting trim value as the system enters into the new stability states. This shows that the proposed control law can guarantee system stability and track ability even if the vehicle model is unstable. Compared with other control methods, the structure of this controller is simple, and the control gains are adaptively obtained in terms of the anticipated control qualities such that the controller is feasible to apply for the real design.

Furthermore, the altitude and velocity control gains can be gotten based on the similar iteration process in line with the guardian maps theories, and these gains are K_{ph} = 0.001, K_{dh} = 0.004 and K_{pV} = 0.01, K_{dV} = 0.005. In the following simulation, the commands are Δh_d = 50 m, ΔV_d = 30 m/s, and the resulting response curves after 100 seconds are demonstrated in Figures 7 and 8.

According to Figures 7 and 8, the real altitude and velocity can track the respective command signals well. This manifests

that the control action relieves the coupling relations between the inputs and the outputs, while ensuring that the flight states and control inputs rapidly reach the new steady values without the presence of the undesirable oscillation. On the further consideration of system robustness, the model uncertainties including 50% errors concerning the lift and drag coefficients are exerted in the simulation. The according response curves are acquired in Figure 9.

Figure 9 tells us that the system outputs can correspond to the given commands even in the presence of the large model uncertainties. Although there are some flutters in the response process, the track errors are small and the changes of the control inputs are smooth. This shows that the control actions are effective to restrain the uncertain disturbances and provide strong system robustness. More importantly, the control gains are obtained automatically by means of the adaptive iteration methods in terms of the given goals. Such design methods based on the guardian maps theories make that the overall system satisfies the anticipated quality demands, thus resulting in suppressing the unstable model dynamics and the uncertain effects.

6. Conclusion

This paper proposes the robust stabilization control methods based on the guardian maps theories for a longitudinal model

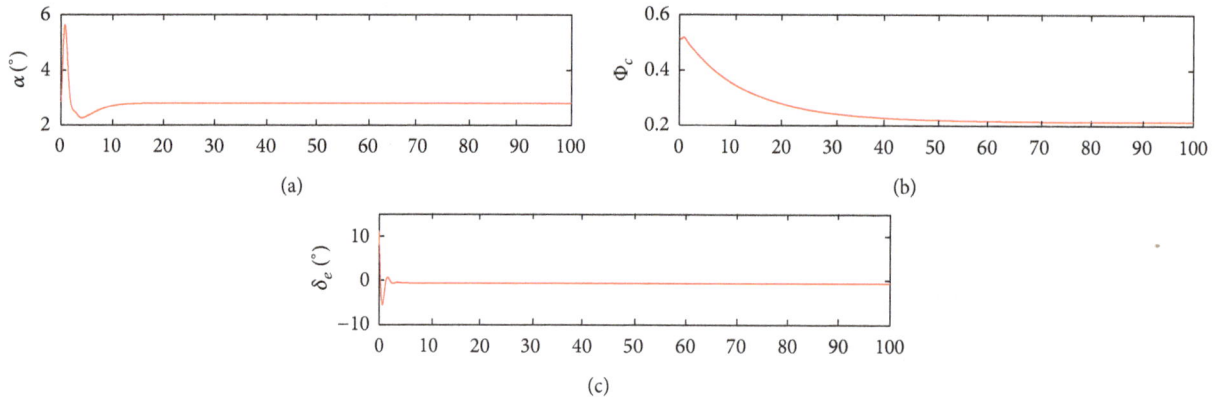

FIGURE 8: Change curves of the control inputs and angle of attack.

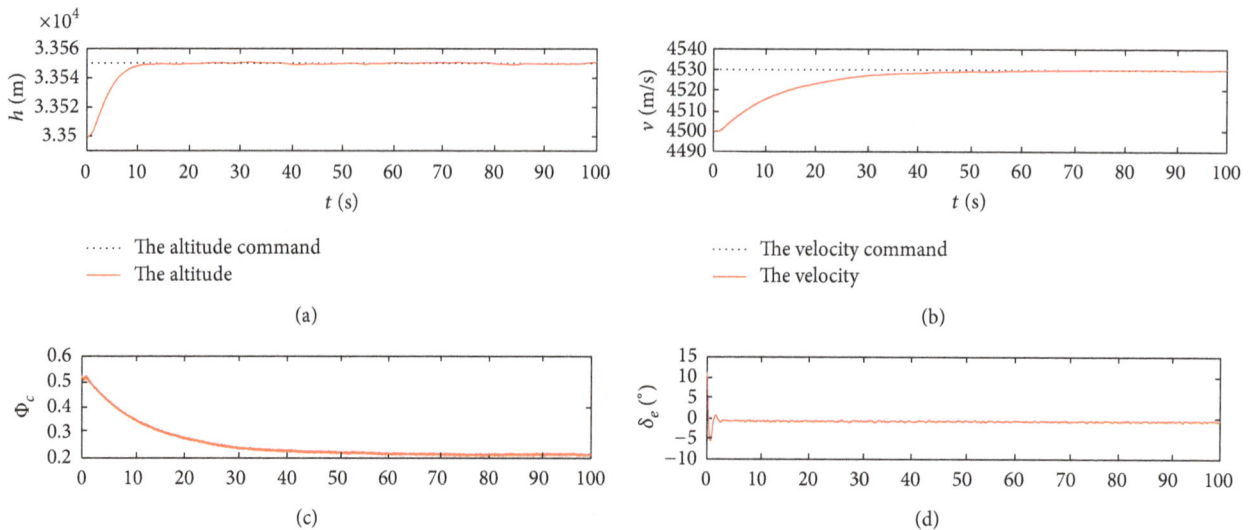

FIGURE 9: Change curves of the control inputs and angle of attack.

of hypersonic vehicle. There are three aspects that need to be considered for the control law design. The first part is to build the general control structure to realize the iteration process of the control gains in line with the guardian maps theories. The second issue is to establish the control-oriented model using the system identification methods, and at the same time the linear model can be acquired according to the small perturbation linearization principles. The last problem is to design the robust stabilization controller for the unstable model of hypersonic vehicle in combination with the general control configuration and the presented iterative process. We believe this work is helpful to design the complicated robust control law and implement the adaptive gain adjustment for hypersonic vehicle in the future.

Conflict of Interests

The authors declare that there is no conflict of interests regarding the publication of this paper.

Acknowledgments

This work is supported by NUAA fundamental research funds under Grant no. NS2014088. The authors thank the editors and the reviewers for their help and improvements to the quality of our presentation.

References

[1] K. Natesan, D.-W. Gu, and I. Postlethwaite, "Design of static H_∞ linear parameter varying controllers for unmanned aircraft," *Journal of Guidance, Control, and Dynamics*, vol. 30, no. 6, pp. 1829–1835, 2007.

[2] K. Peng, G. Cai, B. M. Chen, M. Dong, K. Y. Lum, and T. H. Lee, "Design and implementation of an autonomous flight control law for a UAV helicopter," *Automatica*, vol. 45, no. 10, pp. 2333–2339, 2009.

[3] A. Fujimori, F. Term, and P. N. Nikiforuk, "Flight control design of an unmanned space vehicle using gain scheduling," *Journal of Guidance, Control, and Dynamics*, vol. 28, no. 1, pp. 96–105, 2005.

[4] K. H. Well, "Aircraft control laws for envelope protection," in *Proceedings of the AIAA Guidance, Navigation, and Control Conference*, pp. 1–10, Keystone, Colo, USA, August 2006.

[5] H. J. Gong, Z. Y. Zhen, X. Li et al., "Automatic flight control system design of level change mode for a large aircraft," *International Journal of Advanced Robotic Systems*, vol. 10, no. 96, pp. 1–8, 2013.

[6] L. Xu, Q. Wang, W. Li, and Y. Hou, "Stability analysis and stabilisation of full-envelope networked flight control systems: switched system approach," *IET Control Theory & Applications*, vol. 6, no. 2, pp. 286–296, 2012.

[7] B. Jiang, Y. Guo, and P. Shi, "Adaptive reconfiguration scheme for flight control systems," *Proceedings of the Institution of Mechanical Engineers I: Journal of Systems and Control Engineering*, vol. 224, no. 6, pp. 713–723, 2010.

[8] Q. Zong, J. Wang, B. L. Tian et al., "Quasi-continuous high-order sliding mode controller and observer design for flexible hypersonic vehicle," *Aerospace Science and Technology*, vol. 27, no. 1, pp. 127–137, 2013.

[9] W. D. Gai, H. L. Wang, J. Zhang, and Y. Li, "Adaptive neural network dynamic inversion with prescribed performance for aircraft flight control," *Journal of Applied Mathematics*, vol. 2013, Article ID 452653, 12 pages, 2013.

[10] X. X. Hu, L. G. Wu, C. H. Hu et al., "Adaptive fuzzy integral sliding mode control for flexible air-breathing hypersonic vehicles subject to input nonlinearity," *Journal of Aerospace Engineering*, vol. 26, no. 4, pp. 721–734, 2013.

[11] B. Xu, Z. K. Shi, C. G. Yang, and S. Wang, "Neural control of hypersonic flight vehicle model via time-scale decomposition with throttle setting constraint," *Nonlinear Dynamics*, vol. 73, no. 3, pp. 1849–1861, 2013.

[12] A. J. Calise, S. Lee, and M. Sharma, "Development of a reconfigurable flight control law for tailless aircraft," *Journal of Guidance, Control, and Dynamics*, vol. 24, no. 5, pp. 896–902, 2001.

[13] Y. Shtessel, J. Buffington, and S. Banda, "Tailless aircraft flight control using multiple time scale reconfigurable sliding modes," *IEEE Transactions on Control Systems Technology*, vol. 10, no. 2, pp. 288–296, 2002.

[14] H. B. Duan and P. Li, "Progress in control approaches for hypersonic vehicle," *Science China: Technological Sciences*, vol. 55, no. 10, pp. 2965–2970, 2012.

[15] S. N. Wu, W. Y. Zhou, S. J. Tan, and G. Q. Wu, "Robust H_∞ control for spacecraft rendezvous with a noncooperative target," *The Scientific World Journal*, vol. 2013, Article ID 579703, 7 pages, 2013.

[16] Y. Q. Huang, C. Y. Sun, and C. S. Qian, "Non-fragile switching tracking control for a flexible air-breathing hypersonic vehicle based on polytopic LPV model," *Chinese Journal of Aeronautics*, vol. 26, no. 4, pp. 948–959, 2013.

[17] J. S. Brinker and K. A. Wise, "Stability and flying qualities robustness of a dynamic inversion aircraft control law," *Journal of Guidance, Control, and Dynamics*, vol. 19, no. 6, pp. 1270–1277, 1996.

[18] J. K. Parker, A. Serrani, S. Yurkovich, M. A. Bolender, and D. B. Doman, "Control-oriented modeling of an air-breathing hypersonic vehicle," *Journal of Guidance, Control, and Dynamics*, vol. 30, no. 3, pp. 856–869, 2007.

[19] J. Georgie and J. Valasek, "Selection of longitudinal desired dynamic for dynamic inversion controlled re-entry vehicles," in *Proceedings of the AIAA Guidance, Navigation, and Control Conference and Exhibit*, pp. 1–26, Montreal, Canada, 2001.

[20] A. Marcos and G. J. Balas, "Development of linear-parameter-varying models for aircraft," *Journal of Guidance, Control, and Dynamics*, vol. 27, no. 2, pp. 218–228, 2004.

[21] R. Lind, "Linear parameter-varying modeling and control of structural dynamics with aerothermoelastic effects," *Journal of Guidance, Control, and Dynamics*, vol. 25, no. 4, pp. 733–739, 2002.

[22] W. Xie, "Quadratic stabilization of LPV system by an LTI controller based on ILMI algorithm," *Mathematical Problems in Engineering*, vol. 2007, Article ID 28262, 7 pages, 2007.

[23] D. Saussie, L. Saydy, O. Akhrif, and C. Berard, "Gain scheduling with guardian maps for longitudinal flight control," *Journal of Guidance, Control, and Dynamics*, vol. 34, no. 4, pp. 1–15, 2011.

[24] D. Saussie, O. Akhrif, C. Berard, and L. Saydy, "Longitudinal flight control synthesis with guardian maps," in *Proceedings of the AIAA Guidance, Navigation, and Control Conference and Exhibit*, pp. 1–27, Chicago, Ill, USA, August 2009.

[25] Y. B. Liu, J. Deng, and Y. P. Lu, "Preliminary research on optimal design based on control demands for hypersonic morphing vehicle," *IEEE Aerospace and Electronic Systems Magazine*, vol. 28, no. 5, pp. 23–31, 2013.

[26] H. J. Xu, P. A. Ioannou, and M. Mirmirani, "Adaptive sliding mode control design for a hypersonic flight vehicle," *Journal of Guidance, Control, and Dynamics*, vol. 27, no. 5, pp. 829–838, 2004.

[27] J. Zhao, R. Zhou, and X. L. Jin, "Reentry trajectory optimization based on a multistage pseudospectral method," *The Scientific World Journal*, vol. 2014, Article ID 878193, 13 pages, 2014.

[28] A. A. Rodriguez, J. J. Dickeson, O. Cifdaloz et al., "Modeling and control of scramjet-powered hypersonic vehicles: challenges, trends, and tradeoffs," in *Proceedings of the AIAA Guidance, Navigation and Control Conference and Exhibit*, pp. 1–40, Honolulu, Hawaii, USA, August 2008.

[29] N. Yokoyama, S. Suzuki, T. Tsuchiya, H. Taguchi, and T. Kanda, "Multidisciplinary design optimization of space plane considering rigid body characteristics," *Journal of Spacecraft and Rockets*, vol. 44, no. 1, pp. 121–131, 2007.

[30] L. Saydy, L. T. Andre, and H. A. Eyad, "Guardian maps and the generalized stability of parametrized families of matrices and polynomials," *Mathematics of Control, Signals, and Systems*, vol. 3, no. 4, pp. 345–371, 1990.

[31] M. A. Bolender and D. B. Doman, "Nonlinear longitudinal dynamical model of an air-breathing hypersonic vehicle," *Journal of Spacecraft and Rockets*, vol. 44, no. 2, pp. 374–387, 2007.

[32] T. F. Coleman and Y. Li, "On the convergence of interior-reflective Newton methods for nonlinear minimization subject to bounds," *Mathematical Programming*, vol. 67, no. 1–3, pp. 189–224, 1994.

[33] F. Chavez and D. Schmidt, "Analytical aeropropulsive/aeroelastic hypersonic-vehicle model with dynamic analysis," *Journal of Guidance, Control, and Dynamics*, vol. 17, no. 6, pp. 1308–1319, 1994.

[34] Q. Wang and R. F. Stengel, "Robust nonlinear control of a hypersonic aircraft," *Journal of Guidance, Control, and Dynamics*, vol. 23, no. 4, pp. 577–584, 2000.

[35] D. Saussi, L. Saydy, and O. Akhrlf, "Longitudinal flight control design with handling quality requirements," *The Aeronautical Journal*, vol. 110, no. 1111, pp. 627–637, 2006.

18

Aerodynamic Characteristics of the Ventilated Design for Flapping Wing Micro Air Vehicle

G. Q. Zhang[1] and S. C. M. Yu[2]

[1] Aerospace Engineering Division, School of Mechanical and Aerospace Engineering,
  Nanyang Technological University, Singapore 639798
[2] Academic Division, Singapore Institute of Technology, Singapore 179104

Correspondence should be addressed to G. Q. Zhang; zhangguoqing@ntu.edu.sg

Academic Editors: Z.-H. Han and M. Yamamoto

Inspired by superior flight performance of natural flight masters like birds and insects and based on the ventilating flaps that can be opened and closed by the changing air pressure around the wing, a new flapping wing type has been proposed. It is known that the net lift force generated by a solid wing in a flapping cycle is nearly zero. However, for the case of the ventilated wing, results for the net lift force are positive which is due to the effect created by the "ventilation" in reducing negative lift force during the upstroke. The presence of moving flaps can serve as the variable in which, through careful control of the areas, a correlation with the decrease in negative lift can be generated. The corresponding aerodynamic characteristics have been investigated numerically by using different flapping frequencies and forward flight speeds.

1. Introduction

Micro air vehicles (MAVs) have the potential to revolutionize our sensing and information gathering capabilities in areas such as environmental monitoring and homeland security. In classical stationary wing theory, the tip vortices (TIVs) are seen as wasted energy; in flapping flight, they can interact with the LEV to enhance lift without increasing the power requirements [1, 2]. Just as referred by Ellington et al. [3] in Figure 1(a) (Nature 384: 626–630, 1996), to support the body weight, the wings typically produce 2-3 times more lift than can be accounted for by conventional aerodynamics. Liu and Aono [4] presented computational fluid dynamics (CFD) methods to study the insect hovering aerodynamics in Figure 1(b), which was performed using a biology-inspired dynamic flight simulator that integrated the modeling of realistic wing-body morphology, the modeling of flapping-wing and body kinematics, and an in-house Navier-Stokes solver. The corresponding results can not only give an integrated interpretation on the similarity and discrepancy of the near- and far-field vortex structures in insect hovering,

but also demonstrate that the methods can be an effective tool in the MAVs design. Hueso et al. [5] provided the visualized results that emphasize the presence of vortices in simulated airflow around a motion-captured bat model in Figure 1(c). It shows a vortex in the spanwise plane (streamwise vorticity) captured over the wing in the flight of the bat.

Hu et al. [6] had conducted experiments to explore the potential applications of compact, gearless, and piezoelectric flapping wings with the wing size, stroke amplitude, and flapping frequency within the range of actual insect characteristics for the development of novel insect-sized, flapping-wing-based nanoair vehicles (NAVs). A digital particle image velocimetry (PIV) system was used to achieve phase-locked and time-averaged flow field measurements to quantify the formation and separation processes of the leading-edge vortex (LEV) structures on the upper and lower surfaces of the flapping wing in relation to the phase angle. It was found that the wake vortices in the cross plane at 50% wingspan would form concentrated vortex structures. As they travel downstream, the concentrated anti-clockwise and clockwise wake vortices were found to cross over at first and then align

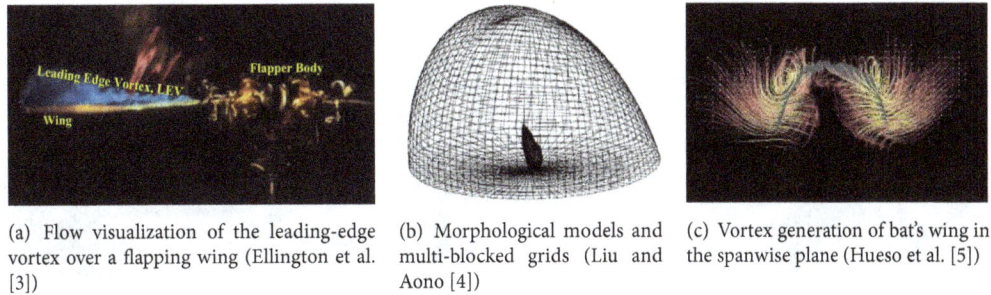

(a) Flow visualization of the leading-edge vortex over a flapping wing (Ellington et al. [3])

(b) Morphological models and multi-blocked grids (Liu and Aono [4])

(c) Vortex generation of bat's wing in the spanwise plane (Hueso et al. [5])

FIGURE 1: The presence of vortices in simulated airflow around the flapping wing model.

themselves in two rows with the clockwise (negative) vortices at above and anticlockwise (positive) vortices below, which is a typical von Karman vortex street.

Nagai et al. [7] also conducted the experimental and numerical studies to investigate the aerodynamic characteristics of a flapping wing of an insect in forward flight. Unsteady aerodynamic forces and flow patterns were measured using a dynamically scaled mechanical model in a water tunnel. The results indicated that these aerodynamic mechanisms had an effect on the aerodynamic characteristics of the flapping wing in forward flight; however, these mechanisms function differently during the up- and downstroke, for different stroke plane angles and for different advance ratios.

Singh and Chopra [8] used an experimental apparatus, with a biomimetic flapping mechanism to measure the thrust generated by a number of insect-based hover-capable flapping wings designs at different pitch settings. The wing mass was found to have a significant influence on the maximum frequency of the mechanism because of a high inertial power requirement. All the wings tested showed a decrease in thrust at high frequencies. In contrast, for a wing held at 90 deg pitch angle, flapping in a horizontal stroke plane with passive pitching caused by aerodynamic and inertial forces, the thrust was found to be larger. Sällström et al. [9] used the stereoscopic particle image velocimetry to investigate the airflow generated by two pairs of flapping Zimmerman planform wings under hovering conditions. The results indicate that the less stiff of the two wings sheds several vortices each half stroke, which may indicate that the wing stalls more rapidly in the beginning of each half stroke than the stiffer wing, in spite of a lower angle of attack. Mahardika et al. [10] conducted the experiments showing that the outer parts of the separated wings are able to deform, resulting in a smaller amount of drag production during the upstroke, while still producing relatively greater lift and thrust during thelinebreak downstroke.

Phillips and Knowles [11] presented the experiment to investigate the effects of varying flapping kinematics on the mean lift produced by an insect-like flapping wing in hover. Results revealed that mean lift scaled with $f(1.5)$ and varied proportionally with Theta. A pitch reversal advanced by up to 5 percent of the flapping period relative to stroke reversal was found to maximize mean lift, and delayed pitch reversals were detrimental to mean lift. Of the parameters tested, mean lift was also maximized for alpha(mid) = 45 degrees and

Theta = 8.6 degrees. Lua et al. [12] were motivated by the works of Dickinson et al. [13] (Science 284: 1954–1960, 1999) and Sun and Tang [14] (J Exp Biol 205: 55–70, 2002) which provided two different perspectives on the influence of wing-wake interaction (or wake capture) on lift generation during flapping motion. They took a more fundamental approach to study the effect of wing-wake interaction on the aerodynamic force generation by carrying out simultaneous force and flow field measurements on a two-dimensional wing subjected to two different types of motion. These results suggest that wing-wake interaction does not always lead to lift enhancement, and it can also cause lift reduction. Mazaheri and Ebrahimi [15] had also investigated the aerodynamic performance of a flexible membrane flapping wing. Results indicated that the thrust increases with the flapping frequency. An increase in the wind tunnel speed and flow angle of attack leads to a reduction in the thrust value and increases the lift component.

To investigate aeroelastic effects of flexible wings (specifically, wing's twisting stiffness) on hovering and cruising aerodynamic performance, a flapping-wing system and an experimental setup were designed and built by Mazaheri and Ebrahimi [16]. Results show how elastic deformations caused by the interaction of inertial and aerodynamic forces with the flexible structure may affect specific power consumption. Based on unsteady numerical simulations, Elarbi and Qin [17] had studied the hovering capability of flapping two-dimensional tandem wing sections inspired by a real dragonfly wing configuration and kinematics. The results suggested that the longer time pitch rotation with the period of 80% of the overall flapping period is closer to the force calculations obtained of a balanced flight.

Using bird flight as the inspiration, the ventilated wing would apply the concept mentioned above and thus focus on "ventilating" the wings during the upstroke motion in hope of reducing any negative lift produced increasing net positive lift for each cycle. As the name suggests, a reduction in wing span would require a more complex mechanism to be built, which is meant to be kept as simple as possible.

2. Experimental Setup and Flapping MAV Design Concept

Based on the new flapping wing rotor design concept, a micro flapping wing rotor test model was designed and built as

(a) Motor and gears

(b) Crankshaft and piston

(c) Completed design

(d) The overall setup in the wind tunnel

(e) The ventilated flapping wing

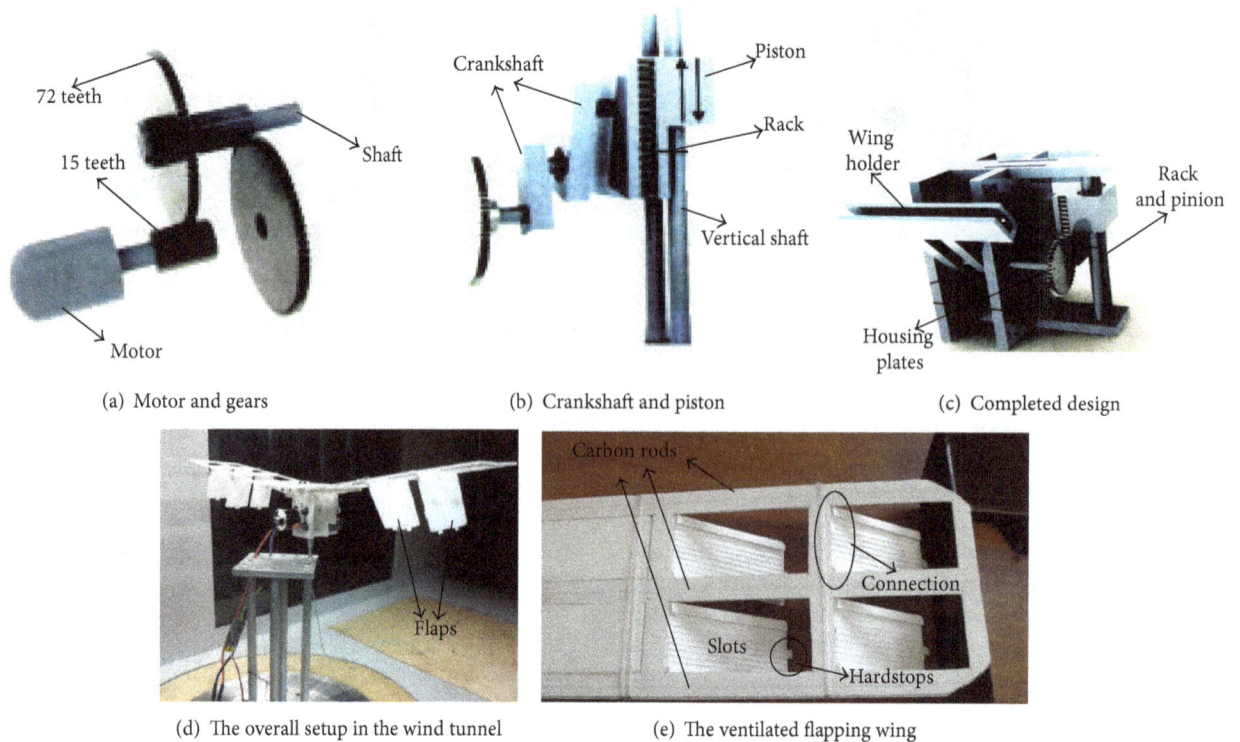

FIGURE 2: Experimental setup and ventilated flapping wing.

shown in Figure 2. The experimental procedure consisted of five primary components:

(1) a high performance Tahmazo ER282610 brushless motor and a Lithium battery, to supply power for the model: this motor is a powerful 540 watt suitable for flying models up to 1.8 kg the speed is at 870 rpm/volt with the battery supplying a minimum of 1 V and up to a maximum of 11.1 V;

(2) a gear mechanism, to decrease the rotation speed: two gears each of module 0.5 with the output/input gear are consisting of 75/15 teeth; the gear ratio is 4.8;

(3) crankshaft, short off-centered bar linked to driving rod to transform the rotation to vertical linear motion as shown in Figure 1;

(4) a rack and pinion actuator is used to convert the linear motion back to rotational motion;

(5) wing holder is designed with a 3 mm slot in which the wings can be attached with screw holes included to firmly attach the wings into the wing holder if necessary.

In the creation of the ventilated wing, a series of slots have to be created on the wings. These slots should have the ability to be opened and closed either through active or passive control mechanisms during the flapping cycle. During the downstroke, the slots should ideally remain fully closed to represent the full wing area and generate maximum positive lift; as the upstroke begins, the slots should open out and allow "ventilation" to occur within the wings by reducing the

exposed wing area, drag, and hence negative lift generated in the process. These slots areas would thus serve as the variable in which, through careful control of the areas, a correlation with the decrease in negative lift can be generated.

The major focus lies both in the design of the mechanism and the lift forces generated at the wings. Therefore, much attention should be paid to the wing design with the preliminary design for the wings shown in Figure 2(e). It is built from polypropylene corrugated sheets which is light in weight yet relatively tough. Additionally, each wing is fitted with three carbon rods, each running through the ends and middle. The carbon rods act as stiffeners to increase the rigidity of the wing and ensure as much as possible zero twist during the flapping maneuver. Thus, to not complicate any lift forces generated, it is essential for the wing to maintain a zero angle of attack with zero twist for both the upward and downward motions. The wing is initially made with a half wing span of 266 mm, areas of $0.0273 \, \text{m}^2$, aspect ratio (AR) of 2.58 with a rectangular shaped throughout and outer edge shaped neatly, and flapping frequency f, 2 to 4.96 Hz.

3. Simulation Setup

3.1. The Ventilated Wing Design. Figures 3(a) and 3(b) show the operating angle of the flapping wing during one flapping cycle. Figure 3(c) shows the three typical wing models, which were used in simulation.

3.2. CFD Method and Boundary Condition. The simulation is conducted using the ANSYS/FLUENT V.6.3.26. When

(a) The operating angles

(b) Flapping angles during one cycle

(c) Three types for the studied flapping wing

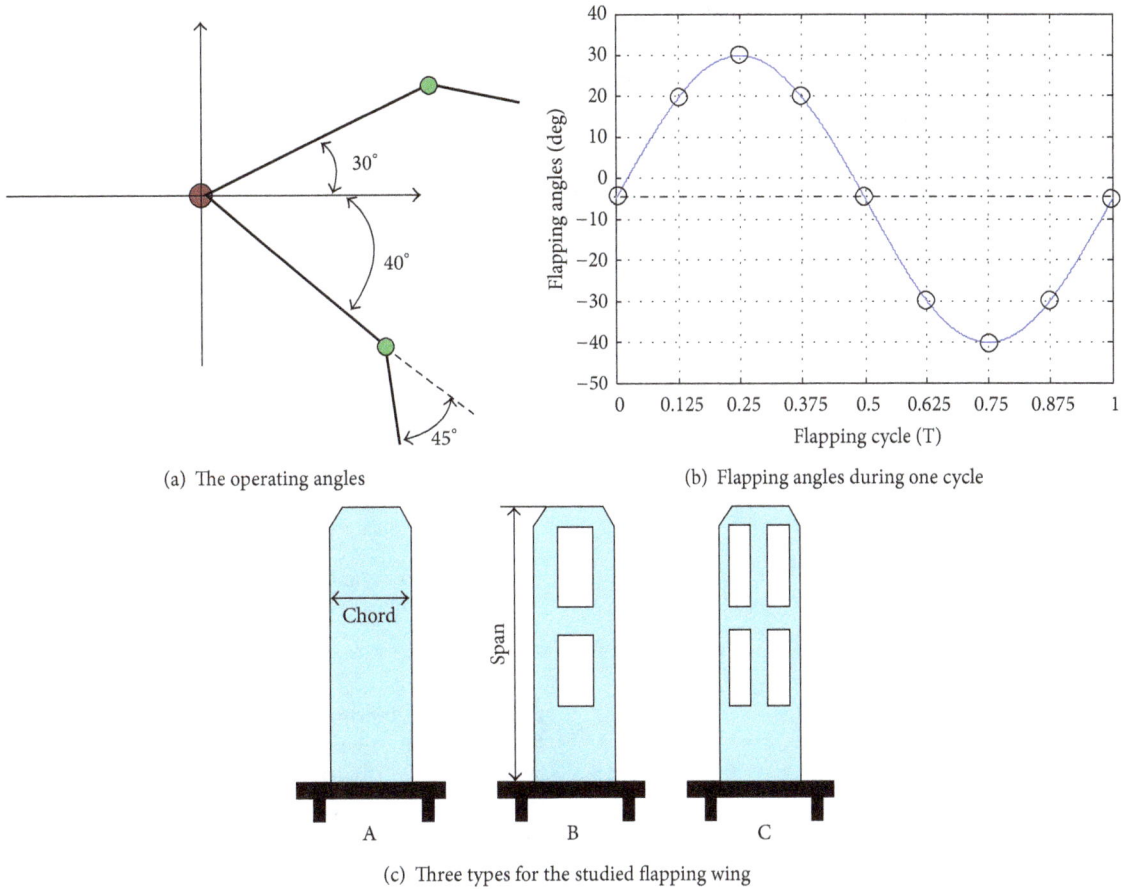

FIGURE 3: The operating process and wing shape type for the flapping mechanism.

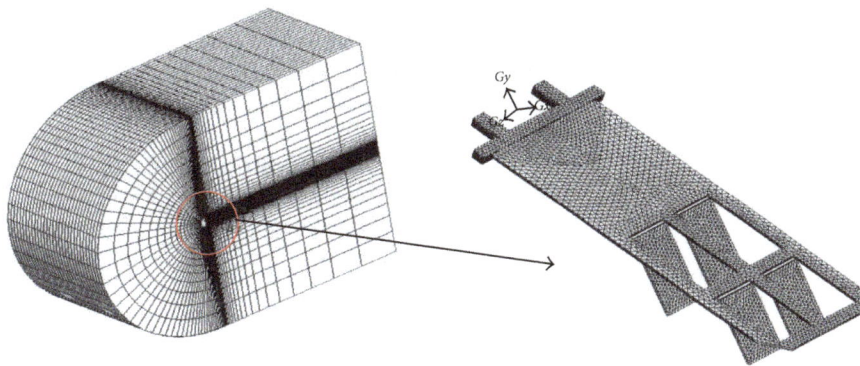

FIGURE 4: Decomposition of the computational domain.

modeling the whole aircraft, the results will be significantly affected by the quality of the mesh. As shown in Figure 4, the number of grid points is $120 \times 56 \times 60$ in the tangential, radial, and spanwise directions, respectively. A nondimensional parameter that is widely used to describe the wing kinematics of flying birds and insects is the Strouhal number, Str = fA/U_∞, which divides flapping frequency (f) and stroke amplitude (A) by the forward flying speed U_∞.

For the present study, as shown in Table 1, the incoming flow velocity or forward flight speed is $U_\infty = 1.4\,\text{m/s}$; the chord length of the piezoelectric flapping wing is $c = 103\,\text{mm}$. The flapping frequency of the piezoelectric flapping wing is $f = 2\,\text{Hz}\sim4.96\,\text{Hz}$. The peak-to-peak amplitude of the piezoelectric flapping wing at middle wingspan was found to be $A = 84.05\,\text{mm}$. Following the work of Triantafyllou et al. [18] and Taylor et al. [19] to use the peak-to-peak flapping

TABLE 1: Specification of flapping wing model.

Case number	Flapping frequency f (Hz)	Free stream velocity (m/s)	Wing type
1	4.96	1.4	A
2	2~4.96	0	A
3	4.96	1.4	A
4	4.96	1.4	B
5	2~4.96	0	C
6	4.96	1.4	C

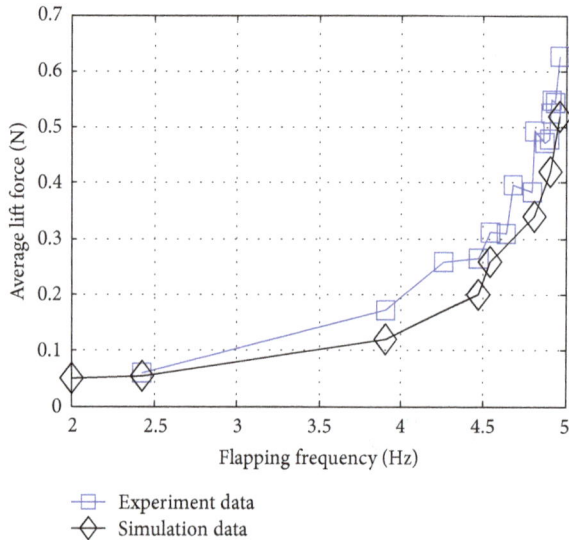

FIGURE 6: The average lift force characteristic for the ventilated wing versus solid wing.

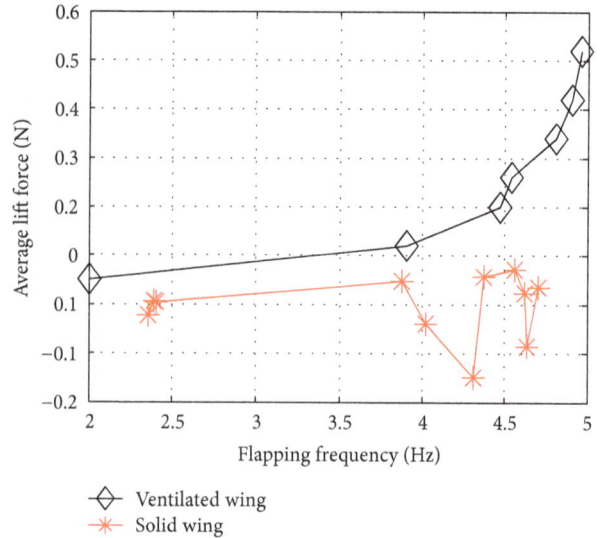

FIGURE 5: Average lift force based on the different flapping frequencies comparing with the experiment.

amplitude at middle wingspan to calculate the equivalent Strouhal number (Str) based on f = 4.96 Hz was found to be 0.298, that is, Str = 0.298, which is within the optimal range of 0.2 < Str < 0.4 usually used by flying birds and insects and swimming fishes.

In the numerical simulation, the relative pressure is adopted instead of absolute pressure. By using velocity inlet and pressure outlet boundary conditions as well as density-based explicit solver, simple iteration is adopted with large Eddy simulation (LES) turbulence model and bounded central differencing solution controls.

4. Results and Discussion

4.1. Comparison of the Simulation and Experimental Results. Figure 5 shows the average lift force produced by the parallel flapping wing (C-type model) as functions of the different flapping frequency (f = 2~4.96 Hz) without airflow introduction. In the results shown in Figure 7, it revealed that the experimental and simulation results can be seen to match in terms of trend. While they both display quadratic patterns, the experimental results can be seen to display more tortuous trend at higher frequencies (f > 4.3 Hz). In terms of magnitude, experimental results always show little bigger

than the simulation. It can be attributed to the presence of four hardstops installed on the moving slots. In order to avoid creating negative volumes during flapping simulation, the corresponding hardstops had been removed. And the ground effect for the experiments should also be taken into consideration. Finally, the simulation has set the flapping wing to be entirely rigid body, ignoring the flexibility effects.

In addition, it can also be found that, in all the studied flapping frequencies, the average lift force produced by the single flapping motion will increase with the increasing flapping frequency monotonically.

As shown in Figure 6, the results for net lift force generated for the solid wing can be seen to fall very close to zero with all results falling within a band ±0.15 N from zero line. There is also no clear trend of net lift generated increasing with increased frequency with negative net lift still being generated at higher frequencies.

However, at the first glance for the case of a ventilated wing, as expected, it can be clearly seen that all the results for net lift force are positive which immediately justifies the effect created by the "ventilation" in reducing negative lift force from the upward stroke. A clear trend is also present whereby net lift force increases with increasing frequency. This increase in net lift can be seen to be linear from 2 to 3.8 Hz after which it starts increasing exponentially with frequency. At the highest frequency tested (4.96 Hz), as high as 0.63 N is generated with the plot predicting even steeper increases in net lift for frequencies above 4.96 Hz.

Table 2 shows the enhancement in net lift using ventilated wing compared to solid wing at selected frequencies. Besides depicting the exponential rise in lift force as frequency increases, the percentage increase shown in the table also serves to provide an initial indication of the frequency in which the ventilated wing mechanism starts activating fully which can be observed at about after 4 Hz (increase by 107.34%).

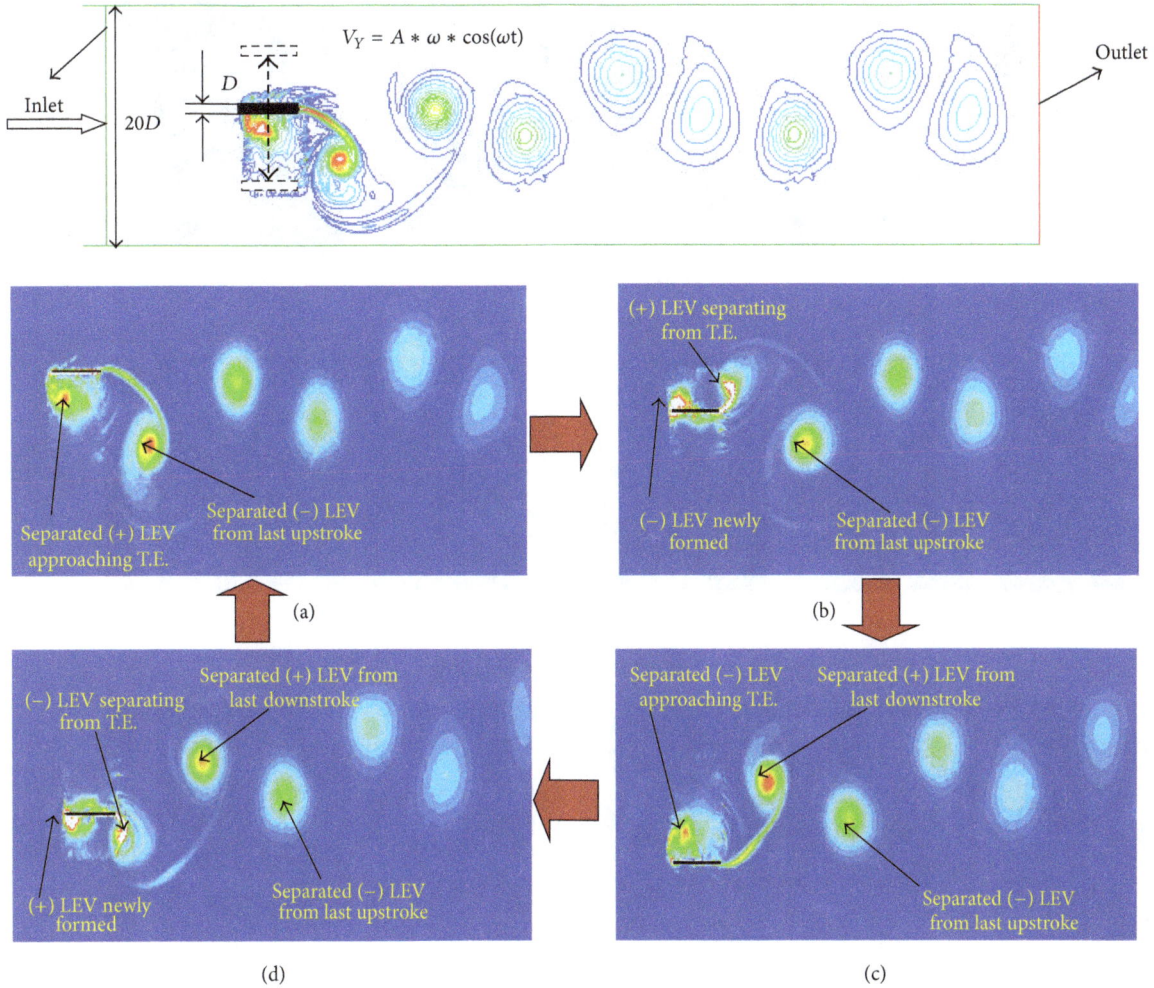

FIGURE 7: 2D vortex shedding on the chordwise direction at the different positions. (a) Wing at the upmost position: downstroke starts. (b) Wing at the neutral position, during downstroke. (c) Wing at the bottom most position: upstroke starts. (d) Wing at the neutral position, during upstroke.

TABLE 2: Enhancement in net lift using ventilated wing compared to solid wing.

Frequency (Hz)	Increase in net lift (N)	Percentage increase (%)
2.5	0.059	—
3	0.0828	40.34
3.5	0.1063	28.38
4	0.2204	107.34
4.7	0.3721	68.83

In order to investigate the mechanism of 3D flapping wing, firstly, we conducted the 2D flapping wing model. Figure 7 shows the vortex shedding at four typical positions: (a) upmost position (i.e., at the end of upstrokes or beginning of downstrokes); (b) neutral position during downstrokes; (c) bottom most position (i.e., at the end of downstrokes or the beginning of upstrokes); (d) neutral position during upstrokes, respectively.

As shown clearly in Figure 7(a), when the flapping wing was at its upmost position to start a downstroke, the anti-clockwise (positive) LEV newly formed on the lower surface of flapping wing by the previous upstroke was found to be intensified rapidly in strength and separated eventually. It would move downstream approaching the trailing edge of the wing. And we also can observe that the separated clockwise (negative) LEV was separated and hung to the trailing edge of the flapping wing. The smaller and weaker wake vortex structures were found to be dissipated rapidly and eventually vanished at further downstream, identical to the typical Karman vortex street configuration. After the upmost position, the flapping wing reached the neutral position during the downstroke. From Figure 7(b), we can see that, during the downstroke, the clockwise LEV was gradually formed on the upper surface near the leading edge of the wing. At mean time, the anticlockwise LEV on the lower surface began to be separated, and it will be pushed to move further downstream gradually. Meanwhile, the newly formed clockwise LEV was found to be strengthened itself

(a) Solid wing

(b) Tandem ventilated flapping wing

(c) Parallel ventilated flapping wing

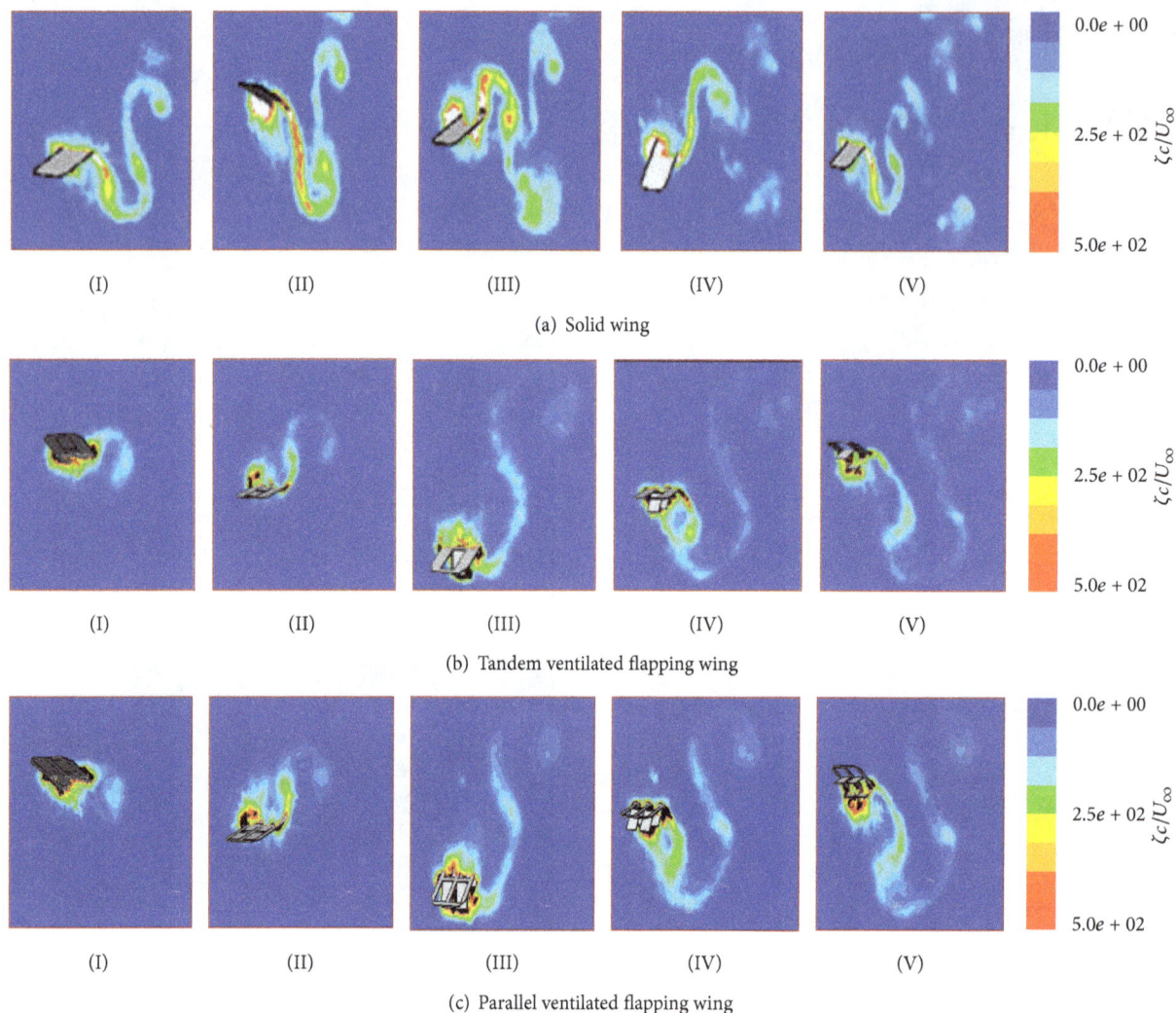

FIGURE 8: Contours of normalized vorticity magnitude $\zeta c/U$ at different time and at zero angle of attack (AOA).

and stayed attached to the leading edge of the flapping wing. And the corresponding separated clockwise LEV formed by the previous upstroke has totally broken away from any edges of flapping wing and stepped into the downstream. As revealed in Figure 7(c), the newly formed clockwise LEV developed its structure and strength; it has covered almost 90% upper surface of the wing at the bottommost position, and it was found to be detached from the upper surface and approaching the trailing edge of the wing. Meanwhile, the separated anticlockwise (positive) LEV created by the previous downstroke was separated and hung to the trailing edge of the flapping wing. At the end of the upstroke of the flapping circle, shown in Figure 7(d), a new anticlockwise (positive) LEV has been formed on the lower surface of the flapping wing, and the separated clockwise LEV on the upper surface would shed from the trailing edge of the wing. Other shed vortexes will become smaller and weaker and be vanished eventually. This process (downstroke and upstroke) would repeat in cycles. Finally, the clockwise and anticlockwise vortexes were shed alternatively in the wake of the flapping wing.

In Figures 8(a), 8(b), and 8(c), the LEV vortex forming and shedding for the solding wing and two typical ventilated wings have been shown. Although they have the same general trend, including the forming and shedding, due to the presence of slots, it can make the flow pattern become more complex. From Figure 8(a), we can see the LEV vortex has formed earlier and easier than the other two ventilated wing cases, and the shed vortex shows more islated with each other. Different from the solid wing case, the LEV vortex formation and shedding for the ventilated wing seem to need more time. Except this, the period is also relatively longer than the solid wing, and the upmost and downmost impacted region extended five times as long as the flapping amplitude A. The shed vortexes are also connected mutually with the long "tails."

The pressure coefficient is negative for pressure less than the free stream, which may occur on the top or bottom

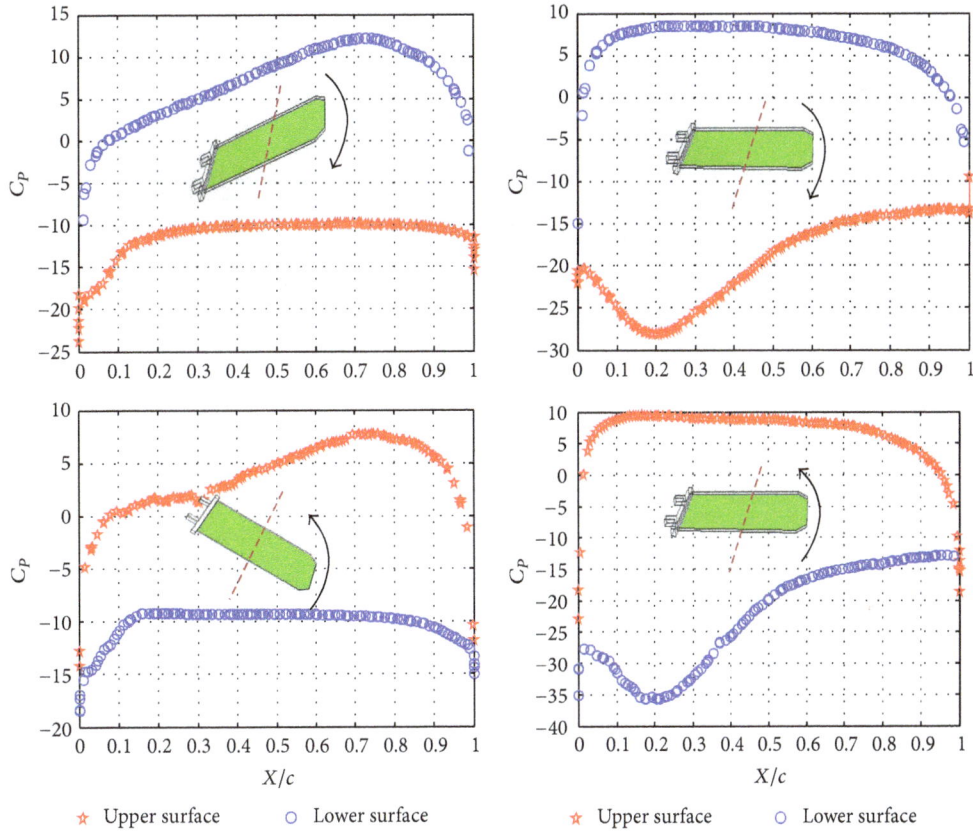

(a) Solid wing

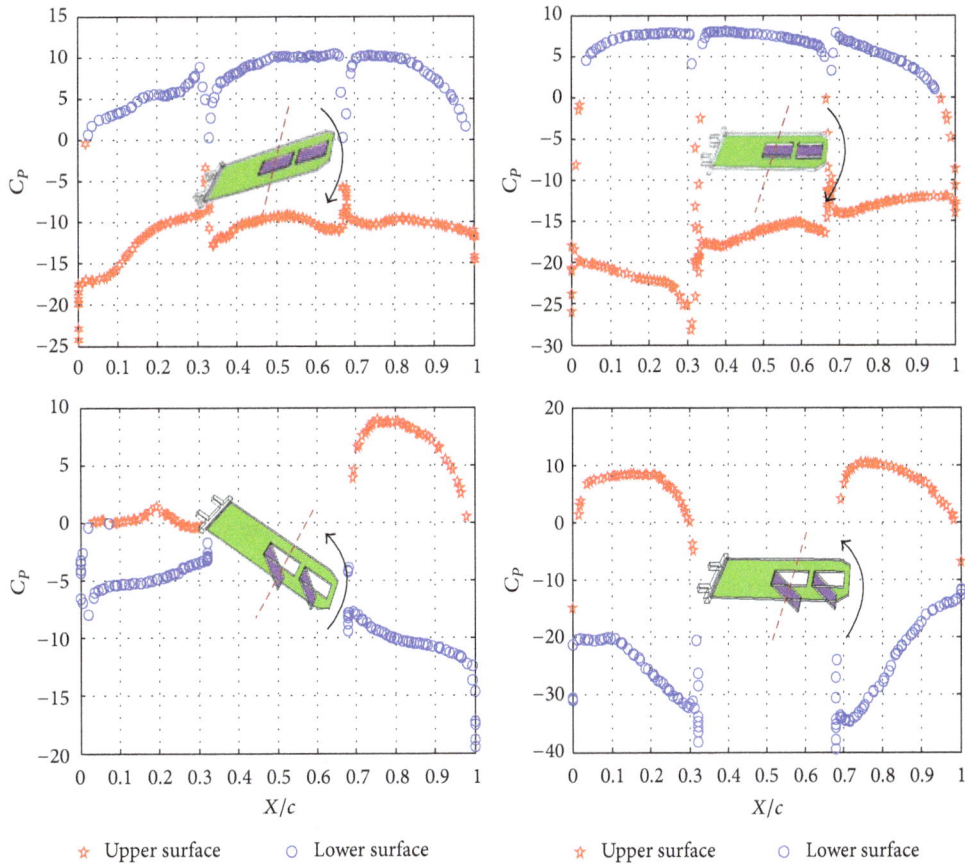

(b) Tandem ventilated flapping wing

FIGURE 9: Continued.

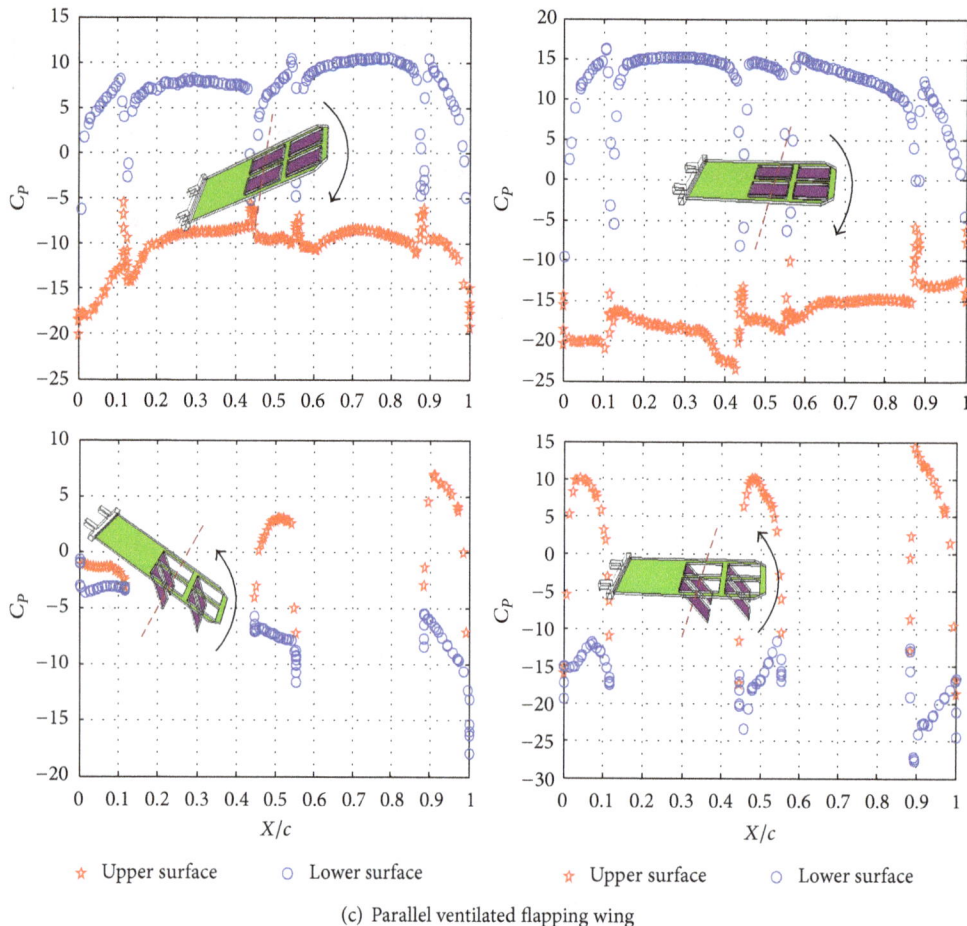

(c) Parallel ventilated flapping wing

FIGURE 9: Pressure coefficient (C_p) at t = 0.25 T, 0.5 T, 0.75 T, and T across the 50% chord length.

surface of the wing or canard. The data from the simulation is expressed with respect to the local chord length c. Figure 9 shows the pressure coefficient (C_p) distribution on the solid wing (a), tandem ventilated wing (b), and parallel ventilated wing (c) at t = 0.25 T, 0.5 T, 0.75 T, and T, respectively, based on the four typical flapping positions.

Comparing with Figures 9(a), 9(b), and 9(c), for the solid wing case, due to the fact that it did not include any slots, so the corresponding C_p curves are showing much smoother than the other two ventilated flapping wings. And the area of pressure curves of solid wing is entire and blank. Comparing with the solid wing, the area of C_p curves of tandem ventilated wing has been divided into three parts: main wing and two slots (seen in Figure 9(b)). And due to the presence of slots, the area of C_p curves of parallel ventilated wings has also been divided into five parts: main wing and four slots (seen in Figure 9(c)).

When the wing starts to flap up, the upper surface of the wing is actually playing the role of "the lower surface." Contrary to the conventional theory, the higher pressure region is on the top surface of the wing (upper surface), and the lower pressure region is on the bottom of the wing (lower surface). So the upstrokes always generate the negative lift force for the flapping wing. However, after the upstroke starts,

both the upper and lower surfaces will become reversed. So the downstrokes always generate the positive lift force. Consequently, there is, a common characteristic, that is C_p curves of the lower surface are always on the top of the upper surface during the downstrokes, and C_p curves of the lower surface are always on the bottom of the upper surface during the upstrokes.

However, we should not ignore the moving slots, because in all the above figures, we only focus on the entire movement, not specially considering the special time of slots movements. For it will open and close during the very beginning upstroke and downstroke, there will exist a short time that the C_p curves of the upper and lower slots surface have a contrary trend with the other common flapping time. When the main wing begins to flap up and down, the slots will also begin to open and close (moving down and up) immediately. In this short time (t = 0.04 T), the pressure on the upper slots surfaces actually should be higher and lower than the lower slots surfaces, respectively. Only after the slots had completed their own motion and began to flap together with the main wing, the C_p distribution would be changed into the above figures trends.

Figure 10(a) shows the lift force history during five flapping cycles. The static angle of attack (AOA) of the flapping

(a) Solid wing

(b) Parallel ventilated flapping wing

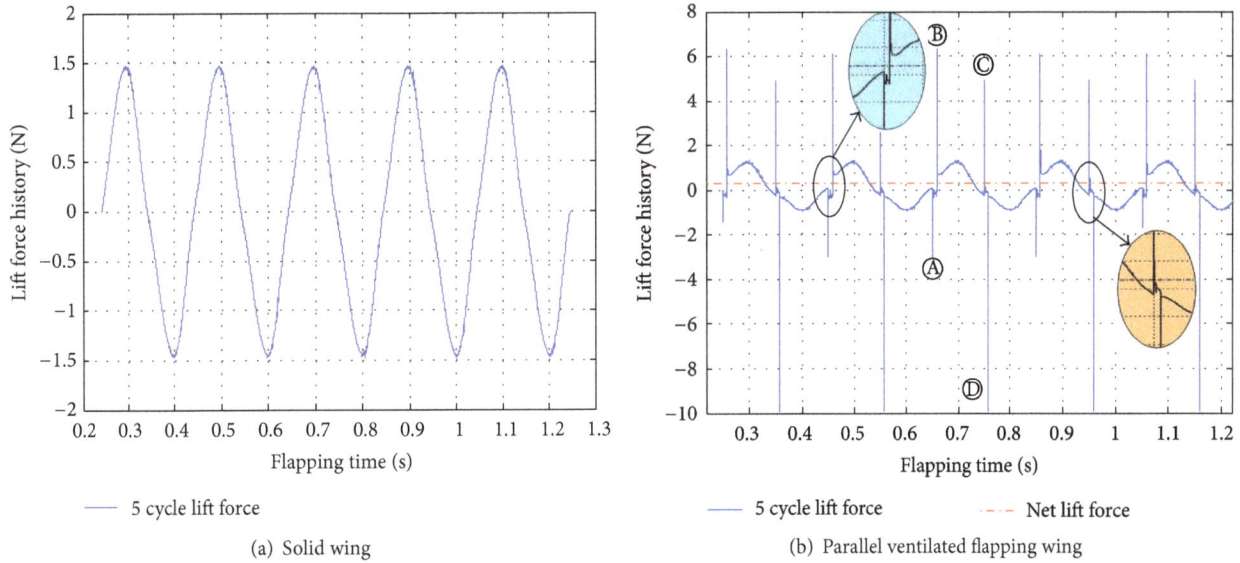

FIGURE 10: Lift force history during five flapping cycles.

wing was set to be zero (i.e., AOA = 0 deg). Due to the downstroke and upstroke motion is repeated. The uniform velocity of the incoming flow was steady, so the maximum was similar to the minimum in values. As a result, the curves have been shown symmetrically, and the corresponding net lift force always fluctuated near the zero value. And compared with the other two typical ventilated flapping wings, the lift curve is much smoother and more regular (will be discussed as follows).

As shown in Figure 10(b), the lift force history for the parallel ventilated wing case had displayed different from the solid wing case; there are some rhythmic vibrations on the curve. The typical four peak values have been found and marked A, B, C, and D, respectively. Actually, these marks have further described the open and close motion with respect to the moving slots. Firstly, when the ventilated flapping wing begin to flap down, the slots rooted on the main wing would not only have to flap down with the main wing, but also begin to close (moving up); although the downstrokes should create the positive lift force, due to the fact that the small slots have accounted for almost 85% of the wing area, the small slots are actually the main lift parts for this kind of ventilated flapping wing. So when the main wing begins to flap down, the slots will begin to close; immediately, the total lift force will be jumped into the negative values, so the peak A was formed at this short moment. After this immediate start, the slots will be moved up slowly ($t = 320$ time step per time step = 2.5×10^{-5} s); when it completed the close motion, the upmotion has suddenly stopped. The negative force will also immediately jump into the positive, so this process has created the mark B. After half of the flapping cycle, the ventilated wing will begin to flap up. However, the fully closed slots will begin to open again at this time. Due to this motion, the negative lift force created by the upstroke has to jump into the positive, so this instant has created the mark C. After the open motion is completed

($t = 320$ time step), the fully opened slot will flap up together with the main wing, so the positive lift force has to jump immediately into the negative, creating the mark D at this moment. The downstrokes and upstrokes process (including the open and close motion of the slots) will repeat in cycles continuously. Finally, four typical marks will alternatively appear, respectively, in the whole flapping cycles.

5. Conclusion

Using bird flight as the inspiration, three types of the flapping wing have been investigated by using the dynamic mesh method. The corresponding conclusions have been drawn as follows.

The results for net lift force generated for the solid wing can be seen to fall very close to zero with all results falling within a band ±0.15N from zero line. There is also no clear trend of net lift generated increasing with increased frequency with negative net lift still being generated at higher frequencies. While for the case of a ventilated wing, all results for net lift force are positive which immediately justifies the effect created by the "ventilation" in reducing negative lift force from the upward stroke. A clear trend is also present whereby net lift force increases with increasing frequency.

For the LEV vortex forming and shedding for the soliding wing and two typical ventilated wings, although they have the same general trend, including the forming and shedding, due to the presence of slots, it can make the flow pattern become more complex.

For the solid wing case, due to the fact that it did not include any slots, so the corresponding C_P curves are showing much smoother than the other two ventilated flapping wings. And the area of pressure curves of solid wing is entire and blank. However, the area of C_P curves of tandem ventilated wings will be divided into three parts: main wing and two slots. The area of C_P curves of parallel ventilated wing will also be divided into five parts: main wing and four slots.

For the solid wing, due to the downstroke and upstroke, motion is repeated. The uniform velocity of the incoming flow was steady, so the maximum is similar to the minimum in values. As a result, the curves have been shown symmetrically, and the corresponding net lift force always fluctuated near the zero value. While the parallel ventilated wing case had displayed different from the solid wing case, the lift curve will not be so smooth and regular; there are some rhythmic vibrations on the curve.

Conflict of Interests

The authors declare that there is no conflict of interests regarding the publication of this paper.

Acknowledgments

The authors' deepest gratitude goes first and foremost to Professor Yu for his constant guidance. Financial support from Nanyang Technological University is gratefully acknowledged. The authors also would like to thank the reviewers for their helpful suggestions.

References

[1] W. Shyy, H. Aono, S. K. Chimakurthi et al., "Recent progress in flapping wing aerodynamics and aeroelasticity," *Progress in Aerospace Sciences*, vol. 46, no. 7, pp. 284–327, 2010.

[2] Y. Hong and A. Altman, "An experimental study on lift force generation resulting from spanwise flow in flapping wings," in *Proceedings of the 44th AIAA Aerospace Sciences Meeting*, pp. 5269–5287, January 2006.

[3] C. P. Ellington, C. D. van Berg, A. P. Willmott, and A. L. R. Thomas, "Leading-edge vortices in insect flight," *Nature*, vol. 384, no. 6610, pp. 626–630, 1996.

[4] H. Liu and H. Aono, "Size effects on insect hovering aerodynamics: an integrated computational study," *Bioinspiration & Biomimetics*, vol. 4, no. 1, Article ID 015002, pp. 1–13, 2009.

[5] E. I. Hueso, Pivkin, S. Swartz, D. H. Laidlaw, G. Karniadakis, and K. Breuer, "Visualization of vortices in simulated airflow around bat wings during flight," http://vis.cs.brown.edu/organization/people/hueso.html.

[6] H. Hu, L. Clemons, and H. Igarashi, "An experimental study of the unsteady vortex structures in the wake of a root-fixed flapping wing," *Experiments in Fluids*, vol. 51, no. 2, pp. 347–359, 2011.

[7] H. Nagai, K. Isogai, T. Fujimoto, and T. Hayase, "Experimental and numerical study of forward flight aerodynamics of insect flapping wing," *AIAA Journal*, vol. 47, no. 3, pp. 730–742, 2009.

[8] B. Singh and I. Chopra, "Insect-based hover-capable flapping wings for micro air vehicles: experiments and analysis," *AIAA Journal*, vol. 46, no. 9, pp. 2115–2135, 2008.

[9] E. Sällström, L. Ukeiley, P. Wu, and P. Ifju, "Flow measurements in the wake of flexible flapping wings," in *Proceedings of the 28th AIAA Applied Aerodynamics Conference*, July 2010.

[10] N. Mahardika, N. Q. Viet, and H. C. Park, "Effect of outer wing separation on lift and thrust generation in a flapping wing system," *Bioinspiration & Biomimetics*, vol. 6, no. 3, Article ID 036006, pp. 1–10, 2011.

[11] N. Phillips and K. Knowles, "Effect of flapping kinematics on the mean lift of an insect-like flapping wing," *Proceedings of the Institution of Mechanical Engineers G*, vol. 225, no. 7, pp. 723–736, 2011.

[12] K. B. Lua, T. T. Lim, and K. S. Yeo, "Effect of wing-wake interaction on aerodynamic force generation on a 2D flapping wing," *Experiments in Fluids*, vol. 51, no. 1, pp. 177–195, 2011.

[13] M. H. Dickinson, F.-O. Lehmann, and S. P. Sane, "Wing rotation and the aerodynamic basis of insect flight," *Science*, vol. 284, no. 5422, pp. 1954–1960, 1999.

[14] M. Sun and J. Tang, "Unsteady aerodynamic force generation by a model fruit fly wing in flapping motion," *Journal of Experimental Biology*, vol. 205, no. 1, pp. 55–70, 2002.

[15] K. Mazaheri and A. Ebrahimi, "Experimental investigation on aerodynamic performance of a flapping wing vehicle in forward flight," *Journal of Fluids and Structures*, vol. 27, no. 4, pp. 586–595, 2011.

[16] K. Mazaheri and A. Ebrahimi, "Experimental study on interaction of aerodynamics with flexible wings of flapping vehicles in hovering and cruise flight," *Archive of Applied Mechanics*, vol. 80, no. 11, pp. 1255–1269, 2010.

[17] E. M. Elarbi and N. Qin, "Effects of pitching rotation on aerodynamics of tandem flapping wing sections of a hovering dragonfly," *Aeronautical Journal*, vol. 114, no. 1161, pp. 699–710, 2010.

[18] G. S. Triantafyllou, M. S. Triantafyllou, and M. A. Grosenbaugh, "Optimal thrust development in oscillating foils with application to fish propulsion," *Journal of Fluids and Structures*, vol. 7, no. 2, pp. 205–224, 1993.

[19] G. K. Taylor, R. L. Nudds, and A. L. R. Thomas, "Flying and swimming animals cruise at a Strouhal number tuned for high power efficiency," *Nature*, vol. 425, no. 6959, pp. 707–711, 2003.

19

Experimental Investigation of a Wing-in-Ground Effect Craft

M. Mobassher Tofa,[1,2] **Adi Maimun,**[1,2] **Yasser M. Ahmed,**[1,2]
Saeed Jamei,[1,2] **Agoes Priyanto,**[1,2] **and Rahimuddin**[2]

[1] *Faculty of Mechanical Engineering, Universiti Teknologi Malaysia, 81310 Skudai, Johor, Malaysia*
[2] *Marine Technology Centre, Universiti Teknologi Malaysia, 81310 Skudai, Johor, Malaysia*

Correspondence should be addressed to Yasser M. Ahmed; yasser@mail.fkm.utm.my

Academic Editors: K.-M. Chung and J. E. Lamar

The aerodynamic characteristics of the wing-in-ground effect (WIG) craft model that has a noble configuration of a compound wing was experimentally investigated and Universiti Teknologi Malaysia (UTM) wind tunnel with and without endplates. Lift and drag forces, pitching moment coefficients, and the centre of pressure were measured with respect to the ground clearance and the wing angle of attack. The ground effect and the existence of the endplates increase the wing lift-to-drag ratio at low ground clearance. The results of this research work show new proposed design of the WIG craft with compound wing and endplates, which can clearly increase the aerodynamic efficiency without compromising the longitudinal stability. The use of WIG craft is representing an ambitious technology that will help in reducing time, effort, and money of the conventional marine transportation in the future.

1. Introduction

Wing in ground effect is quite a new concept of designing fast ships, which has vast relevance in numerous areas such as transportation of cargo, tourism, rescue operations, and military functions. WIG craft gives an alternate solution to gain higher speed [1]. The ground effect (GE) is a phenomenon where the lift-to-drag ratio of a body will increase while it is cruising at a very close distance to the surface of water or ground [2]. Volkov and Russetsky [3] and Hooker [4] widely discussed useful characteristics of WIG craft. The development of WIG crafts originated from observations made of the landing performance of aircraft in 1920's. Later USA and the USSR became interested in attempting to exploit the potential benefits of ground effect. The USA abandoned efforts to produce ground effect craft in the mid 1960's as they were more interested in surface effect ship development. Germany began work in the late 1960's using the designs of Alexander Lippisch. However, USSR was the undisputed leader in research and development of WIG crafts up to the late 1980's [5]. Under these circumstances, the Ministry of Science Technology and Innovation (MOSTI) Malaysia provided funds to develop a WIG craft here.

The lift and drag that are produced by a wing define the performance and general attributes of the craft that it supports. A wing that is moving through the air produces a resultant force. Lift is the component of the resultant force perpendicular to the velocity vector of the wing. The component that is resultant force parallel to the velocity vector of the wing is defined as induced drag. There are other forms of drag, which are collectively known as parasite or profile drag; this drag is created by the friction of the object moving through the air. The total drag of an object moving through the air, can be achieved by summing up induced drag and parasite drag. Both lift and drag are functions of a number of variables such as the density of the air, the velocity of the object through the air and the geometry of the object. Figure 1 depicts the formation of lift (L) and induced drag (D_i) from the resultant force (R) created by the wing's movement through the air. In addition, the previous figure shows that the position of the wing as it moves through the air is characterized by the geometric angle of incidence (α), where the geometric angle of incidence is the angle between the chord line of the wing section and the velocity vector of the wing [6]. If the wing is designed properly, the lifting surface near ground generates higher lift at smaller ground

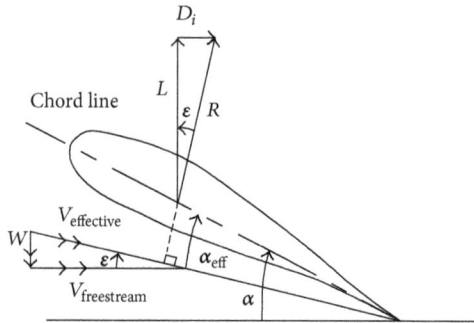

FIGURE 1: Lift and drag of a wing section [6].

clearances. However, the drag decreases with decreasing ground clearance for span-dominated GE.

A properly designed lifting system should increase lift-to-drag ratio and reduce drag for constant lift as the ground clearance reduces [5]. Many experimental and numerical simulations have been conducted by several researchers to study the phenomenon of GE. Jung et al. [7] conducted experimental investigation of wing with NACA 6409 section. Three different types of endplates were tested in those experiments at various ground clearances, angle of attacks, and aspect ratios. Endplates help to increase the performance of the wing though moving its center of pressure forward to the leading edge. Furthermore, it was found that the freestream velocity has insignificant effect on the wing aerodynamic coefficients. Pressure distribution through the symmetrical airfoil surface under different ground clearances and angle of attack were studied by Ahmed and Sharma [8]. NACA4412 airfoil section with flap in extreme ground effect were numerically investigated by Ockfen and Matveev [9]. The study consists of a steady-state, incompressible, finite volume method using Spalart-Allmaras turbulence model for the turbulent flow. It was found that favorable trailing-edge flap configuration improves aerodynamic characteristics of NACA4412 wing section. Chun and Chang [10] numerically analyzed the turbulent flow around two-dimensional WIG with respect to two ground boundary conditions, that is, moving and fixed bottom boundary. According to this study, the lift force and moment are not influenced by different bottom conditions, though the drag force simulated by the moving bottom is greater than that by the fixed one. The influence of the endplates on the aerodynamic characteristics of small aspect ratio (AR) wing was studied by Park and Lee [11]. According to their study, the reduction of the tip vortex will develop a substantial rise in the lift and lift-to-drag ratio, which can be developed by endplate. They claimed endplates also made a small deviation of height stability at different ground clearances and angle of attacks, which can reduce the height stability. Park and Lee [12] demonstrated the optimal profiles of two-dimensional wings to reach a moderate stability and high efficiency. In an another research work by Yang et al. [13], it was shown that for three-dimensional wing, end plates actually enhance the height stability at low ground clearance. Kornev and Matveev [14] found that the profiles of tail wing and main wing are the main factors of static height stability. They suggested that for acceptable stability of a WIG craft, the

centre of gravity should be close to the height of aerodynamic center (X_h), and it should be between the height of aerodynamic center and pitch aerodynamic center (X_α). Irodov [15] recommended a height static stability criterion as follows:

$$
\begin{aligned}
\text{HS} &= \frac{CM_\alpha}{CL_\alpha} - \frac{CM_z}{CL_z} < 0, \\
X_\alpha &= \frac{CM_\alpha}{CL_\alpha}, \\
X_h &= \frac{CM_z}{CL_z},
\end{aligned}
\tag{1}
$$

where CM_α, CL_α, CM_z, and CL_z are derivatives of lift and moment coefficient with respect to pitching angle and height. In a stable WIG craft, these derivatives usually are $CL_z > 0$, $CM_\alpha < 0$, $CL_\alpha > 0$, and $CM_z > 0$ [15].

A new configuration of the compound wing was developed in UTM by Jamei et al. [16] to improve the aerodynamic performance of the wing. The compound wing was divided into three parts: the middle part of rectangular wing shape and the side parts with reverse taper wings with an anhedral angle. The compound wings could create a greater reduction of downwash velocity and modify the pressure distribution on the lower side. The high increment of lift-to-drag ratio for the present wing in extreme ground effect recognizes a good efficiency for wing-in-ground (WIG) craft. Furthermore, it has been found that the performance of the wing improves noticeably for a certain total span of compound wing when the span of the middle part becomes smaller. It can be seen from the literature review that improving the aerodynamic characteristics of the wing in proximity to the ground by different means, such as the endplates, different configuration of wing is one of the important parameters in designing a WIG craft. The improvement of wing can increase the range and endurance of flight and decrease the fuel consumption and CO_2 emission. Therefore, there are many research articles dealt with the wing in ground effect, but most of these articles focused on the aerodynamic characteristics of the wing only and not the whole WIG craft.

In this research work, the aerodynamic characteristics and static stability of the WIG craft with the compound wing configuration [14] developed at UTM was experimentally investigated with and without endplates.

2. UTM Wind Tunnel

The aerodynamic characteristics, specifically the ground effect, of the WIG craft with a compound wing were investigated in a low speed wind tunnel at the Universiti Teknologi Malaysia (UTM-LST). This wind tunnel was able to deliver a maximum air speed of 80 m/s (160 knots or 288 km/hr) inside the test section. The size of test section was 2.0 meters wide, 1.5 meters height, and 5.5 meters long. The flow inside the wind tunnel was of good quality, with a flow uniformity <0.15%, temperature uniformity <0.2, flow angularity uniformity <0.15, and turbulence <0.06%. UTM-LST had high-quality facilities that allow for accuracy and repeatability of experiment results [17].

FIGURE 2: New compound wing configuration designed in UTM [16].

The wind tunnel is equipped with a 6-component balance for load measurements. The balance is a pyramidal type with virtual balance moment at the centre of the test section. The balance has a capability to measure aerodynamic forces and moment in the 3-dimensional. The aerodynamic loads can be tested at various wind direction by rotating the model via turntable. The accuracy of the balance is within 0.04% based on 1 standard deviation. The maximum load range is ±1200 N for axial and side loads [17].

3. WIG Craft Model

The experiments were carried out on a WIG craft model that used a new compound wing configuration [16] as shown in Figures 2 and 3. The compound wing was composed of three parts: a rectangular wing in the middle and two reverse taper wings with an anhedral angle at the sides [16]. The NACA 6409 airfoil section was selected as a section of the compound wings. The principal dimensions of the WIG craft are summarized in Table 1.

4. Experimental Procedures and Set-Up

In the wind tunnel, aerodynamic force measurements were carried out for a range of ground clearances (h/c) and different angles of attacks (α), from $h/c = 0.18$ to $h/c = 0.25$ and from $\alpha = 4°$ to $\alpha = 6°$. Ground clearance (h/c) was defined as the distance ratio between the wing trailing edge centre and ground surface (h) to root chord length (c) of the wing.

In this study, the floor of wind tunnel was used as a fixed flat ground as shown in Figure 4. The WIG craft model was mounted on the test section of the wind tunnel with a strut as shown in Figure 5. The position of the strut was at the 40% of chord length from the leading edge of the compound wing. The strut was adjustable and then was fixed at any height. The WIG craft model could be rotated about an axis at the strut position.

The frontal area ratio of the WIG craft model and the test section was small; therefore, the blockage ratio for the wings related to side and roof walls of the wind tunnel can be neglected. In this study, all experiments were performed with freestream velocity of 25 m/s.

5. Results and Comparisons

The WIG craft model aerodynamic coefficients and centre of pressure were obtained in this study using the following formulas:

$$C_L = \frac{L}{0.5\rho U^2 S},$$

$$C_D = \frac{D}{0.5\rho U^2 S},$$

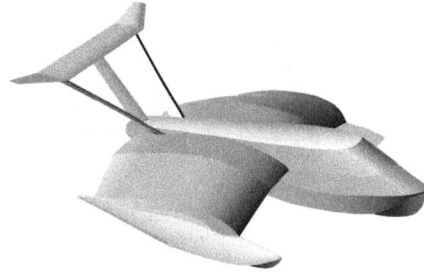

WIG craft model with endplates

(a)

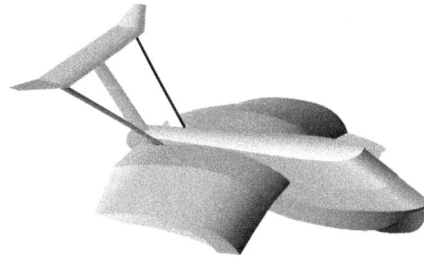

WIG craft model without endplates

(b)

FIGURE 3: WIG craft model with compound wing.

TABLE 1: Principle dimensions of WIG craft.

Scale factor	1 : 6
Wing span	83.4 cm
Wing root chord	66.7 cm
Middle wing span	41.4 cm
Aspect ratio	1.25
Anhedral angle	13°
Length of WIG craft	1.2 m
Breadth of WIG craft	0.13 m
Tail wing span	0.78 m
Tail wing chord	0.15 m

$$C_M = \frac{M}{0.5\rho U^2 Sc},$$

$$X_{cp} = 0.4 + \frac{C_M}{C_L \cos \alpha + C_D \sin \alpha}.$$

(2)

The lift coefficients of the WIG craft model with and without endplates are shown in Figure 6. As expected the lift coefficient of the WIG craft model with endplates was higher than the lift coefficient of the WIG craft model without endplates. The lift coefficients of the WIG craft model increase up to 42% and 50% for angles of attack of 4° and 6°, respectively. From the previous figure, it is clear for higher angle of attack and lower ground clearance endplates significantly enhanced the lift coefficient.

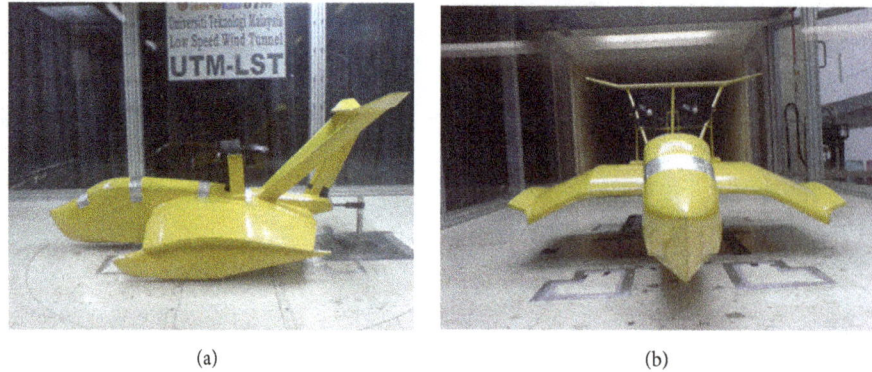

(a)   (b)

FIGURE 4: Wind tunnel test for WIG craft model: (a) model with endplates and (b) model without endplates.

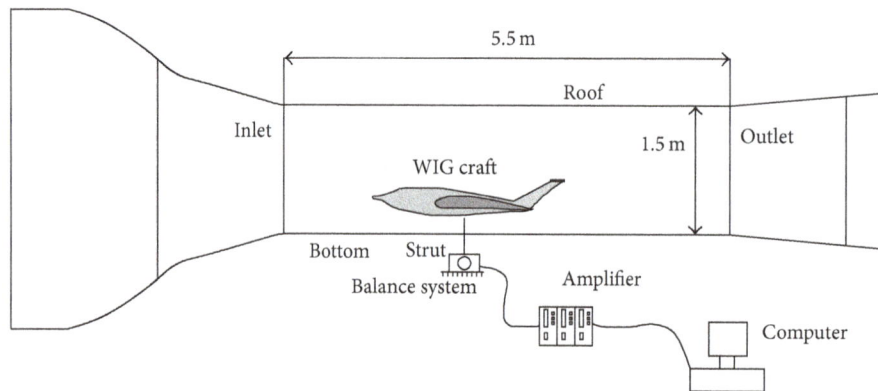

FIGURE 5: Experimental setup in the wind tunnel at the Universiti Teknologi Malaysia.

Figure 7 depicts the drag coefficient (C_D) of the WIG craft model with and without endplates. The drag coefficients of the WIG craft model with and without endplates do not show high discrepancy for each angle of attack, probably because of reduction of induced drag due to the effect of endplates. Thus, the increment of drag can be related only to angle of attack and ground clearance. For higher angle of attack and ground clearance drag coefficient increases up to 11% and 5%, respectively, as shown in the figure.

The performance of an aircraft is defined by its lift to drag ratio (L/D). The comparison of lift to drag ratio between the WIG craft model with and without endplates is shown in Figure 8. For lower ground clearance endplates enhance the lift-to-drag ratio up to 45%; the lift-to-drag ratio of WIG craft model with endplates decreases sharply compared to the model without endplates as the ground clearance increases. Augmentation of lift to drag ratio can also be attributed to the angle of attack, where larger angle of attack amplifies the lift to drag ratio. In addition, L/D decreased when the ground clearance was smaller.

The moment coefficients (C_M) of the WIG craft model with and without endplates are shown in Figure 9. The pitching moments were measured about the point which was 40% of the chord length from the leading edge. The CM_z being positive for the wing with end plates is consistent with the prior discussion about stability for a WIG.

The second parameter that has main effect on the stability of WIG craft is its center of pressure. The distance between the leading edge and center of pressure on the wing is defined as X_{cp}. Figure 10 shows the center of pressure for the WIG craft model with and without endplates. In general, the position of center of pressure of the WIG craft model shifted towards leading edge due to effect of endplates. This effect is expected as pitching moment reduced due to endplates.

As discussed earlier, static stability depends on the rate of change of moment and lift coefficient with respect to angle of attack and ground clearance. The height of aerodynamic center (X_h) and pitch aerodynamic center (X_α) of the WIG craft model with endplates were upstream than that of the WIG craft model without endplates as depicted in Figures 11 and 12. The height of static stability (HS) of the WIG craft model for both with and without endplates was negative with respect to ground clearance (Figure 13) that makes the craft statically stable for both cases [15].

6. Conclusion

This study experimentally investigated the aerodynamic characteristics of a WIG craft model with a new configuration of a compound wing. Effect of endplates on the craft aerodynamic characteristics was studied and it was found that endplates increase the lift and drag ratio of the WIG craft model. Finally,

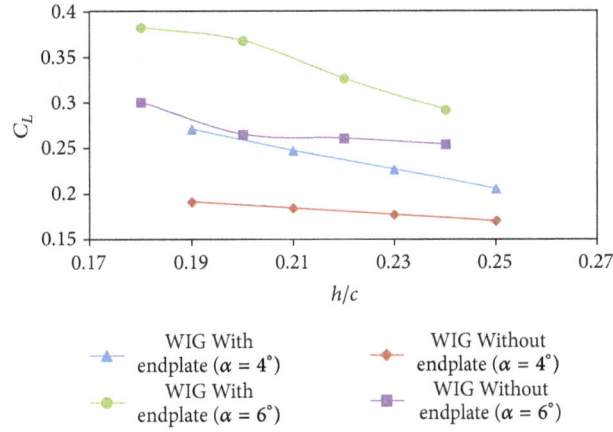

FIGURE 6: Lift coefficient (C_L) of WIG craft model with and without endplates versus ground clearance (h/c) at different angles of attack.

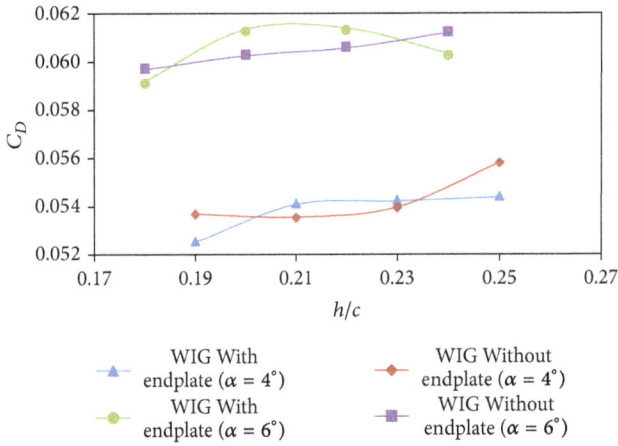

FIGURE 7: Drag coefficient (C_D) of WIG craft model with and without endplates versus ground clearance (h/c) at different angles of attack.

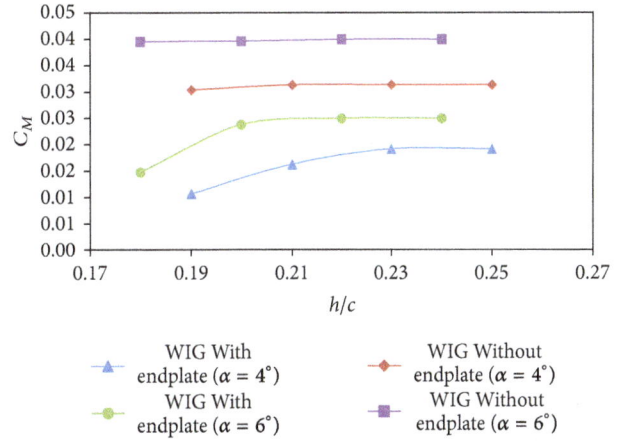

FIGURE 9: Moment coefficient (C_M) of WIG craft model with and without endplates versus ground clearance (h/c) at different angles of attack.

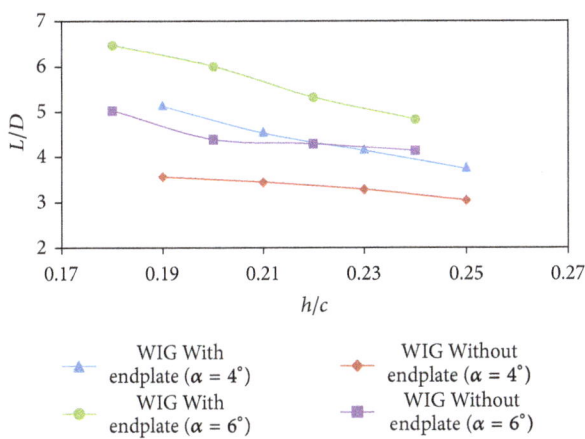

FIGURE 8: Lift-to-drag ratio (L/D) of WIG craft model with and without endplates versus ground clearance (h/c) at different angles of attack.

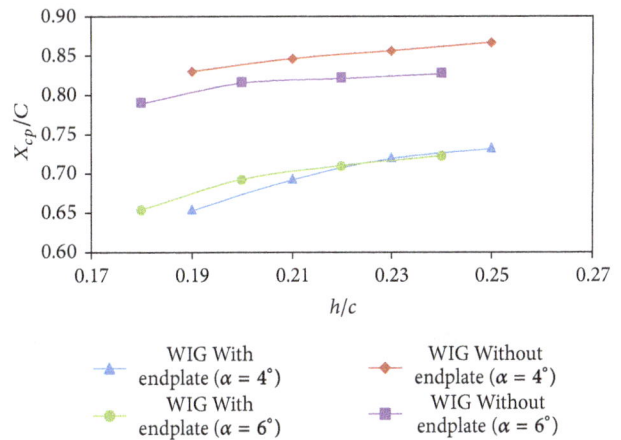

FIGURE 10: The position of the center of pressure (X_{cp}) of WIG craft model with and without endplates versus ground clearance (h/c) at different angles of attack.

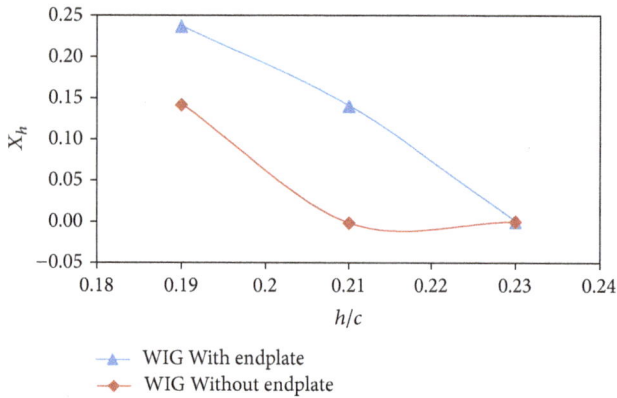

FIGURE 11: Height of aerodynamic center (X_h) of WIG craft model with and without endplates versus ground clearance (h/c) at angle of attack of 4°.

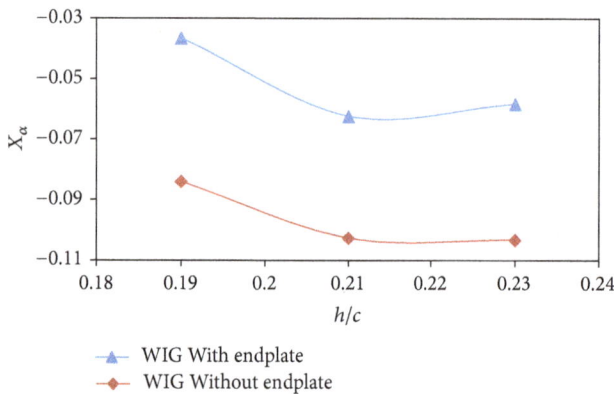

FIGURE 12: Pitch aerodynamic center (X_α) of WIG craft model with and without endplates versus ground clearance (h/c) at angle of attack of 4°.

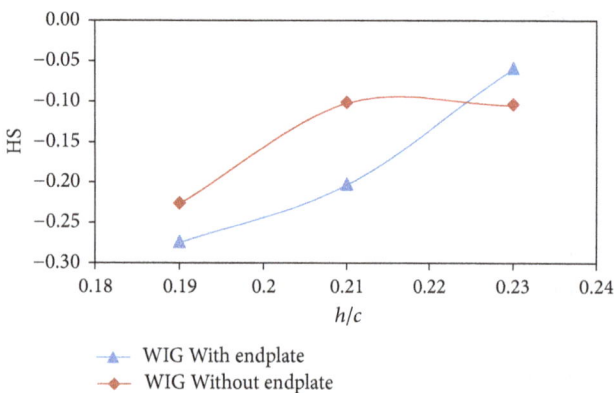

FIGURE 13: Height of static stability (HS) of WIG craft model with and without endplates versus ground clearance (h/c) at angle of attack of 4°.

the following points can be concluded from this research work.

(i) A new design of WIG craft was presented by combining compound wing with new configuration and endplates.

(ii) The aerodynamic performance of the new design was investigated by wind tunnel tests and aerodynamic coefficients are presented for the new design.

(iii) The static stability of the new design was investigated and it was found that the new design is statically stable enough.

Nomenclature

c: Chord length
C_L: Lift coefficient
C_D: Drag coefficient
C_M: Moment coefficient
D: Drag corce
h: Height of trailing edge above the ground
h/c: Ground clearance
HS: Height of static ctability
L: Lift force
L/D: Lift-to-drag ratio
S: Planform area of wing
X_{cp}: Center of pressure
U: Freestream velocity
α: Angle of attack, geometric angle of incidence.

Conflict of Interests

The authors declare that there is no conflict of interests regarding the publication of paper.

Acknowledgment

The authors would like to thank the Ministry of Science, Technology and Innovation (MOSTI) Malaysia, for funding this research project. The authors are grateful to Dr. S. Mansor and Eng. Abd Basid of UTM aeronautics lab for providing necessary help and guidance needed to perform this test.

References

[1] K. V. Rozhdestvensky, "Ekranoplans—the GEMs of fast water transport," *Transactions of the Institute of Marine Engineers*, vol. 109, no. 1, pp. 47–74, 1997.

[2] J. M. Reeves, "The case for surface effect research, platform applications and development opportunities," in *Proceedings of the NATO-AGARD Fluid Mechanics Panel Symposium in Long Range and Long Range Endurance Operation of Aircraft (FMP '93)*, pp. 24–27, The Hague, The Netherlands, 1993, Session 1A.

[3] L. D. Volkov and A. A. Russetsky, "Ekranoplans: problems and perspectives," *Sudostroenie Journal*, pp. 1–6, 1995.

[4] S. Hooker, *Wingships: Prospect for High-Speed Oceanic Transport*, Jane's All the World's Surface Skimmers, Jane's Information Group, Coulsdon, UK, 1982.

[5] K. V. Rozhdestvensky, "Wing-in-ground effect vehicles," *Progress in Aerospace Sciences*, vol. 42, no. 3, pp. 211–283, 2006.

[6] M. Halloran and S. O'Meara, "Wing in ground effect craft review," Contract Report CR-9802, The Sir Lawrence Wackett Centre for Aerospace Design Technology, Royal Melbourne Institute of Technology, Melbourne, Australia, 1999.

[7] K. H. Jung, H. H. Chun, and H. J. Kim, "Experimental investigation of wing-in-ground effect with a NACA6409 section," *Journal of Marine Science and Technology*, vol. 13, no. 4, pp. 317–327, 2008.

[8] M. R. Ahmed and S. D. Sharma, "An investigation on the aerodynamics of a symmetrical airfoil in ground effect," *Experimental Thermal and Fluid Science*, vol. 29, no. 6, pp. 633–647, 2005.

[9] A. E. Ockfen and K. I. Matveev, "Aerodynamic characteristics of NACA 4412 airfoil section with flap in extreme ground effect," *International Journal of Naval Architecture and Ocean Engineering*, vol. 1, no. 1, pp. 1–12, 2009.

[10] H. H. Chun and C. H. Chang, "Turbulence flow simulation for wings in ground effect with two ground conditions: fixed and moving ground," *International Journal of Maritime Engineering*, pp. 211–227, 2003.

[11] K. Park and J. Lee, "Influence of endplate on aerodynamic characteristics of low-aspect-ratio wing in ground effect," *Journal of Mechanical Science and Technology*, vol. 22, no. 12, pp. 2578–2589, 2008.

[12] K. W. Park and J. H. Lee, "Optimal design of two-dimensional wings in ground effect using multi-objective genetic algorithm," *Ocean Engineering*, vol. 37, no. 10, pp. 902–912, 2010.

[13] W. Yang, Z. Yang, and C. Ying, "Effects of design parameters on longitudinal static stability for WIG craft," *International Journal of Aerodynamics*, vol. 1, no. 1, pp. 97–113, 2010.

[14] N. Kornev and K. Matveev, "Complex numerical modeling of dynamics and crashes of wing-in-ground vehicles," AIAA, 2003-600, 2003, http://authors.library.caltech.edu/21291/1/Kornev_N_Complex_Numerical_modeling.pdf.

[15] R. D. Irodov, "Criteria of longitudinal stability of ekranoplan," *Ucheniye Zapiski TSAGI*, vol. 1, no. 4, pp. 63–74, 1970.

[16] S. Jamei, A. Maimun, S. Mansor, and N. Azwadi, "Numerical investigation on aerodynamic characteristics of a compound wing in ground effect," *Journal of Aircraft*, vol. 49, no. 5, pp. 1297–1305, 2012.

[17] Aeronautic Laboratory Universiti Teknologi Malaysia, 2013, http://aerolab.fkm.utm.my/?id=home&pid=523.

20

Exploration of a Capability-Focused Aerospace System of Systems Architecture Alternative with Bilayer Design Space, Based on RST-SOM Algorithmic Methods

Zhifei Li, Dongliang Qin, and Feng Yang

College of Information System and Management, National University of Defense Technology, Changsha, Hunan 410073, China

Correspondence should be addressed to Zhifei Li; lee.nudt@gmail.com

Academic Editors: P.-C. Chen, J. Y. Fu, and C. Mohan

In defense related programs, the use of capability-based analysis, design, and acquisition has been significant. In order to confront one of the most challenging features of a huge design space in capability based analysis (CBA), a literature review of *design space exploration* was first examined. Then, in the process of an aerospace system of systems design space exploration, a bilayer mapping method was put forward, based on the existing experimental and operating data. Finally, the feasibility of the foregoing approach was demonstrated with an illustrative example. With the data mining RST (rough sets theory) and SOM (self-organized mapping) techniques, the alternative to the aerospace system of systems architecture was mapping from P-space (performance space) to C-space (configuration space), and then from C-space to D-space (design space), respectively. Ultimately, the performance space was mapped to the design space, which completed the exploration and preliminary reduction of the entire design space. This method provides a computational analysis and implementation scheme for large-scale simulation.

1. Introduction

Recently, capability-based analysis, design, and acquisition have had a significant impact in defense related programs. The paradigm shift to capabilities-based acquisition is causing a fundamental shift in the way defense-related systems are both engineered and purchased. New mission needs and technological advancements have led to novel directives that are causing defense acquisition planning to utilize a capability-based approach. In particular, advancements in communication and transportation, combined with new and diverse enemies, have led to a call for increased joint operations, more integrated operations, and a better method of designing and acquiring systems and SoS (system of systems) to support these needs.

This capability-based mentality shares a natural link with architecting, in that capabilities are achieved through a series of activities. These activities can be represented as an operational architecture. Through the architecting process, they can be mapped to candidate solutions, which can then be evaluated and compared. These solutions provide the *ways* and *means* by which a capability is achieved. This kind of approach has been suggested to help address high level capability needs and help avoid the stove piping that has often plagued defense acquisition [1].

The challenge presented by the sheer number of possible alternatives is compounded in SoS problems. In fact, not only is the number of alternatives extremely large, but the alternatives also vary in their specifications, including alternatives across all aspects of the DOTMLPF (doctrine, organization, training, materiel, leadership, people, and facilities) spectrum. It is difficult to gather enough information early on to make an informed decision, but it is also difficult to even determine the criteria by which two extremely different solutions can be compared. Even justifying the acquisition of a new system can be difficult, because it must be shown that the same mission level cannot be achieved with a new arrangement or new uses of existing systems. To further illustrate this challenge, consider a simple mission, which is comprised of completing 10 activities. Then consider that these activities can be performed in two different sequences, thus creating two operational alternatives. Furthermore, each activity can be performed by one of three candidate systems. Three possible organizations could be responsible for conducting this mission and, last, consider that there are two types of networks being considered for enabling communication in the architecture. There are then 2 organizational alternatives $\times 3^{10}$ system alternatives $\times 3$ organizational alternatives $\times 2$ network alternatives, resulting in a total of 708,588 alternatives.

Thus, there are several criteria for a design space exploration method for CBA. First, it must be able to capture and define the large number of architectural alternatives available for consideration during the early phases of acquisition and systems engineering. Next, it must provide a way to filter through the design space and find only the promising alternatives for evaluation, while eliminating those that are either unrealistic or are not expected to meet mission goals. Finally, because even the filtering processes will still leave large numbers of alternatives to be evaluated, there must be a way to quickly and accurately evaluate the remaining alternatives.

2. Literature Review

Currently, the research of aerospace system of systems architecture alternatives for design space exploration focuses mainly on the *design of the experiment*, the *approximation model*, and *optimization algorithms*.

2.1. Design of Experiment. Design of the experiment [2] is a mathematical method of statistical analysis that allows for the study of the development of a reasonable alternative using data space technology. DOE has become an indispensable tool in computer-aided design optimization [3]. The main DOE methods include Monte Carlo sampling (MCS) [4], Latin hypercube sampling (LHS) [5], orthogonal array sampling (OA) [6], D-optimal design (DO) [7], and uniform design (UD) [8].

2.2. Approximation Model. In order for large-scale computing to simplify the design space and to generate a full understanding of space exploration, especially for large-scale multidisciplinary design space exploration and optimization, the *approximation model* was introduced into the design process. The main approximation models are the *response surface model* (RSM) [9], the *radial basis function neural network* (RBFNN) [10], and the *kriging model* [11].

2.3. Optimization Algorithm. In engineering design, optimization algorithms are often used to search among global optimal solutions in the design space; the method can be divided into two categories: *exact methods* and *approximation methods*. The exact methods include *branch and bound* [12], *mathematical programming* [13], and *coordination decomposition* [14]. The exact methods can be proven to be the optimal global solution but are only capable of solving smaller problems. The approximation methods can obtain a solution quickly in large-scale problems but cannot ensure that the resulting solution is optimal [15].

2.4. Comparative Analysis. DOE is an essential basic experimental approach in engineering design optimization, which represents the performance of the design space through different distributions of sampling points. However, while the DOE method is capable of sampling within the developed design space and then analyzing on the sampled points, it cannot explore the design space through the sampling itself nor can it divide or reduce the scope of the design space.

As mentioned earlier, design space exploration is one of the application directions of the *approximation model*. Approximate models, however, require repeated sampling when used in design space exploration problems, which will increase the load of computation. At the same time, there are no design space exploration methods that are suitable for the aerospace system design process.

Optimization algorithms of design space exploration, which belong to the latest developments in design optimization, can be used to explore and optimize the design space to find the global optimal solution or a feasible solution. The costs and computational load of the *optimization algorithm* for large-scale design space exploration are very high and inappropriate for an aerospace system of systems design and optimization in the early phase.

Above all, we can see that there is a lack of effective methods to utilize various existing experimental and historical data, as well as data from aerospace SoS, leaving a need for knowledge-based design space exploration methods as a guide for system design optimization. For one thing, since a large amount of computer technology and simulation software in engineering applications is required for the process of aerospace SoS design, when there are large numbers of simulations and experiments, there will be massive amounts of data stored in the data warehouse. It is important to take advantage of this useful data for subsequent SoS design optimization and to then support aerospace SoS design space exploration. Secondly, the existing design space exploration methods are used to approximate and explore directly within the aerospace system design space. In the early phase, however, there is typically a lot of uncertainty and a definite lack of knowledge. The existing methods have a too large computational load and cannot hold up to the design practices and processes. It is imperative to guide the designer to focus on the design space area of concern.

3. Proposed Approach

3.1. The General Framework of the Method. Traditional aerospace SoS optimization is a process that flows from the design space to the performance space, called "*Forward Mapping.*" However, successful experiences and experimental data are difficult to use in the design and development process. Additionally, acquisition staffs tend to pay more attention to the overall SoS performance, hoping to map the route from the performance space (the actual SoS performance and performance evaluation results) to the design space, in order to complete the design space exploration, which can help accurately locate the design space area of concern. Limiting the aerospace SoS design optimization to a smaller space saves time spent searching in an unnecessary area, making

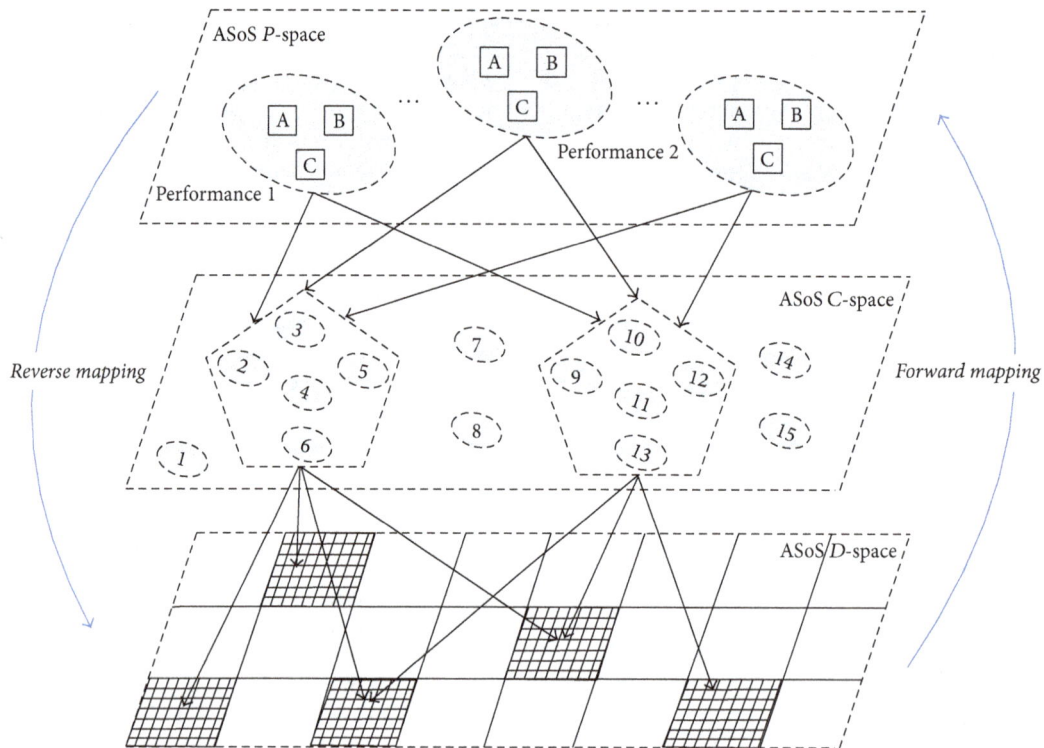

FIGURE 1: The general framework of the method.

the whole design optimization more targeted. Mapping from the "performance space" to the "design space," referred to here as "*reverse mapping*," complies with the general rules of aerospace equipment acquisition, as shown in Figure 1.

3.2. Bilayer Exploration Process

Layer 1: RST-Based Mapping from P-Space to C-Space. As shown in Figure 2, this paper studied the aerospace system of systems design space exploration methods of the architecture alternatives, primarily learning from previous design experience to better guide the overall design optimization with use of RST reasoning, based on the analysis of similar cases.

Similar, relevant cases are first selected, according to the capability gap and required operational activities, in order to determine the initial aerospace system configuration, which provides foundational data for subsequent derivation of configuration rules. Secondly, it must be determined whether or not the parameter attributes are complete. Thirdly, if the attribute data of the configuration program is complete, then the configuration rules from the complete configuration decision table are derived, using RST. If incomplete data is included, then reasoning with corresponding use of RST in the incomplete configuration decision table is utilized.

In the process of complete rule reasoning, the selected attributes are first analyzed and the continuous data is discretized, using the FCM (fuzzy C-means) algorithm, which preprocesses data for the use of RST. Secondly, in accordance with the selected configuration, similar cases are

collected from the corresponding performance estimates, along with a variety of configuration attribute data, constituting a configuration decision table. Again, the simplest related configuration rules from the configuration decision table are acquired with RST. Finally, when the performance space and the configuration space are positioned corresponding to configuration rules, the mapping from P-space to C-space can be completed.

In the incomplete configuration reasoning process, discretized continuous data must first be put into an incomplete configuration scheme. In accordance with the selected configuration, similar cases can be collected in the corresponding performance estimates, along with a variety of configuration attribute data, marking any uncertainties or missing data in the configuration alternatives with an "∗." The configuration decision table can then be compiled. Again, due to the incomplete data, there will be uncertain causality. The optimal configuration rules can thus be determined with the similarity function in Section 3.3. The optimized configuration rules should be assessed. If the rules meet the design specifications and system requirements of the design staff, the performance space and the configuration space can be positioned according to the configuration rules, completing the mapping from P-space to C-space. If not, the requirements of attribute decision can be relaxed, and the optimized generalized configuration rules can then be calculated with the optimization of general configuration rule functions, as defined in Section 3.3. Once again, the new optimal general configuration rules must be assessed to determine whether they meet the design specifications

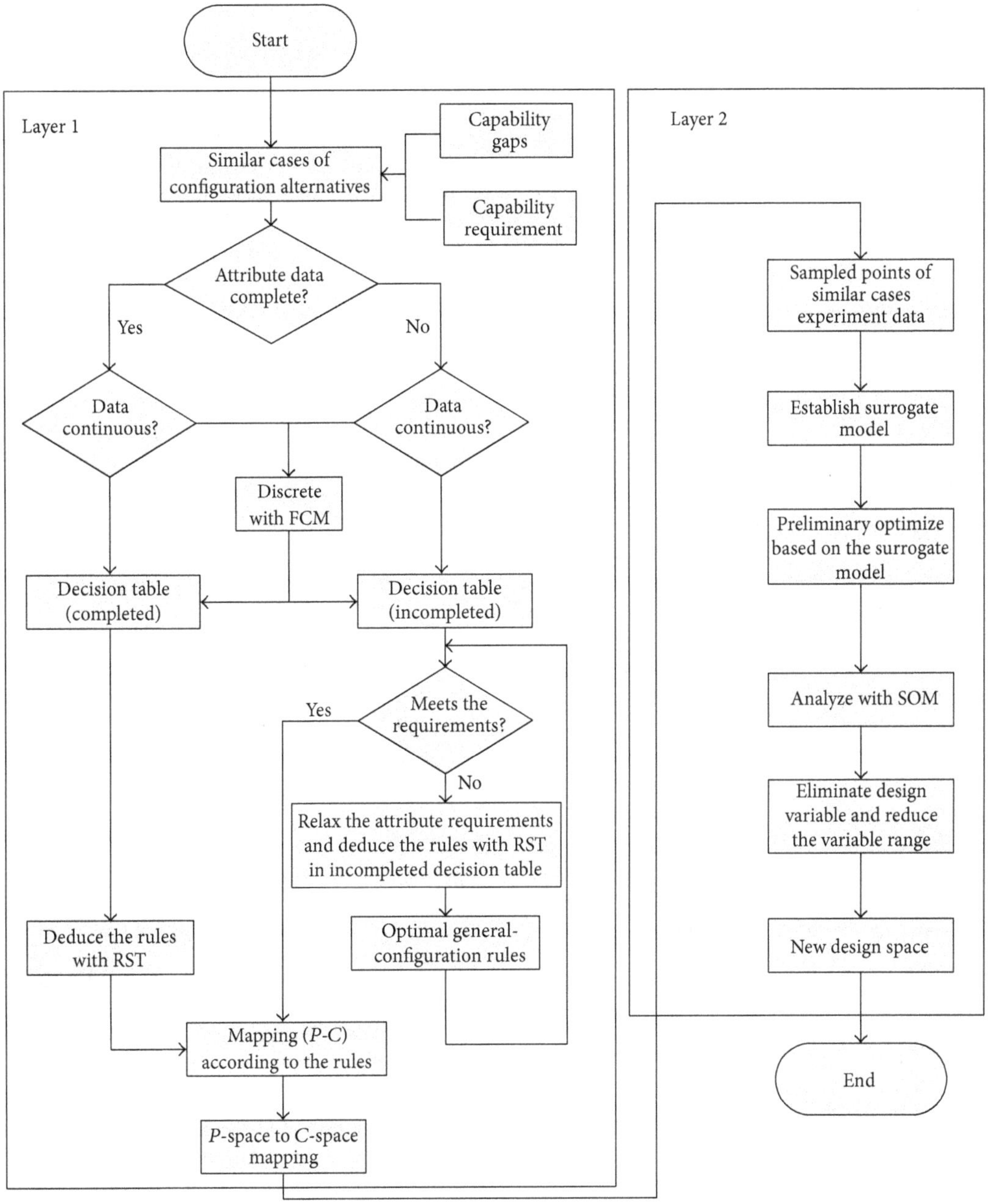

FIGURE 2: The Bilayer exploration process.

and requirements. If so, the iteration is terminated. Finally, according to the configuration rules, positioning the performance space, and the configuration space area according to configuration rules, the mapping from P-space to C-space can be completed.

Layer 2: SOM-Based Mapping from C-Space to D-Space. Upon completion of the preliminary configuration of aerospace SoS, relevant experimental data or the actual running information can primarily be selected from similar cases, according to the given aerospace systems within the configuration. Secondly, the relevant *surrogate models* can be established,

using relevant information and data, and then preliminary optimization can be made based on the model. Again, the design variables and the objective functions were analyzed using the SOM. A detailed study of the relationship between design variables and the objective function can then be made, using the color changes of a two-dimensional hexagonal grid, eliminating the unimportant design variables and reducing the associated interval of design variables. Finally, the dimensions and the design variables of concern can be determined for the design space and then a new design space can be constructed with a smaller design optimization range than the original, including local and global optimums.

The smaller range of a more targeted and relatively transparent design space optimization can improve efficiency, saving design time, and cost.

3.3. RST-Based Exploration Algorithm

3.3.1. Aerospace System C-Space Modeling. Aerospace system configuration can be defined as

$$S(U, A, V, f), \tag{1}$$

where U is a nonempty set of alternatives, A is a set of nonempty attributes of a selected configuration, V is the range of α, $\alpha \in A$, and f is an information function, $f: U \to V_\alpha$, giving each attribute of each object an information value, where $\alpha \in A$, $x \in U$, and $f(x, a) \in V_\alpha$.

The decision table for aerospace SoS C-space and P-space is defined as follows:

$$S = (U, A \cup \{d\}, V, f), \tag{2}$$

where U, A, V, and f have the same meaning within the configuration space model and $\{d\}$ is a decision attribute. The entire aerospace SoS performance space is divided through the actual aerospace system and the user evaluation. Therefore, designers can get $\{d\}$ attribute values from the performance space.

3.3.2. The Definition of Upper and Lower Approximation in the C-Space. In the aerospace SoS configuration model, each attribute subset $M \subseteq A$, IND(M) expresses metarelations between any two configuration alternatives, called indiscernible relations, which are defined as follows:

$$\text{IND}(M) = \{(x, y) \in U \times U \mid \forall \alpha \in M, \alpha(x) = \alpha(y)\}, \tag{3}$$

where $M \subseteq A$ (M is a subset of the entire attribute A) and X is a subset of all optional configurations, U.

For X, the upper and lower approximation of M is defined as

$$\underline{M}X = \cup \left\{ Y \in \frac{U}{\text{IND}(M)} \mid Y \subseteq X \right\},$$
$$\overline{M}X = \cup \left\{ Y \in \frac{U}{\text{IND}(M)} \mid Y \cap X \neq \emptyset \right\}. \tag{4}$$

As seen from the definitions, for the selected configuration X, the lower approximation represents the minimum optional configuration set similar to M and the upper approximation represents the maximum optional configuration set similar to M.

3.3.3. The Definition of the Division Matrix and Division Function in the Configuration Space. The division matrix of selected attributes M in the configuration decision tables is defined as follows:

$$\left(C_{ij} \right) = \left\{ \alpha \in M \mid \alpha(x_i) \neq \mid \alpha(x_j) \right\} \quad \text{for } i, j = 1, 2, \dots, n. \tag{5}$$

The division function is defined as follows:

$$f(M) = \prod \sum_{(x,y) \in U \times U} \alpha(x, y). \tag{6}$$

The division matrix and division function are used to infer the smallest reduction, which is a small subset of the attributes that can reflect implicit relationships in the selected configuration decision tables.

With the introduction of new technology or new systems, the relevant information is incompletely or vaguely stored, which leads to incomplete configuration space information. At this time, any attribute value field, V_α, may contain unknown or missing attribute values, represented with an "*."

3.3.4. The Similarity of the Configuration Alternatives. In the configuration alternatives decision table, SIM(M) is defined as

$$\text{SIM}(M) = \{(x, y \in U \cup U \mid \forall a \in M, f_a(x)$$
$$= f_a(y) \text{ or } f_a(x) = * \text{ or } f_a(y) = *)\}, \tag{7}$$

where SIM(M) is a compatible relationship; there is no distinction between any two configuration collections through a variety of attribute values.

$S_M(x)$ represents a set of configuration alternatives, similar to a configuration:

$$S_M(x) = \{y \in U \mid (x, y) \in \text{SIM}(M)\}. \tag{8}$$

Generalized decision function $\partial_A(x)$ is as follows:

$$\partial_A(x) = \{f_d(y) \mid y \in S_M(x)\}. \tag{9}$$

In the incomplete configuration decision table, the role of $\partial_A(x)$ is to relax the evaluation rating requirements of the performance of the configuration alternatives, which might include multiple decision attributes.

3.3.5. Calculation of Determined Rules of the System Configuration Optimization. Any configuration rules where $t \to d$ (where t is a conditional attribute value and d is the decision attribute value) are called the determination rules, only if $t \to d$ is unambiguous in S and $\|t\| \subseteq \|d\|$.

For any configuration in S, $t \to d$ is determined, leaving no other condition attribute subset to determine the decision attribute value d in values t, which is to say the configuration rule $t \to d$ is determined.

For any configuration alternatives $x \in U$ and $I_A(x) \subseteq I_{\{d\}}(x)$, $\Delta_U(x)$ is a division function only if $\Delta_U(x) = \prod_{y \in Y} \sum \alpha(x, y)$, where $Y_U = U/I_{\{d\}}(x)$ and $\Delta_U(x)$ is a function of the determination of the configuration rules. We can get the optimization determination of configuration rules of the decision table through the establishment and reduction of the function.

3.3.6. Calculation of Generalized Rules of the System Configuration Optimization. $\Delta_g(x)$ is a determined division function of x ($x \in U$), only when $\Delta_g(x) = \prod_{y \in Y} \sum \alpha(x, y)$.

Exploration of a Capability-Focused Aerospace System of Systems Architecture Alternative with Bilayer Design Space, Based on RST-SOM Algorithmic Methods

199

Where $Y_g = U \setminus \{y \in U \mid d(y) \in \partial_{AT}(x)\}$ and $\Delta_g(x)$ is a function of the optimization of generalized configuration rules. We can get the optimization of generalized configuration rules from the decision table through the establishment and reduction of the function.

3.3.7. Fuzzy C-Means Algorithm. We use the *fuzzy C-means* method to discrete the continuous data. The definition of FCM is summarized as follows:

$X = \{x_1, x_2, \ldots, x_n\}$, sampling set of an attribute,

$x_j = (x_{j1}, x_{j2}, \ldots, x_{jk})$, jth k-dimensional vector of each attribute,

c, the number of clusters that are specified,

v_i, the center of the ith cluster,

$$v_i = \frac{\sum_{j=1}^{n}\left(u_{ij}\right)^q x_j}{\sum_{j=1}^{n}\left(u_{ij}\right)^q}, \tag{10}$$

$V = (v_1, v_2, \ldots, v_c)$, center vector composed of a cluster center,

q, real number greater than 1,

u_{ij}, weight index which control the fuzziness of the attribute clustering,

ε, termination condition determined by the engineering staff,

$\|x_j - v_i\|^2$, Euler distance of jth attribute and the cluster center.

The definition of the membership function of each attribute vector to each attribute cluster is

$$u_{ij} = \frac{\left[1/\|x_j - x_i\|^2\right]^{1/(q-1)}}{\sum_{k=1}^{c}\left[1/\|x_j - x_k\|^2\right]^{1/(q-1)}}. \tag{11}$$

In the process of discretization of continuous data, the minimal value of the following objective function is required:

$$J_q\left(u_{ij}, v_k\right) = \sum_{j=1}^{n}\sum_{i=1}^{c}\left(u_{ij}\right)^q\|x_j - v_i\|^2; \quad c \le n. \tag{12}$$

The application procedures are summarized as follows.

Step 1. Determine the target that needs to be analyzed and the related attributes that need to be discretized.

Step 2. Determine a set of sampling points of the configuration attributes $X = \{x_1, x_2, \ldots, x_n\}$ and jth k-dimensional vector of each attribute's sampling point.

Step 3. After discretization of the configuration attributes, allocate the value of c, q, and ε.

Step 4. Initialize the membership function matrix u_{ij}^0, which represents the distance of each configuration attribute point to the initial cluster center.

Step 5. Use u_{ij}^0 and v_i to upgrade the center of each configuration property cluster.

Step 6. Calculate $u_{ij}^{(L+1)}$, which represents the relationship of each configuration attribute point to its center.

Step 7. If $\max[\|u_{ij}^{(L)} - u_{ij}^{(L+1)}\|] \le \varepsilon$, then stop iteration; otherwise return to Step 5.

3.4. SOM-Based Exploration Algorithm. After the C-space area of concern is determined, using the SOM method, the configuration space is mapped to part of the design space, and the subsequent optimization is then capable of meeting the design specifications and requirements only in the area of concern.

SOM is an unsupervised learning neural network, which is a type of data clustering and high-dimensional data visualization method. The purpose of visualization is to project data onto a graphical representation to provide a qualitative idea of its properties. Typically, the multidimensional data is mapped to the two-dimensional space with hexagonal grids. Therefore, SOM further maps the configuration space region to the smaller design space area, which is the area of concern in the design space. Unlike conventional geographical methods, SOM cannot provide any geographical features, coordinates, distances, and so on, but it can describe closeness or distribution of the input design variables. After the initial aerospace system configuration is determined, the input layer of the n-dimensional design variables and the m-objective function as an input vector can be determined, where n and m are positive integers. The $n + m$ neurons can then be assigned. In the output layer, the $n + m$ dimensional weight vector $= \{v_1, v_2, v_3, \ldots, v_{n+m}\}$ is randomly assigned to neurons.

In SOM, unsupervised learning clusters similar patterns together, while preserving the topology of the input space and maintaining a full connection of the input vectors to neurons in the output layer. There are two main goals to be achieved. The first is that the output layer searches for the winning unit with a closer weight vector to each input vector.

The second is that, in order to be closer to the input design variables and objective function vectors, weight vectors of the winning unit and its neighboring neurons will be updated. As a result, the $n + m$-dimensional input vectors are projected onto a sequence of neighboring neurons in the two-dimensional hexagonal grid. From the color of the neurons in the output layer, we can compare the change trends of design variables or the correlation between design variables and objective functions.

The detailed steps of SOM application are summarized as follows.

Step 1. Assign the weight vector $V = \{v_1, v_2, v_3, \ldots, v_{n+m}\}$.

Step 2. Select n design variables and m-objective functions as the input vectors.

Step 3. Get the neuron that has the least distance from input vectors.

Step 4. Update the weight vectors of the winning unit and its neighboring neurons.

Step 5. If the predefined iterative requirement is satisfied, stop. All the design variables and objective functions are projected onto the two-dimensional hexagonal grid. Otherwise, go to Step 2.

4. Case Study

4.1. Problem Description. In order to better demonstrate this method, a simple example problem will be used. This illustration is adapted from an example previously published by Griendling [16]. Note that the example is not designed to reflect reality, in order to avoid publication restrictions. The SEAD mission demonstrated the need for CBAs to explore a broad range of operational and materiel solutions. The considered alternatives included variations on operations, systems, organizational responsibilities, network structure, interoperability level, and force structure. Since the total alternative space had over 700,000,000 feasible architectures, it was decided to first group the alternatives by their system portfolios and eliminate portfolios with overall poor performance.

4.2. Parameter Settings. The following several alternatives were selected from numerous architecture alternatives as the basis for the aerospace SoS configuration. After processing the corresponding attribute values, the list was compiled, as shown in Table 1.

4.3. Experimental Results. Using the standard rough set theory for data mining, the continuous data should be discretized. In order to facilitate attribute processing, the attribute set C is divided into three categories. Among them, the first category includes C_1 (cost) and C_2 (time), the second category C_3 (risk) and C_4 (support level), and the third category C_5 (P-success). The first class of continuous attributes is discrete with equal interval division, the attribute values of the second class use the range standardized management approach to discrete data and the third class attribute values are directly converted to discrete data.

Therefore, the attribute C_1 is divided by 20 for each interval, C_2 is discretized by 25 for each interval, and in C_3, 1 represents general and 2 represents high. For attribute C_5, 1 represents a success rate of 0.5 or more and 2 represents a success rate below 0.5. For attribute C_4, 1 represents class I and 2 represents class II.

A sample attribute classification is shown in Table 2.

Calculated by the software *Rosetta*, the reduction of C by D can be obtained with $\{C_1, C_2\}$; the key of C is $\{C_1, C_2\}$.

The decision rules deduced from Table 2 are as follows.

Rule 1. If $C_1 = [80, 100)$ *and* $C_2 = [95, 120)$, *then* evaluation results = 1.

Rule 2. If $C_1 = [100, 120)$ *and* $C_2 = [95, 120)$, *then* evaluation results = 1.

Rule 3. If $C_1 = [80, 100)$ *and* $C_2 = [145, 170)$, *then* evaluation results = 1.

Rule 4. If $C_1 = [100, 120)$ *and* $C_2 = [95, 120)$, *then* evaluation results = 2.

Rule 5. If $C_1 = [120, 140)$ *and* $C_2 = [120, 145)$, *then* evaluation results = 2.

Rule 6. If $C_1 = [80, 100)$ *and* $C_2 = [170, 195)$, *then* evaluation results = 3.

Rule 7. If $C_1 = [140, 160)$ *and* $C_2 = [170, 195)$, *then* evaluation results = 2.

Rule 8. If $C_1 = [60, 80)$ *and* $C_2 = [45, 70)$, *then* evaluation results = 4.

Rule 9. If $C_1 = [60, 80)$ *and* $C_2 = [70, 95)$, *then* evaluation results = 5.

Rule 10. If $C_1 = [80, 100)$ *and* $C_2 = [70, 95)$, *then* evaluation results = 5.

Rule 11. If $C_1 = [80, 100)$ *and* $C_2 = [45, 70)$, *then* evaluation results = 6.

Rule 12. If $C_1 = [60, 80)$ *and* $C_2 = [70, 95)$, *then* evaluation results = 5.

Among which

Rule 1 and Rule 2 can be merged together:

if $C_1 = [80, 120)$ *and* $C_2 = [95, 120)$, *then* evaluation results = 1;

Rule 9, Rule 10, and Rule 12 can be merged together:

if $C_1 = [60, 100)$ *and* $C_2 = [70, 95)$, *then* evaluation results = 5.

Uncertainty rules are as follows.

Rule 13. If $C_1 = [100, 120)$ *and* $C_2 = [70, 95)$, *then* evaluation results = 3, and rule certainty factor is 0.5.

Rule 14. If $C_1 = [100, 120)$ *and* $C_2 = [70, 95)$, *then* evaluation results = 4, and the rule certainty factor is 0.5.

In the first mapping layer, the rules list which attributes have the greatest impact on the performance of the aerospace SoS.

Configuration rules show that cost and time are the core attributes of the decision table that influence the evaluation results.

In the process of aerospace SoS design or selection, the designer can select the satisfactory alternatives based on the extracted configuration rules, narrowing the range of options for candidate configuration alternatives.

In practical applications, decisions can be made according to the above rules of certainty and uncertainty.

Exploration of a Capability-Focused Aerospace System of Systems Architecture Alternative with Bilayer Design Space, Based on RST-SOM Algorithmic Methods

201

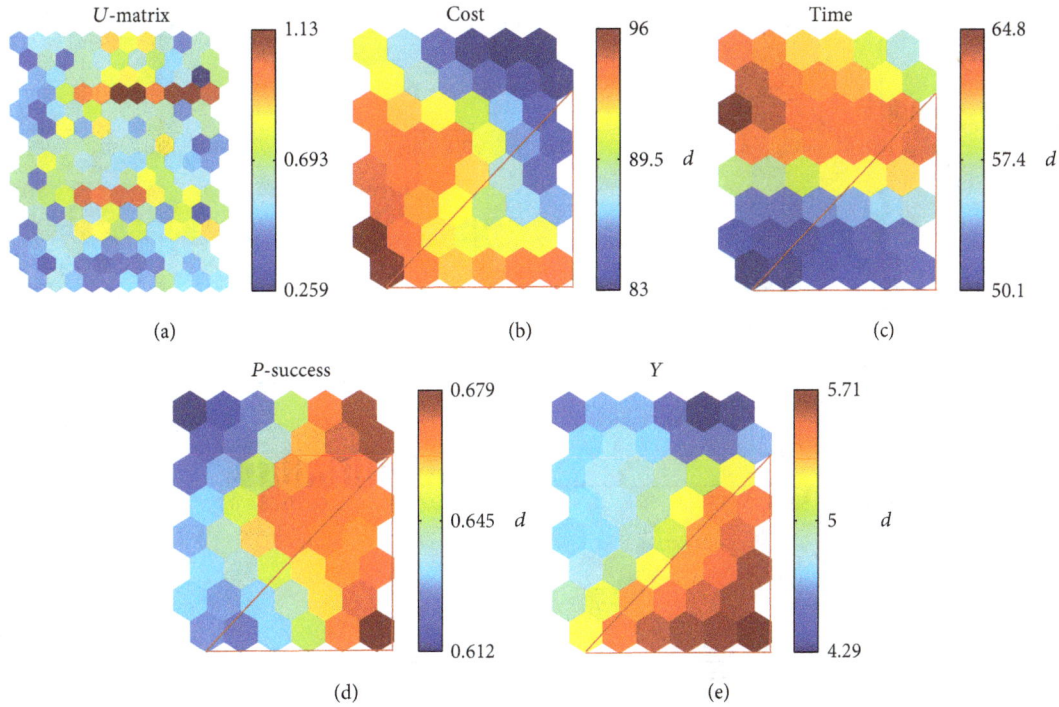

FIGURE 3: The SOM result I.

TABLE 1: The similar cases and corresponding data.

Alternative	Cost	Time	Risk	Support level	P-success	Evaluation results
1	99	112	High	I	0.67	1
2	110	110	High	I	0.55	1
3	95	150	General	I	0.71	1
4	108	108	General	II	0.52	2
5	125	125	General	II	0.49	2
6	86	190	High	II	0.67	3
7	146	192	High	II	0.68	2
8	108	90	General	II	0.71	3
9	60	65	General	II	0.72	4
10	74	79	General	II	0.80	5
11	102	80	General	II	0.66	4
12	94	94	High	II	0.54	5
13	80	45	High	I	0.61	6
14	66	78	General	II	0.59	5

After the first mapping, suppose that the designer needs to get the alternatives with evaluation results of 6. He can then choose configuration alternatives according to the rules $C_1 = [80, 100]$ and $C_2 = [45, 70]$ and $C_5 = [0.6, 0.7]$, meaning that the costs should be between 80 and 100, time should be no more than 70 but not less than 45, and the task success rate will be between 0.6 and 0.7.

Before analysis with the SOM, a surrogate model must be established to approximately express the relationship between the variables and objective functions. Sampling 100 sets of data from the existing simulation database using the *Latin hypercube* experimental method, a neural network surrogate model must be established, using SOM to analyze the relationships between the objective function and design variables.

Figures 3 and 4 show the results of the analysis, using the SOM method. Y represents an optimized objective function (the highest evaluation value).

Objective function Y focuses on the right bottom of the graph; the costs graph is concentrated in the left corner.

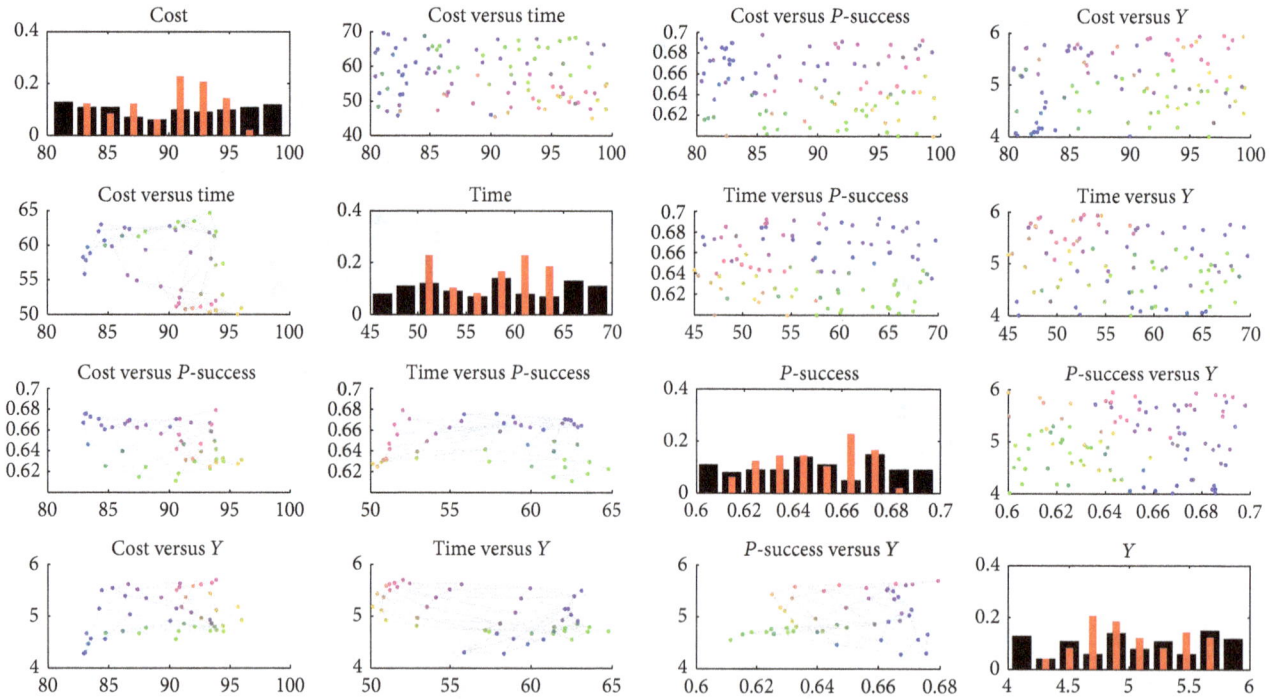

FIGURE 4: The SOM result II.

TABLE 2: The classification of sample attributes.

S	C_1	C_2	C_3	C_4	C_5	D
S_1	2	3	2	1	1	1
S_2	3	3	2	1	2	1
S_3	2	5	1	1	1	1
S_4	3	3	1	2	2	2
S_5	4	4	1	2	2	2
S_6	2	6	2	2	1	3
S_7	5	6	2	2	1	2
S_8	3	2	1	2	1	3
S_9	1	1	1	2	1	4
S_{10}	1	2	1	2	1	5
S_{11}	3	2	1	2	1	4
S_{12}	2	2	2	2	2	5
S_{13}	2	1	2	1	1	6
S_{14}	1	2	1	2	2	5

For the sake of a bigger value of Y, as the red triangle in Figure 3(e), more attention should be paid to the corresponding red triangle in Figures 3(b), 3(c), and 3(d).

In this way, the value range of P-success should be (0.645, 0.679), rather than (0.612, 0.679), the cost of area is reduced to (81, 93), and the value range of time is reduced to (45, 60).

In Figure 4, simple scatter plots and histograms of all variables are shown. Original data points are in the upper triangle, map prototype values are in the lower triangle, and histograms are on the diagonal: black for the data set and red

for the map prototype values. The variable values have been denormalized.

Therefore, compared with the initial design space, the interval of design variables has largely narrowed.

5. Conclusions

In this paper, we studied capability-focused aerospace system of systems architecture alternative design space exploration problems with bilayer mapping. Our results suggest that the RST method can effectively map aerospace system performance space to the configuration space, while a different configuration space is mapped to different regions, efficiently narrowing the design range and providing new ideas for the quick selection of alternatives. At the same time, the SOM method can effectively map the configuration space of aerospace system of systems to the design space and reduce the design dimension or range. This allows the focus to remain on the areas of concern. The optimized efficiency of aerospace system of systems design is fundamentally improved and, as mentioned above, the proposed method effectively explores the design space, reducing the design space range. Starting with the initial stage of the aerospace system of systems design, the method is optimized in the conceptual design phase, sufficiently solving the problem of computing complexity and search difficulty.

Conflict of Interests

The authors declare that there is no conflict of interests regarding the publication of this paper.

Exploration of a Capability-Focused Aerospace System of Systems Architecture Alternative with Bilayer Design Space, Based on RST-SOM Algorithmic Methods

203

Acknowledgments

This research was supported in part by the National Natural Science Foundation of China under Grant nos. 61273198 and 71031007. The authors are grateful to the anonymous reviewers for their valuable comments and suggestions to improve their work.

References

[1] Chairman of the Joint Chiefs of Staff Instruction, "Joint capabilities integration and development system," Defense Acquisition University CJCSI 3170.01G, 2009.

[2] L. Eriksson, E. Johansson, N. Kettaneh-Wold, C. Wikström, and S. Wold, *Design of Experiments: Principles and Applications*, Umetrics Academy, Umeå, Sweden, 2000.

[3] F. Pukelsheim, *Optimal Design of Experiments*, John Wiley & Sons, New York, NY, USA, 1993.

[4] N. Metropolis and S. Ulam, "The Monte Carlo method," *Journal of the American Statistical Association*, vol. 44, no. 247, pp. 335–341, 1949.

[5] M. Stein, "Large sample properties of simulations using latin hypercube sampling," *Technometrics*, vol. 29, no. 2, pp. 143–151, 1987.

[6] R. C. Bose and K. A. Bush, "Orthogonal Arrays of Strength two and three," *The Annals of Mathematical Statistics*, vol. 23, no. 4, pp. 508–524, 1952.

[7] K. T. Fang and C. Ma, *Orthogonal and Uniform Design Experimentation*, Science Press, Beijing, China, 2001.

[8] K. T. Fang, "The uniform design: application of number-theoretic methods in experimental design," *Acta Mathematicae Applicatae Sinica*, vol. 3, no. 4, pp. 363–372, 1980.

[9] D. C. Montgomery, *Design and Analysis of Experiments*, John Wiley & Sons, New York, NY, USA, 1991.

[10] S. Chen, C. F. N. Cowan, and P. M. Grant, "Orthogonal least squares learning algorithm for radial basis function networks," *IEEE Transactions on Neural Networks*, vol. 2, no. 2, pp. 302–309, 1991.

[11] A. D. Cliff and J. K. Ord, "Model building and the analysis of spatial patterns in human geography—with discussion," *Journal of the Royal Statistical Society: Series B (Methodological)*, vol. 37, no. 3, pp. 297–348, 1975.

[12] P. Afentakis and B. Gavish, "Optimal lot-sizing algorithms for complex product structures," *Operations Research*, vol. 34, no. 2, pp. 237–249, 1986.

[13] K. Saitou, K. Izui, S. Nishiwaki, and P. Papalambros, "A survey of structural optimization in mechanical product development," *Journal of Computing and Information Science in Engineering*, vol. 5, no. 3, pp. 214–226, 2005.

[14] M. Danilovic and T. R. Browning, "Managing complex product development projects with design structure matrices and domain mapping matrices," *International Journal of Project Management*, vol. 25, no. 3, pp. 300–314, 2007.

[15] P. Pongcharoen, C. Hicks, and P. M. Braiden, "The development of genetic algorithms for the finite capacity scheduling of complex products, with multiple levels of product structure," *European Journal of Operational Research*, vol. 152, no. 1, pp. 215–225, 2004.

[16] K. Griendling, *ARCHITECT: the architecture-based technology evaluation and capability tradeoff methodology [Ph.D. thesis]*, Georgia Institute of Technology, Atlanta, Ga, USA, 2011.

Permissions

The contributors of this book come from diverse backgrounds, making this book a truly international effort. This book will bring forth new frontiers with its revolutionizing research information and detailed analysis of the nascent developments around the world.

We would like to thank all the contributing authors for lending their expertise to make the book truly unique. They have played a crucial role in the development of this book. Without their invaluable contributions this book wouldn't have been possible. They have made vital efforts to compile up to date information on the varied aspects of this subject to make this book a valuable addition to the collection of many professionals and students.

This book was conceptualized with the vision of imparting up-to-date information and advanced data in this field. To ensure the same, a matchless editorial board was set up. Every individual on the board went through rigorous rounds of assessment to prove their worth. After which they invested a large part of their time researching and compiling the most relevant data for our readers. Conferences and sessions were held from time to time between the editorial board and the contributing authors to present the data in the most comprehensible form. The editorial team has worked tirelessly to provide valuable and valid information to help people across the globe.

Every chapter published in this book has been scrutinized by our experts. Their significance has been extensively debated. The topics covered herein carry significant findings which will fuel the growth of the discipline. They may even be implemented as practical applications or may be referred to as a beginning point for another development. Chapters in this book were first published by Hindawi Publishing Corporation; hereby published with permission under the Creative Commons Attribution License or equivalent.

The editorial board has been involved in producing this book since its inception. They have spent rigorous hours researching and exploring the diverse topics which have resulted in the successful publishing of this book. They have passed on their knowledge of decades through this book. To expedite this challenging task, the publisher supported the team at every step. A small team of assistant editors was also appointed to further simplify the editing procedure and attain best results for the readers.

Our editorial team has been hand-picked from every corner of the world. Their multi-ethnicity adds dynamic inputs to the discussions which result in innovative outcomes. These outcomes are then further discussed with the researchers and contributors who give their valuable feedback and opinion regarding the same. The feedback is then collaborated with the researches and they are edited in a comprehensive manner to aid the understanding of the subject.

Apart from the editorial board, the designing team has also invested a significant amount of their time in understanding the subject and creating the most relevant covers. They scrutinized every image to scout for the most suitable representation of the subject and create an appropriate cover for the book.

The publishing team has been involved in this book since its early stages. They were actively engaged in every process, be it collecting the data, connecting with the contributors or procuring relevant information. The team has been an ardent support to the editorial, designing and production team. Their endless efforts to recruit the best for this project, has resulted in the accomplishment of this book. They are a veteran in the field of academics and their pool of knowledge is as vast as their experience in printing. Their expertise and guidance has proved useful at every step. Their uncompromising quality standards have made this book an exceptional effort. Their encouragement from time to time has been an inspiration for everyone.

The publisher and the editorial board hope that this book will prove to be a valuable piece of knowledge for researchers, students, practitioners and scholars across the globe.

List of Contributors

D. G. Aggelis
Department of Mechanics of Materials and Constructions, Vrije Universiteit Brussel, Pleinlaan 2, 1050 Brussels, Belgium
Department of Materials Science and Engineering, University of Ioannina, 45110 Ioannina, Greece

D. Kleitsa and T. E. Matikas
Department of Mechanics of Materials and Constructions, Vrije Universiteit Brussel, Pleinlaan 2, 1050 Brussels, Belgium

Wen-Qin Wang and Huaizong Shao
School of Communication & Information Engineering, University of Electronic Science and Technology of China, Chengdu, China

Michael J. Krasowski, Norman F. Prokop, Lawrence C. Greer and Philip G. Neudeck
NASA Glenn Research Center, 21000 Brookpark Road, Cleveland, OH 44135, USA

Liangyu Chen and Joseph M. Flatico
Ohio Aerospace Institute, NASA Glenn Research Center, 21000 Brookpark Road, Cleveland, OH 44135, USA

Phillip P. Jenkins
U. S. Naval Research Laboratory, 4555 Overlook Avenue SW, Washington, DC 20375, USA

Danny C. Spina
Jacobs Technology, NASA Glenn Research Center, 21000 Brookpark Road, Cleveland, OH 44135, USA

Li Zhimeng, He Chuan, Qiu Dishan, Liu Jin and Ma Manhao
Science and Technology on Information Systems Engineering Laboratory, National University of Defense Technology, Changsha 410073, China

Ali Abdul-Aziz and Mark Woike
NASA Glenn Research Center, Cleveland, OH 44135, USA

GaigeWang and Luo Liu
Changchun Institute of Optics, Fine Mechanics and Physics, Chinese Academy of Sciences, Changchun 130033, China
Graduate School of Chinese Academy of Sciences, Beijing 100039, China

Lihong Guo and Heqi Wang
Changchun Institute of Optics, Fine Mechanics and Physics, Chinese Academy of Sciences, Changchun 130033, China

Hong Duan
School of Computer Science and Information Technology, Northeast Normal University, Changchun 130117, China

Akshoy Ranjan Paul and Anuj Jain
Department of Applied Mechanics, Motilal Nehru National Institute of Technology Allahabad, Allahabad 211004, India

Shrey Joshi and Aman Jindal
Department of Mechanical Engineering, Motilal Nehru National Institute of Technology Allahabad, Allahabad 211004, India

Shivam P. Maurya
Department of Chemical Engineering, Motilal Nehru National Institute of Technology Allahabad, Allahabad 211004, India

He Chuan, Qiu Dishan and Liu Jin
Science and Technology on Information Systems Engineering Laboratory, National University of Defense Technology, Changsha 410073, China

GaigeWang and Mingzhen Shao
Changchun Institute of Optics, Fine Mechanics and Physics, Chinese Academy of Sciences, Changchun 130033, China
Graduate School of Chinese Academy of Sciences, Beijing 100039, China

Lihong Guo, Heqi Wang and Luo Liu
Changchun Institute of Optics, Fine Mechanics and Physics, Chinese Academy of Sciences, Changchun 130033, China

Hong Duan
School of Computer Science and Information Technology, Northeast Normal University, Changchun 130117, China
Graduate School of Chinese Academy of Sciences, Beijing 100039, China

Mark Woike and Ali Abdul-Aziz
National Aeronautics and Space Administration, Glenn Research Center, Cleveland, OH 44135, USA

Nikunj Oza and Bryan Matthews
National Aeronautics and Space Administration, Ames Research Center, Moffett Field, CA 94035, USA

Konstantinos G. Dassios, Evangelos Z. Kordatos and Theodore E. Matikas
Department of Materials Science & Engineering, University of Ioannina, 45110 Ioannina, Greece

Dimitrios G. Aggelis
Department of Mechanics of Materials and Constructions, Vrije Universiteit Brussel, Pleinlaan 2, 1050 Brussels, Belgium

Ibrahim Yilmaz, Ece Ayli and Selin Aradag
Department of Mechanical Engineering, TOBB University of Economics and Technology, Sogutozu Cad., No. 43, 06560 Ankara, Turkey

Kung-Ming Chung
Aerospace Science and Technology Research Centre, National Cheng Kung University, Tainan 711, Taiwan

Po-Hsiung Chang and Keh-Chin Chang
Institute of Aeronautics and Astronautics, National Cheng Kung University, Tainan 711, Taiwan

Wen-Qin Wang
The School of Communication and Information Engineering, University of Electronic Science and Technology of China (UESTC), Chengdu 611731, China

Zhenqing Wang, Xiaojun Tang and Hongqing Lv
College of Aerospace and Civil Engineering, Harbin Engineering University, Harbin 150001, China

Yanbin Liu, Mengying Liu and Peihua Sun
College of Astronautics, Nanjing University of Aeronautics and Astronautics, Nanjing 210016, China

G. Q. Zhang
Aerospace Engineering Division, School of Mechanical and Aerospace Engineering, Nanyang Technological University, Singapore 639798

S. C. M. Yu
Academic Division, Singapore Institute of Technology, Singapore 179104

M. Mobassher Tofa, Adi Maimun, Yasser M. Ahmed, Saeed Jamei and Agoes Priyanto
Faculty of Mechanical Engineering, Universiti Teknologi Malaysia, 81310 Skudai, Johor, Malaysia

Rahimuddin
Marine Technology Centre, Universiti Teknologi Malaysia, 81310 Skudai, Johor, Malaysia

Zhifei Li, Dongliang Qin and Feng Yang
College of Information System and Management, National University of Defense Technology, Changsha, Hunan 410073, China

www.ingramcontent.com/pod-product-compliance
Lightning Source LLC
Chambersburg PA
CBHW080658200326
41458CB00013B/4899